Computer Vision

Song-Chun Zhu • Ying Nian Wu

Computer Vision

Statistical Models for Marr's Paradigm

 Springer

Song-Chun Zhu
Beijing Institute for General Artificial
Intelligence
Peking and Tsinghua Universities jointly
Beijing, China

Ying Nian Wu
Department of Statistics
University of California, Los Angeles
Los Angeles, CA, USA

ISBN 978-3-030-96532-7   ISBN 978-3-030-96530-3   (eBook)
https://doi.org/10.1007/978-3-030-96530-3

Paper in this product is recyclable.

This Springer imprint is published by the registered company Springer Nature Switzerland AG
The registered company address is: Gewerbestrasse 11, 6330 Cham, Switzerland

Preface

The highest activity a human being can attain is learning for understanding, because to understand is to be free.

—Baruch Spinoza, *Philosopher, 1632–1677*

Story of David Marr

This book is intended for researchers and graduate students in statistics, computer science, and engineering. Based on contributions from multiple authors in the past 20+ years in the Department of Statistics and the Department of Computer Science at the University of California, Los Angeles, it may be used as a reference in the fields of computer vision and pattern recognition. As the first book of a three-part series, this book is offered as a tribute to pioneers in vision, such as Béla Julesz, David Marr, King-Sun Fu, Ulf Grenander, and David Mumford. In this book, the authors hope to provide a mathematical framework and, perhaps more importantly, further inspiration for continued research in vision.

An overarching goal of the three-book series is to provide a mathematical framework for research on vision, cognition, and autonomy in artificial intelligence. With the rise of deep learning, applications of neural networks continue to grow but still very much at the expense of understanding how these models truly work or generate their solutions. It is all too common to refer to neural networks as "black boxes," representing a common lack of understanding in the community. In an effort to promote understanding of neural networks, and to promote the unification of various artificial intelligence theories under a common framework, the vision models in this book take inspiration from many authors in artificial intelligence and other fields, such as statistics, physics, neuroscience, and psychology. Models in this book may lead to more explicit understandings of neural networks, or in some cases may help to modify or generalize existing neural networks, so that more explicit and more efficient models, in terms of both data efficiency and computational efficiency, may emerge.

David Marr is well known for pioneering a resurgence of interest in computational neuroscience and for integrating the fields of psychology, neurophysiology, and artificial intelligence in his research on visual processing. After he passed away at the age of 35, his work was published posthumously in 1982 in the book *Vision: A computational investigation into the human representation and processing of visual information* [169]. Although Marr published early works on the cerebellum in 1969, neocortex in 1970, and hippocampus in 1971, he is most well-known for his research on vision. He laid foundations for continued studies of vision for various fields, such as computational neuroscience and computer vision.

Referred to as Marr's tri-level hypothesis, Marr viewed vision as an information-processing system that should be understood at three distinct, but complementary, levels: the computational level, the algorithmic or representational level, and the implementational or physical level. At the computational or mathematical level, one seeks to understand the problems the visual system solves and, in a similar sense, why it solves them. At the algorithmic or representational level, one seeks to understand how the visual system solves its problems, i.e., the representations it uses and the algorithmic processes it employs to manipulate those representations. At the implementational or physical level, one seeks to understand how the visual system is physically realized. In the same spirit as Marr's tri-level hypothesis, in this book, vision problems are addressed with respect to each of the computational, representational, and implementational levels. Crucially, vision problems may be studied at the computational and representational levels, independent of the implementational level, which is often realized through neural networks [170].

Marr believed that a deep understanding of the brain entails an understanding of the problems it encounters, i.e., the input, and how it solves them, i.e., the steps taken to produce the output. As his interests gradually evolved from the brain to visual processing, he began to treat vision similarly. He described vision as a computational process that takes as input a two-dimensional array on the retina and outputs a three-dimensional description of the world. His three stages of vision, depicted in Fig. 1, include a primal sketch, a 2.5D sketch, and a 3D model. A 2.1D sketch was also proposed by Nitzberg and Mumford [185].

A primal sketch extracts key components of a scene, such as simple edges and regions. Textures and textons are constituent parts of a primal sketch, which correspond to the early stages of human visual perception, i.e., the first visual phenomena noticed by humans when viewing an object. A 2.5D sketch reflects textures and depth. The 2.1D and 2.5D sketches correspond to mid-level vision. In a complete 3D model, a scene is visualized in a continuous 3D map. The 3D model corresponds to high-level vision and provides an object-centered perspective, while the primal 2.1D and 2.5D sketches provide view-centered perspectives.

Various research influenced Marr's book *Vision*, such as psychology experiments by Béla Julesz on textures and textons, neuroscience discoveries in edge detection and filters, the random dot stereopsis by Julesz, the shape-from-shading theory from Berthold K. P. Horn, and research on generalized cylinders and 3D representations. Various works facilitated the transitions from primal sketch to 2D sketch to 2.5D sketch to a complete 3D model. For a primal sketch formulation, work on textures and textons paired with work on edge detection and filters provided crucial foundations. For the 2.5D sketch, the random dot stereopsis from Julesz and the shape-from-shading theory by K. P. Horn facilitated the modeling of

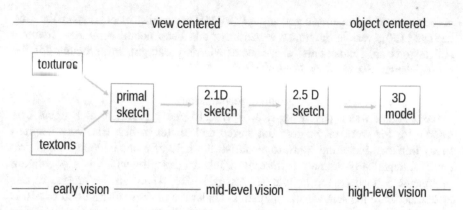

Fig. 1 Stages of vision according to Marr's paradigm

textures and depth. Lastly, for the 3D model, work on generalized cylinders and 3D representations made it possible to model scenes in continuous, 3D maps.

As a whole, Marr's work proposed a holistic framework for understanding vision and touched on broader questions about how cognition may be studied. Thirty years later, the main problems that occupied Marr remain fundamental, open problems in the study of vision. This book may be understood as a bridge between Marr's theory of vision and the modern treatment of computer vision with statistical models. It explores connections between Marr's paradigm and neuroscience discoveries while solidifying such discoveries with mathematical models.

The mathematical framework studied in this book also counteracts a "big data for small tasks" paradigm that dominates the machine learning community today. This paradigm refers to the practice of exploiting massive amounts of data for highly specific tasks and essentially repeating this process for each new task. A great deal of research readily falls susceptible to such a "task trap," such that solutions often do not contribute to a greater, unified framework for vision.

Given the wide variety of artificial intelligence methods used today on similar problems, it is clear much effort is devoted by the community as a whole to often entirely different research approaches. Undoubtedly, ingenuity and novelty contribute greatly to research advancement, but the community as a whole could work more in harmony and hence more efficiently if it adopted a common ground, or overarching framework, for general-purpose research in vision. This type of maturation for the field of vision is comparable to a process other academic fields have undergone and is arguably long overdue.

Beyond David Marr's Paradigm

This book provides mathematical frameworks and models for many of Marr's concepts. Notably not available during Marr's time, they include Markov random fields; the FRAME (Filters, Random fields, and Maximum Entropy) model [282], i.e., a predecessor of the energy-based model (EBM); generative models; sparse coding models [190]; various inference algorithms; and deep neural networks. Textures and textons are fundamental elements of Marr's paradigm, in particular for the primal sketch. For textures, Markov random fields may be used to mathematically formulate models for textures. For textons, generative models and sparse coding models provide essential mathematical formulations.

Béla Julesz was a psychologist at Bell Labs and a professor at Rutgers best known for his work on random dot stereo and texton theory. His work inspired many thinkers, including Marr, to establish the field of computational vision. We can mathematically define a texture by a Julesz ensemble, which is a population of images defined on the infinite 2D image lattice, where all the images in this ensemble share the same spatial statistics. Under the uniform distribution over this ensemble of infinitely large images, the probability distribution of images on a finite image patch follows a Markov random field model.

In general, a visual concept can be defined by an ensemble or population of images, whose distribution can be mathematically defined by a probability distribution or a probability density function. The goal of learning is to estimate or approximate this probability distribution based on a finite number of examples randomly sampled from this population.

For modeling textons, sparse coding models and active appearance models (AAM) have found some success. Sparse coding states that objects may be represented by the strong activation of a relatively small set of nodes, or neurons. It is motivated in part by research in biological vision. For example, Huber and Wiessel in the 1960s performed experiments on cats to record the activations of cortical cells in the V1 section of the mammalian brain. It was observed that bars of light oriented mostly vertical activated a particular set of neurons in the V1 section of the brain, while the same bar of light oriented in other directions failed to activate the same neurons. In another neurophysiological experiment, a neuron was recorded to selectively fire only to images of Jennifer Aniston, inspiring a sparse coding scheme called grandmother cell coding. Furthermore, some neurons referred to as mirror neurons fire not only when performing an action but also when observing other subjects perform the same action. In general, one of the striking observations about physiological recordings from sensory cortical areas of the brain is the difficulty of finding stimuli that effectively activate some given neurons. These difficulties reflect the narrow functionality of cortical neurons, which, given the incredibly large number of them, suggests that a sparse coding system may be in place in the brain.

Field [54] performed experiments on the primary visual cortex to suggest that basis functions limited in both space and frequency domains, such as Gabor functions, maximize sparseness when applied to natural images. Olhausen and Field [187] gave examples of sparse coding in other brain regions. In [190], Olhausen and Field defined an explicit objective function that promoted both high sparseness and low reconstruction error. The minimization of this function on natural images leads to a set of basis functions that resemble localized receptive fields of simple cells in the primary visual cortex. Sparse coding is closely related to independent component analysis [15].

Various inference algorithms and deep neural networks, also methods not available during Marr's time, are widely used throughout this book. Of course, the vision community has seen incredible success in recent years due to advances in deep learning, but as mentioned before, research efforts in vision are often devoted to entirely different approaches. In an effort to recapture the main theme of vision, this book carefully establishes mathematical frameworks for general-purpose research in vision by examining theories in the literature, beginning with Marr's paradigm and progressing up to the most recent uses of neural networks.

Introducing the Book Series

The book series consists of three parts.

The first book, introduced here, covers David Marr's paradigm and various underlying statistical models for vision. The mathematical framework herein integrates three regimes of models (low-, mid-, and high-entropy regimes) and lays the ground for research in visual coding, recognition, cognition, and reasoning. Concepts in this book are first explained for understanding and then supported by findings in psychology and neuroscience, after which they are established by statistical models and further linked to research in other fields. A reader of this book will gain a unified, cross-disciplinary view of artificial intelligence research in vision and will accrue knowledge spanning from psychology to neuroscience to statistics.

The second book defines stochastic grammar for parsing objects, scenes, and events, posing computer vision as a joint parsing problem. It summarizes research efforts over the past 20 years that have worked to extend King-Sun Fu's paradigm of syntactic pattern recognition. Similar to David Marr, King-Sun Fu was a pioneer and influential figure in the vision and pattern recognition community.

The third book discusses visual commonsense reasoning, including subjects such as functionality, physics, intentionality, causality, and values. The third book connects vision to cognition and artificial intelligence.

The authors would like to thank many current and former Ph.D. students at UCLA for their contributions to this book: Erik Nijkamp, Eric Fischer, Jonathan Mitchell, Linqi Zhou, Mitchell Hill, Yaxuan Zhu, Ruiqi Gao, Yuxin Qiu, Chi Zhang, Peiyu Yu, Sirui Xie, Dehong Xu, Deqian Kong, Adrian Barbu, and Tianfu Wu. Erik Nijkamp and Eric Fischer, especially, have worked extensively on editing the manuscript.

Beijing, China
Los Angeles, CA, USA

Song-Chun Zhu
Ying Nian Wu

The original version of the book has been revised. A correction to this book can be found at https://doi.org/10.1007/978-3-030-96530-3_13

Contents

About the Authors

S.-C. Zhu

Y. N. Wu

Song-Chun Zhu and Ying Nian Wu studied at the University of Science and Technology of China (USTC), Hefei, in the late 1980s, majoring in computer science. They first met in April of 1992 at the front gate of the US consulate in Shanghai, while applying for their F1 student visas to attend Harvard University. At the Harvard Graduate School of Arts and Sciences, Zhu studied computer science under Dr. David Mumford and Wu studied statistics under Dr. Donald Rubin, and they became collaborators. Their first joint project was the random field model for texture and the minimax entropy learning method, published in 1997. After graduation, Zhu and Wu continued to carry out research on textures, textons, and sparse coding models. They became colleagues in 2002 as tenured faculty at the University of California, Los Angeles (UCLA), continuing work on the primal sketch model, information scaling phenomena, and connecting these classical models and others to deep neural network models.

Zhu started his search for a theory of intelligence while in college, after encountering the book *Vision: A Computational Investigation into the Human Representation and Processing of Visual Information*, written by MIT computational neuroscientist David Marr. At the time in the late 1980s, Zhu was a student helper at a cognitive science lab at USTC. Marr's book is considered the first attempt to outline a computational paradigm for vision, and it inspired Zhu to study computer vision with Dr. David Mumford, who also had an interest in pursuing a mathematical framework for vision and intelligence. The 1990s were characterized by a transition from models based on logic and symbolic reasoning to models based on statistical modeling and computing. During this time and since then, Wu has brought expertise in statistics to the team. Since the late 1990s, Zhu and Wu have been working on statistical modeling and learning for computer vision.

Notably, the pursuit of a unified framework for computer vision dates back to Ulf Grenander, a pioneer who launched General Pattern Theory in the 1960–70s while at the Division of Applied Math at Brown University. Zhu studied pattern theory as a postdoc at Brown and benefitted greatly from interactions with the Brown group. On this note, an underlying goal of this book is to apply the methodology of general pattern theory to formulating concepts in Marr's paradigm.

At the Center for Vision, Cognition, Learning, and Autonomy at UCLA, Zhu has supervised many Ph.D. students and postdocs who have both contributed their thesis work as examples in this book and assisted with editing the book.

Chapter 1
Introduction

The primary aim of this book is to pursue knowledge representation for vision. In a world in which technology generates immense amounts of data from a wide spectrum of sources, it is of growing importance to establish a common framework for knowledge representation, learning, and discovery. A current problem in artificial intelligence is how to acquire truly massive amounts of knowledge, akin to the faculty of common sense in humans, from raw sensory signals and, moreover, use this knowledge for inference and reasoning. In this book, the vision models presented for acquiring such knowledge are based most fundamentally on statistical properties discovered for natural images over the past several decades. Similar to models in physics, models are advanced based on empirical grounds, such as discovering additional patterns in data. In this way, models stay free of bias.

Discriminative modeling approaches are not discussed in this book in depth. In statistical classification, two main approaches exist: discriminative and generative modeling. In discriminative modeling, the conditional probability of a target y is modeled, given an observation x, i.e., $p(y|x)$. Often, the value of a target variable y is determined by training a model on thousands of examples. A discriminative classifier may perform well at discriminating between, e.g., chairs and non-chairs, but does it truly understand the concept of a chair? Consider adversarial attacks in the literature, in which an image that a discriminative classifier can otherwise correctly classify with a high degree of certainty is injected with a small amount of noise, in many cases barely perceptible to a human, and the predicted class is entirely different, e.g., a bus instead of a chair. This behavior demonstrates that a discriminative model does not actually learn the concept of a chair as a human does—it only learns to discriminate between chairs and non-chairs using whatever obscure image features it may prefer.

In cases in which discrimination is not the ultimate goal, generative modeling approaches, explored in this book, facilitate the construction of models more representative of human learning. In generative modeling, given an observable variable x and a target variable y, the joint probability distribution $p(x, y)$ is modeled. From this, the conditional probability $p(y|x)$, modeled directly in discrim-

© Springer Nature Switzerland AG 2023
S.-C. Zhu, Y. N. Wu, *Computer Vision*, https://doi.org/10.1007/978-3-030-96530-3_1

Fig. 1.1 One can observe, as an example of texture, the tree leaves in the background and, as an example of textons, the edges created by tree trunks in the foreground

inative modeling, is computed. Generative modeling is not only more indirect but also more probabilistic in comparison to discriminative modeling, allowing more domain knowledge and probability theory to be applied. In this way, generative models better represent human learning. For example, with prior domain knowledge encoded, a generative model can generalize knowledge to new but related tasks. And this is akin to human learning; a human may be able to intuit, for example, how to make orange juice with the prior knowledge of how to make lemonade.

After discussing statistical properties discovered for natural images in Chap. 2, the most basic units of visual perception, textures and textons, are introduced in Chaps. 3 and 4, respectively. A texture could be sand, or the thousands of leaves of a tree viewed from afar, as in Fig. 1.1, for which individual components in pre-attentive vision are not disentangled. Pre-attentive vision is characterized as human vision before focusing on any specific region of some visual stimuli. A texton, or a "token" as referred to by Marr, can be thought of as a most basic element, like a bar, edge, corner of an eye, or trunk of a tree as in Fig. 1.1. Textons form the structural part of an image and object boundaries. Note that the same object may be perceived as textons or texture, depending on potential viewing distance or focal point, and in natural scenes, these two entities are seamlessly interwoven. Following David Marr's insight, these two modeling components may be integrated to form a generative image representation called primal sketch, which he referred to as a "symbolic" image representation in terms of image primitives. Primal sketch is discussed further in Chap. 6, after a discussion of Gestalt laws and perceptual organization, which naturally follows from Chap. 4 on textons.

Marr defined different stages of vision, starting from a 2D visual array on the retina to a 3D description of the world. His stages of representation include a primal sketch, a 2.1D sketch, a 2.5D sketch, and a 3D model. The first two stages perform separation between four main factors known to influence perceived visual intensity: geometry (shape and position), reflectance of visible surfaces, illumination, and viewpoint. A primal sketch extracts fundamental components of the scene, like edges and regions, and it looks like a pencil sketch. In a primal sketch, geometric structures, pixel intensity changes, and illumination effects are detected. A primal sketch captures spatial layout by using textons, such as edges, bars, and blobs. A 2.1D sketch introduces layered representations of the input image and is discussed in Chap. 7. A 2.5D sketch, discussed in Chap. 8, represents orientation, depth, and textures, considering the distance from the viewer and discontinuities in depth and surface orientation. Both primal and 2.5D sketches are viewer-centered perceptions. In a 3D model, the scene is visualized in a continuous 3D map, as an object-centered perception. The 3D representation describes shapes and their organization using a hierarchical organization of volumetric and surface primitives. Marr's stages of visual perception serve as a basis for further analyses in the book.

In Chap. 9, information projection is introduced as a framework for learning a statistical model as an approximation to the true data distribution. In Chap. 10, information scaling and regimes of models are discussed, including entropy, metastability, and energy landscapes. In Chap. 11, image models with multilayer neural networks, such as deep FRAME/energy-based model and the generator model, are presented. In Chap. 12, three main families of machine learning models, i.e., discriminative, generative, and descriptive models, are further examined.

1.1 Goal of Vision

In the parse graph in Fig. 1.2, there can be several types of nodes including scenes and objects, minds and intents, hidden objects, actions, imagined actions, attributes, and fluents, i.e., how objects change over time. Nodes are organized in a tree-like hierarchical structure with potential connections between sibling nodes. Parse graphs for knowledge representation in vision will be discussed in greater detail in the second book, but for now, it may be understood that it involves parsing scenes and events in a picture, or a video sequence, into nodes such as in Fig. 1.2.

These nodes may be represented with words one might use to describe them in natural language, e.g., "backpack" or "vending machines," or they may be represented with "words" that are not really words at all—perhaps a symbol, a number, some other character, an expression in sign language, or a facial expression. Either way, for simplicity, the descriptors of these nodes are referred to as words. They can be thought of as *visual* words in the sense that even a primate, for example, could have a word for a node, even though it is not endowed with a capacity for

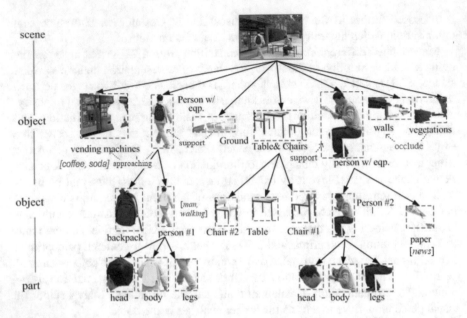

Fig. 1.2 Scenes and events may be understood in terms of a parse graph. From a single image, a dense 3D scene may be reconstructed, by estimating camera parameters, materials, and illumination. The scene may be parsed hierarchically in terms of relations, intents, beliefs, attributes, and fluents. The actions of agents may be predicted over time and hidden object states, e.g., water boiling inside a metal kettle, which humans can naturally infer from the image, may even be recovered

language. Sometimes in this book, words are referred to as symbols or concepts, but this is only to aid with explanation depending on the context.

A common misconception is that concepts are encoded explicitly in pictures. The raw pixel values of a given picture do not directly correlate to, or represent, any concepts. Consider, for instance, a grayscale image of 1000×1000 pixels, which may be represented as a 1000×1000 table, a portion of which would look similar to Fig. 1.3. For a color image, each table entry would contain three numbers, indicating the intensities of the colors red (R), green (G), and blue (B). Nowhere in this collection of numbers can an explicit representation of a concept such as a person, table, or chair be found. They have to be inferred in a sophisticated manner from the collection of numbers representing the image. Visual computation is a daunting task for computers, just the same as it is for humans. In fact, half of the human brain is devoted to visual computation, and most of the brain's activities that involve cognition and reasoning are based on visual stimuli.

At this stage, researchers have yet to grasp how the so-called "signal to symbol" transition is realized, i.e., how concepts are inferred and ultimately learned from visual stimuli. In order to answer this question, the relationship between concepts and image patches, such as the one shown in Fig. 1.3, needs to be better understood,

Fig. 1.3 An image is merely a collection of numbers indicating the intensity values of the pixels

as well as how different concepts relate to one another. The relationship between images and concepts as it relates to modeling is explored in depth in Chap. 2.

In *Vision*, David Marr stated, "Vision is the process of discovering from images what is present in the world, and where it is." The parse graph in Fig. 1.2 depicts "what" objects are present in the image and "where" they are spatially relative to other objects. The distinction between "what" and "where" in vision is noteworthy. The ventral pathway in the brain is thought to account for "what" objects humans see and the dorsal pathway for "where" humans see them. The notion of a division between a ventral and a dorsal visual stream has been an essential principle of visual processing since David Milner and Melvyn Goodale published the two-streams hypothesis in 1992.

The two-streams hypothesis argues that humans have two distinct visual systems. After visual information exits the occipital lobe, it follows two pathways or streams. The ventral pathway, or "vision-for-perception" pathway for "what" humans see, leads to the temporal lobe, which is involved with object identification and

Fig. 1.4 The dorsal pathway is responsible for the spatial aspect of vision, i.e., "where" humans see objects. The ventral pathway is responsible for the main content of vision, i.e., "what" objects humans see

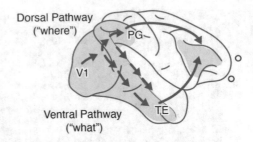

recognition. This pathway is believed to mainly identify and discriminate between shapes and objects. The dorsal pathway, or "vision-for-action" pathway, leads to the parietal lobe, which is involved with processing objects' spatial locations. This pathway has been tied to movements such as reaching and grasping, which are based on evolving spatial locations, shapes, and orientations of objects. Understanding the difference between the two visual pathways, the ventral pathway for perception and the dorsal pathway for action, is useful knowledge for vision research and for understanding the inspiration behind theories presented later on (Fig. 1.4).

Another consideration for vision research that must be accounted for is the fact that human vision is often task-driven. After viewing an instructional video about how to make watermelon juice, one could most likely reenact the process, including chopping watermelon, blending it, and so forth. The physics and the functionality of the objects, and potentially hundreds of other subtasks involved in making the juice, may be easily understood. Humans additionally understand the causality and how to, e.g., switch the order of the steps without affecting the final result. From a small amount of data, i.e., the instructional video in this case, humans can absorb a massive amount of information, forming what is actually just referred to as common sense. Humans possess this remarkable ability because vision is incredibly task-driven, absorbing specific information from complex visual stimuli.

Thus, the faculty of common sense, which forms part of the prior knowledge each human possesses, ultimately entails an understanding of physics, functionality, causality, intentionality, and utility, among other properties of a scene. Pursuing holistic knowledge and learning reminiscent of prior knowledge and learning in humans may be referred to as the "small data for big tasks" paradigm. This is where efforts are concentrated in this book. Many avenues in machine learning today, unlike human learning, depend on massive amounts of data to learn highly specific tasks, such as the task of making watermelon juice or identifying a person from his or her face in facial recognition. This assumes no prior knowledge and this type of learning is not generalizable. As mentioned previously, this contrasting and all-to-common paradigm in machine learning may be referred to as the "big data for small tasks" paradigm.

As a given task or purpose very much influences the interpretation of visual stimuli to form knowledge, similar behavior should be exhibited by vision models. Accordingly, in this book the optimal model and knowledge representation of some visual data depends not only on the data but also on the relevant task. Here, a contrast may be made with the field of physics. In vision, there are subjective, in addition to the objective, considerations that guide the formation of models. As a vision model seeks structure in the data, i.e., an image space, the optimal structure or architecture of the model depends on its given purpose. As an example, the hierarchical structure of an And–Or graph depends on its purpose, such that it evolves as the purpose of the model evolves.

As mentioned, vision models should also generalize well. Although excellent for a specific task, discriminative modeling approaches clearly lack generalization, which does not reflect human learning. Again, a human could likely infer how to make orange juice after learning how to make lemonade or watermelon juice. Or a human could infer how to crush a walnut with a book after observing someone crushing it with a hammer. Humans have the ability to generalize knowledge to the novel but similar tasks, often after learning from just one example. The generative modeling approach offers the tools to begin to mimic this incredible learning ability.

Humans generalize knowledge well primarily due to prior knowledge and imagination, among other faculties. From birth, humans continually gain a better understanding of space, time, causality, functionality, and other features of the world, utilizing that knowledge each time a new task is learned. As humans learn, they add to their prior knowledge, in this way creating a continual accumulation of it. With imagination, humans have the ability to extrapolate prior knowledge to imagined future or hypothetical tasks, a powerful tool for generalizing knowledge.

1.2 Seeing as Bayesian Inference

Clearly, the way in which humans form and represent knowledge from visual stimuli is complex and relies on more than just vision—it is dependent on prior knowledge, imagination, and in general the mind, or agent, that perceives. In fact, every image has literally infinite interpretations, and humans must derive only one or two meaningful interpretations for each one. Vision, in this sense, is under-constrained and necessitates some form of guessing by the agent, using prior knowledge. Images, by themselves, lend minimal information. This is an important point and essentially what it means for vision to be ill-posed and mostly an illusion. Interpretation of visual stimuli is largely independent of the visual stimuli itself—it depends on the agent's prior knowledge and imagination.

It has been shown that a form of top-down processing performed by the visual system also contributes greatly to visual interpretation. Top-down processing makes use of global information propagation, exploiting high-level knowledge possessed by the agent. The agent considers high-level contextual information and performs top-down inferential reasoning based on prior knowledge, both innately acquired through evolution and learned through experience. With top-down processing, the agent may form an interpretation of visual stimuli, even if the stimuli were only partially perceived or perceived poorly, e.g., due to poor lighting conditions or quick movement. To give an example, top-down processing may help one intuit the object or scene of a partially completed puzzle, although only portions of what typically represents the object or scene are present. With the role of top-down processing in the visual system, top-down architectures have become paramount to the design of learning and inference algorithms in vision.

Given prior knowledge, a basic assumption since the time of Helmholtz's research in the 1860s is that a given visual input may be represented as the most probable computed interpretation of an image. To give a Bayesian formulation for vision, let $\mathbf{I}$ be an image and pg be a semantic representation of the world, such as a parse graph. The most probable representation pg^* of the image is defined as $pg^* = \arg\max_{pg\in\Omega} p(pg|\mathbf{I}) = \arg\max_{pg\in\Omega} p(\mathbf{I}|pg) p(pg)$. To obtain probable interpretations, it is necessary to sample from the posterior $p(pg|\mathbf{I})$ and obtain candidates $(pg_1, pg_2, ..., pg_k) \sim p(pg|\mathbf{I})$. This is a crucial point—sampling from the posterior probability lends the possible visual interpretations. The result of this process, pg^*, represents the most probable semantic representation of the image and the visual interpretation of the agent.

The quality of the prior knowledge of a model may be judged using a method referred to as analysis by synthesis, based on synthesized examples. Analysis by synthesis can be compared to a Turing test as a way to judge the humanlike thinking capability, or intelligent behavior, of an artificial intelligence model. In a Turing test, a subject asks questions to two different agents, one a machine and the other a human, and the machine is determined to be more humanlike the more uncertain the subject is in identifying the human from the machine, based on each of their responses. Thus, the machine does not need to have correct responses; it merely needs to respond like a human, to perhaps fool the subject into thinking it is, in fact, the human. The analysis by synthesis method may also be used to judge humanlike thinking capability, but in a way that does not depend on the input, such as a subject asking questions. This eliminates the possibility of any questionable input. The analysis by synthesis method thus provides an unrestricted way to judge what a machine knows, i.e., its prior knowledge—in a way, this is similar to dreaming in humans. Dreaming, or the imagination, is also unrestricted in the sense that these are perceptions, often of great visual detail, that also receive no input visual stimuli. Interestingly, then, the analysis by synthesis method may be viewed as a way to determine what a machine "imagines" or "dreams." If one were to draw several random samples meant to represent a dream from a completely untrained model that possesses no prior knowledge of the world, the random samples would be akin to white noise. A well-informed model, however, would produce a detailed, interesting dream.

The fact that vision relies so heavily on prior knowledge and imagination signifies that it is highly probabilistic. Accordingly, probability and Bayesian inference play large roles in vision research. The probabilistic nature of vision is most obvious with problems exhibiting imperceptibility, in which human vision jumps between image interpretations often entirely different. In the bikini versus martini example, a well-known example demonstrating ambiguity in visual perception, human vision oscillates between interpretations of a bikini bathing suit and a martini drinking glass, depending on how the visual system interprets the image. Mathematically, the fact that perception may jump between interpretations in some structured state space poses a significant challenge for the design of learning and inference algorithms. Ambiguities in the visual inference that occur with human vision are, in general, difficult to model. Imperceptibility in vision is more formally introduced in Chap. 9, which discusses information scaling regimes.

Imperceptibility may be formally defined as the complexity of an image subtracted from the complexity of the world. As a measure of complexity, it may be aptly modeled with entropy. Given a generative model such that pg $\sim$ $p(\text{pg})$ and $\mathbf{I} = g(\text{pg})$, in which pg is a semantic representation of the world and $\mathbf{I}$ is an image, imperceptibility $H(p(\text{pg}|\mathbf{I})) = H(p(\text{pg})) - H(p(\mathbf{I}))$. Entropy measures complexity or uncertainty, e.g., the image complexity $H(p(\mathbf{I})) = E[-\log p(\mathbf{I})] = -\int p(\mathbf{I}) \log p(\mathbf{I}) d\mathbf{I}$. Hence, as world complexity increases relative to the image complexity, imperceptibility grows. When the posterior probability $p(\text{pg}|\mathbf{I})$ exhibits high imperceptibility, it means that certain variables in pg, the semantic representation of the world, cannot be inferred as the uncertainty is too high; hence, they are imperceptible. In this case, the model's representation of the world would need to be reduced in complexity.

As the concept of imperceptibility explains ambiguities in inference, it has been utilized in analyses of abstract art. Artists intentionally create imperceptible aspects of a piece of artwork to induce ambiguous visual interpretations. Viewers may proceed down different perceptual paths to interpret the artwork in various ways.

1.3   Knowledge Representation

To begin to understand the way knowledge may be represented in vision, first consider the space of all image patches of natural scenes and of a fixed size, e.g., 10×10 pixels. These image patches reside in an image space whose dimension is the total size of the image patches, in this case, $10 \times 10 = 100$. Each image patch can be treated as one point in the image space. Hence, the image patches together form a population of points in the image space. One may consider an analogy between this population and the 3D universe, as illustrated in Fig. 1.5. The distribution of mass within the universe is highly uneven. There are high densities of mass at stars, but there are low densities of mass across nebulas (clouds of dust and gas in outer space), in which mass is spread out. The distribution of the population of natural image patches in the image space is also highly uneven. To model this space, it becomes necessary to identify, map out, and catalog high-density clusters such as stars, as well as low-density regions such as nebulas.

Just as a representative in the United States Congress represents a subset of the population, mathematically, a concept represents a subset or sub-population of image patches within the entire universe of image patches. All the image patches in this subset are perceived as the same pattern and hence are described by the same symbol or "visual word." The subset, or sub-population, of image patches that correspond to a concept can be represented mathematically by a probability

Fig. 1.5 Consider the complex distribution of mass in the universe, e.g., at stars (high-density low-volume clusters) and over nebulas (low-density high-volume regions). Left: the universe with galaxies, stars, and nebulas. Right: a zoomed-in view of a small part of the image

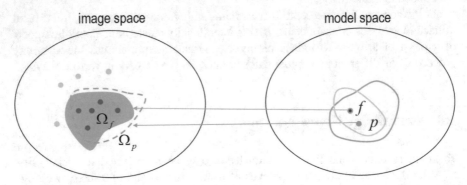

Fig. 1.6 The hope is that Ω_p, the learned estimate of Ω_f, approximates as closely as possible Ω_f. The better the approximation, the more closely the learned concept p approximates, and thus accurately represents, the real concept f

distribution or a probability density function, which can be parametrized by a statistical model. Hence, concepts can ultimately be represented by statistical models, which are quite powerful.

Thus, there are two spaces: the image space of image patches (signals) and the model space of concepts (symbols), depicted in Fig. 1.6. Each concept, which corresponds to a set of image patches in the image space, is represented as just one point in the model space. For example, in Fig. 1.6, the sub-population of image patches Ω_f in the image space corresponds to one concept f, a point in the model space. All the image patches of a concept, despite having diverse pixel intensities, correspond to the same concept because they are perceived as the same pattern or object by the visual system.

Suppose a number of image patch examples are randomly sampled from the sub-population of image patches Ω_f, and the goal is to learn a concept f in the model space based on these examples. In other words, the goal is to recover the sub-population Ω_f to best learn the concept f. First and foremost, it is impossible to fully recover the sub-population Ω_f, since it is impossible to access all image patches in the universe that belong to Ω_f. All the image patches that correspond to the concept of sand, for example, are clearly not somewhere in a dataset. However, it is still possible to come up with an approximated estimate of Ω_f. Call it Ω_p, which corresponds to a concept p in the model space. The hope is that the set Ω_p is close to the set Ω_f in the image space so that the concept p will be as close as possible to the concept f in the model space. For example, one might hope p is within a distance ϵ of f, measured by some predefined metric. In machine learning, identifying Ω_f based on randomly sampled examples from Ω_f is referred to simply as learning, and these examples are training examples.

Concepts can differ greatly in their complexities. Some concepts may appear to be very simple or regular, such as a line segment, a triangle, or even a human face, while other concepts can appear quite complex or random, such as stochastic textures like grasses or other foliage. Some concepts lie in between, such as the face of a tiger. Ultimately, concepts may be defined as textures, textons, or some composition of the two. Compositions of textures and textons are discussed in the second book in the series. Figure 1.7 is a simple illustration of concepts with different complexities. For each concept, such as a texture or the face of a tiger, a number of examples are collected, which may then be aligned in the case of, e.g., faces. The principal component analysis is performed and the eigenvalues are plotted in descending order. For a simple concept such as a face, the eigenvalues drop to 0 very quickly, indicating that the images lie in a very low-dimensional space. For a complex concept such as a texture, the eigenvalues stay high for an extended range, indicating that the images lie in a very high-dimensional space. For the face of a tiger, the eigenvalues lie in between.

Fig. 1.7 Observe a plot of eigenvalues, in decreasing order, both for concepts of low dimension (blue curve) and high dimension (red curve)

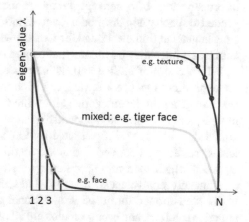

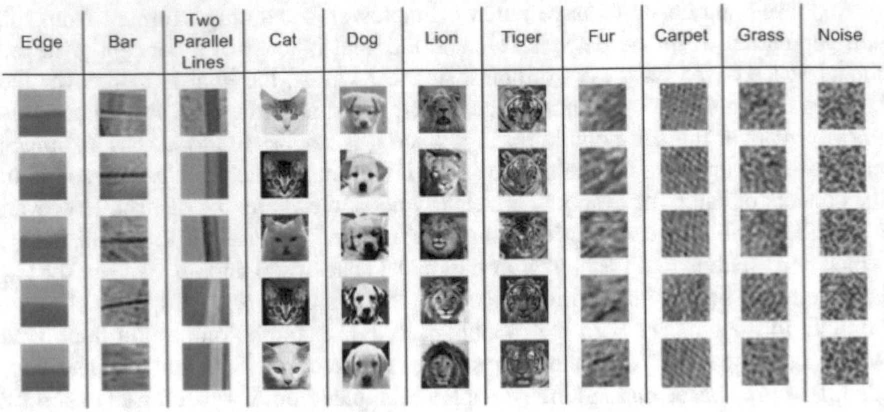

Edge	Bar	Two Parallel Lines	Cat	Dog	Lion	Tiger	Fur	Carpet	Grass	Noise

Fig. 1.8 Simple geometric patterns like edges and bars have low entropy, while stochastic textures like fur and carpet have high entropy

The complexity of a concept f in the model space can be measured by the log volume of the corresponding subset Ω_f in the image space if we assume a uniform distribution. This log volume is called entropy in statistical physics and information theory, and it is called intrinsic dimension in mathematics and coding theory. As one can see in Fig. 1.8, entropy may be used as an axis on which to map concepts. Geometric patterns belong to the low-entropy regime, while texture patterns belong to the high-entropy regime. Many object patterns lie in the mid-entropy regime. Textons are of low entropy, textures are of high entropy, and many compositional structures in between are of mid-entropy. Textures, textons, and compositions provide three ways to characterize a sub-population in the image space that represents a concept in the model space. While patterns in the low-entropy regime tend to be simple and patterns in the high-entropy regime tend to be random, patterns in the mid-entropy regime tend to be quite informative. The informativeness of a concept f can be measured by the number of parameters needed to specify it in the model space. Central to the goal of this book is to establish a mathematical framework and form of knowledge representation under which these entropy regimes may be further studied and defined.

With the aim of unified knowledge representation, a natural question is how to relate different types of knowledge. Vision models that lend knowledge representation can generally be categorized into three paradigms: logic models, probabilistic models, and discriminative models, as shown in Fig. 1.9. In the 1960s and 1970s, knowledge was mostly represented by logic formulas, i.e., propositions and predicates, called well-formed formulas. In this paradigm, a concept is represented by a set, which is in turn specified by well-formed formulas. As such, a concept is ultimately equivalent to a set of states, from a joint state space, that satisfies all the well-formed formulas. Propositional calculus, first-order predicate calculus, situation calculus, and event calculus are all used to represent knowledge. In general, the logic paradigm for knowledge representation is alive and well, and it is used

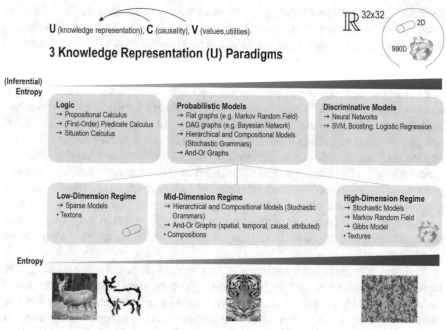

Fig. 1.9 At the highest level of vision and cognition, one can discriminate between three types of models: Knowledge Representation (U) models, Causal (C) models, and Value and Utility (V) models. The last two types are discussed in the third book in the series, along with many other more advanced artificial intelligence concepts. Knowledge representation models aim to describe the world as intuitively and as accurately as possible. Causal models represent cause and effect in the physical world. Value models compute how much value is placed on certain objects or ideas; they involve morality. Causal (C) and Value (V) models together guide the creation of the world. For example, chairs are created because they have a certain value (V) and satisfy some causality (C) in physics, allowing an individual to rest when he or she sits down. Causal and Value models, as they uncover knowledge, feed Knowledge Representation models. There are 3 paradigms for knowledge representation: logic models, probabilistic models, and discriminative models. Probabilistic models have low- (e.g., textons), mid-, and high-dimension (e.g., textures) regimes, each increasing in entropy. At the top right, one can see a 32×32 image space, i.e., 1024 dimensions, in which part of the space is only 2D, while another part of the space is 990D. This conveys that dimensionality often varies greatly in just one image, lending instances of each of the low-, mid-, and high-dimension regimes

today to solve many problems in vision. For example, situation calculus is used for causal reasoning in robot planning. A clear advantage of logic for knowledge representation is that it is rigorous and enables pure reasoning; however, a clear disadvantage is that it is inherently fragile in that it is not grounded on observational data. Truths found by exercising logic alone are called *a priori* truths; they are justified by reasoning that proceeds only from theoretical deduction, completely independent of empirical data.

Beginning in the 1980s, based partly on a desire to ground knowledge representation on empirical data, there was more research into probabilistic models, which provided imperfect knowledge representation and the ability to make inferences from observed signals with some degree of uncertainty. Note that "imperfect" knowledge representation does not imply that knowledge is infused by noise; it signifies that knowledge can be represented in a non-absolute way, with some degree of uncertainty. The Bayesian Belief Network described by Judea Pearl in 1984 is an example of such a probabilistic model. In probabilistic models, concepts are no longer regarded as deterministic sets but rather probabilistic distributions. Two concepts may, for example, overlap in the set of instances they define. There are two regimes of probabilistic graphical models: flat and hierarchical. The flat regime includes descriptive or declarative models, such as constraint satisfaction models, Markov random fields, and Gibbs models. The hierarchical regime includes generative or compositional models, such as Markov trees, stochastic context-free grammars, and sparse coding models. Integrating context into hierarchical models, there are models such as primal sketch, And–Or graphs (spatial, temporal, causal, and attributed), and stochastic context-sensitive grammars. All of these are reviewed in this book.

The last paradigm for knowledge representation in vision, to go along with logic and probabilistic modeling, is discriminative modeling. Discriminative models include neural networks (e.g., convolutional neural networks, recurrent neural networks), boosting, logistic regression, and support vector machines. Discriminative models often have different pipelines for each task in a problem, such as pipelines for object classification, pose estimation, and attribute recognition. This actually leads to non-unified knowledge representation, which is unhelpful for the aim of this book. So, although discriminative models can be used to represent knowledge, they are often used in such a way that does not contribute to a unified representation. Discriminative models also require a huge number of parameters (on the order of $O(10^7)$) and a very large amount of data for supervised training (on the order of $O(10^6)$). Generally, among the three paradigms of knowledge representation, discriminative models are very complex while logic is simple; probabilistic models lie somewhere in between, simple enough to understand thoroughly and complex enough to capture complex data distributions. A goal of this book is to integrate these three paradigms providing knowledge representation: logic models, probabilistic models, and discriminative models.

1.4 Pursuit of Probabilistic Models

So, broadly speaking, why are probabilistic or statistical models useful in computer vision, and what is the origin of these models? Some assume that statistical models and probability have a role in computer vision primarily due to the noise and distortion present in natural images. This is truly a misunderstanding. With the abundance of high-quality cameras that exist nowadays, there is rarely a considerable amount of noise or distortion in images anymore. Rather, probability and statistical models actually help capture more details in images, as opposed to helping represent fewer details, i.e., noise or distortion. Accordingly, statistical models, which provide an intrinsic representation of visual knowledge, actually help capture image *regularities*.

As discussed previously, a population of natural image patches can be viewed as a cloud of points in the image space. The distribution of the cloud of points can be described by a probability density function. Image patches of a certain pattern, such as a face, form a sub-population in this cloud of points. If one randomly samples an image patch from this face sub-population, then the density function one assigns to the random image patch, naturally, should be the density function of the sub-population from which it was sampled. In this way, each pattern, or concept, corresponds to a probability distribution defining a sub-population of the image space, and it can be defined by a probability density function and can be parametrized by a statistical model.

In the literature, Grenander [84] and Fu [68] pioneered using statistical models for various visual patterns. In the late 1980s and early 1990s, statistical models become popular and then indispensable when the computer vision community recognized that problems, typically shape-from-X problems, are intrinsically ill-posed. If interested, the reader is invited to explore the historical context more. Nowadays, it is known that extra information is needed to account for the regularities in natural images, and statistical models can help to encode or represent these regularities. Crucially, statistical models also assist in learning and recognizing patterns or objects in the first place.

Figure 1.10 illustrates two methods of pursuing a statistical model of a concept, i.e., of learning a concept f by approaching Ω_f with a sequence of models. In the first method, starting from the entire image space, at each step, a new constraint is added to shrink the image space. With more constraints continually added, Ω_f is captured from the outside. This method of pursuing a statistical model characterizes descriptive or declarative models, which are flat. Descriptive models represent one regime of probabilistic graphical models and include constraint satisfaction models, Markov random fields, Gibbs models, Julesz ensembles, and other contextual models. In the second method, starting from a single point or small ball inside Ω_f, at each step, some dimensions are expanded to gradually fill in Ω_f from the inside. This method of pursuing a statistical model characterizes generative or compositional models, which are hierarchical. Generative or compositional models represent another regime of probabilistic graphical models and include Markov trees, stochastic context-free grammars, and sparse coding models. Integrating contextual and hierarchical information is, e.g., primal sketch models, And–Or graphs, and stochastic context-sensitive grammars.

The reason for choosing one of the two model pursuit strategies, descriptive or generative, for a given problem is intuitive. Some sets may be of very high dimensionality, and hence it is more efficient to capture them with a descriptive model using constraints, gradually reducing the volume of the search space. These sets are modeled through the description, hence the name "descriptive" models. Other sets may be of much lower dimensionality, and hence it will be more efficient to capture them with a generative model using expansion, gradually increasing

$$q = p_0 \to p_1 \to \cdots \to p_k \quad to \quad f$$

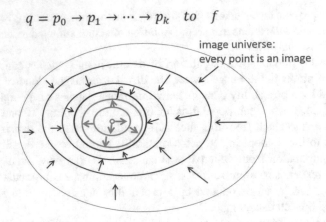

Fig. 1.10 This illustrates two strategies for pursuing a statistical model for a concept. First, in descriptive or declarative models, a set Ω_f can be pursued through a sequence of models that continually reduce the entire image space to capture Ω_f from the outside. The image space is gradually reduced by adding constraints. Second, in generative or compositional models, a set Ω_f can be pursued through a sequence of models that continually expand from a single point or small ball to gradually fill in Ω_f from the inside. In this case, the image space is gradually expanded by adding dimensions

the volume of the search space by adding dimensions. These sets are modeled by pursuing latent variables, which are meant to encode and explain training examples and may be used to generate new examples, hence the name "generative" models. Some sets may be of more middle dimensionality and hence require a more complex combination of a generative and descriptive model.

By analogy, if a teacher is grading a final exam that has a full score of 100, for a very strong student the teacher may start from 100 points and subtract points occasionally for incorrect answers. For a very weak student, the teacher may start from 0 and add points occasionally for correct answers. Accordingly, students at both ends of the spectrum are easy to grade, but students near the exam average require more work. Images for most vision problems are near the middle of an analogous spectrum, making it necessary to capture challenging sets that form complex, multimodal distributions of varying dimensionality.

Fig. 1.11 Scanning from left to right, low-dimension, mid-dimension, and high-dimension regimes may be observed for two images. There is a gradual transition from images that should be fully modeled by a generative model, on the very left, to images that should be fully modeled by a descriptive model, on the very right

Figure 1.11 illustrates use cases for the different regimes of models, using only two images, one of the red maple leaves and the other of green leaves. Both images are simply shown at five varying levels of magnification. On the very left of the figure, the low-dimension regime may be observed. As the images are of low dimension, it is easiest to generate or compose them through expansion with a generative model, increasing the volume of the search space by adding dimensions. On the very right of the figure, the high-dimension regime may be observed. As the images are of high dimension, it is easiest to describe them with a descriptive model, reducing the volume of the search space by introducing constraints. From the left to the right of the figure, there is a gradual transition from images that should be fully modeled by a generative model to images that should be fully modeled by a descriptive model. In between, a combination of the two should be used to optimally model the leaves. This is a powerful example because it shows that, even for the same picture but at varying levels of magnification, there is a need for generative and descriptive models and their combination.

Thus far, relatively small image patches have been discussed, e.g., 10×10 pixels. The concepts that arise from small image patches may be considered "atomic concepts," i.e., the simplest symbols, or descriptors, at the lowest layer of perception. Atomic concepts can be further composed into larger patterns and more abstract concepts. This leads to a hierarchy of concepts at multiple layers, as illustrated in Fig. 1.12.

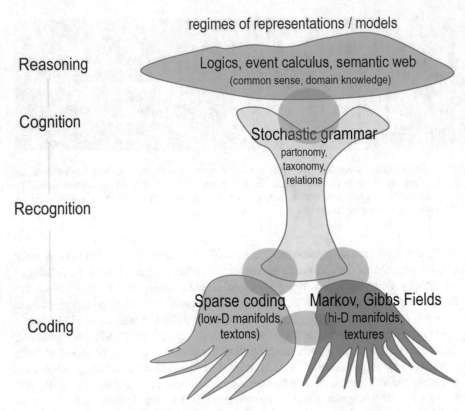

Fig. 1.12 Concepts become increasingly abstract, moving up through the hierarchy from coding to recognition to cognition to reasoning. However, concepts at all levels can be learned and described by Spatial–Temporal–Causal And–Or graphs

At the bottom layer, there are concepts for image patches. High-entropy concepts can be described by, e.g., Markov random fields, and low-entropy concepts can be described by, e.g., sparse coding models. At the top layer, there is reasoning based on logic and event calculus. In the middle layers, there are recognition and cognition, which can be described by stochastic grammar allowing for recursive compositions. As will be discussed, stochastic grammars in the form of Spatial–Temporal–Causal And–Or graphs are actually well suited to learning and describing concepts at all layers.

Chapter 2
Statistics of Natural Images

The population of natural images may be defined as all the images of the natural world that have been observed by humans. Images and videos captured by humans with cameras in modern civilization represent only a very small subset of this population. Accordingly, the size of the natural image population far exceeds the size of the set of images captured by cameras, although it is much smaller in comparison to the size of the entire, unconstrained image space. Natural images contain an overwhelming variety of structures and patterns resulting from a myriad of physical processes. They exhibit hierarchical compositions of objects at a broad continuum of scales.

It may be surprising to note that natural images, with their diverse array of patterns and orientations, consistently share any statistical features. Indeed, they do and so much so that they can be easily distinguished from non-natural images based solely on these features. For example, most natural images contain simple subcomponents such as flat surfaces, well-defined edges, and regular textures. The common appearance of these structures suggests that concise but flexible models can efficiently describe natural images at the most basic level by repeated instances of similar patterns.

Barlow in 1961 [10] and Gibson in 1966 [79] were among the earliest researchers to emphasize the role of ecology in visual perception, i.e., visual concepts are learned and mentally categorized based on examples in the environment. Computer vision, like human vision, consists of an agent that learns abstract representations from many observed examples. The learned representations of vision models, similar to concepts for humans, should be sufficiently generalizable so that the model can adapt to new, unseen examples just as a human can.

Throughout this book, statistical phenomena associated with natural images are discussed. This begins the process of establishing a common vocabulary and grammar for vision supported by a unified mathematical foundation. In this particular chapter, some initial statistical properties discovered for natural images are examined, such as the $1/f$-power law, high kurtosis, and scale invariance. Manifolds in the image universe, image scaling in the image space, and impercep-

© Springer Nature Switzerland AG 2023
S.-C. Zhu, Y. N. Wu, *Computer Vision*, https://doi.org/10.1007/978-3-030-96530-3_2

tibility are discussed. There are many motivations for understanding natural image statistics, but a few include potential optimizations in image/video compression due to redundancy reduction and improved image/video coding, an understanding of ecologic influences on neural receptive fields (e.g., how neurons in visual cortical areas *adapted* to visual environments over time), and the knowledge to exploit image regularities and prior models to solve ill-posed problems, such as image restoration (e.g., denoising, inpainting), estimating surface from stereo, motion, texture, etc., and the development of concepts and generative models (as prior knowledge) for scenes, objects, actions, events, and causality from images and video.

2.1 Image Space and Distribution

Beginning with a concrete description of image space, consider the population of all grayscale image patches of size 10×10 pixels with each pixel intensity in the range $[0, 1]$. Each possible image patch is a point in a high-dimensional space $[0, 1]^{100}$. This means that an image can be represented by a single vector with 100 elements for the 100 pixel intensities. If the images are color instead of grayscale, each of the three color channels (red, green, blue) has its own 10×10 image patch and, accordingly, the image space is $[0, 1]^{300}$. The population of natural image patches forms a cloud of points, which is only a subset of the entire image space. The distribution of this cloud of points exhibits a high-dimensional geometry reflecting the various structures in natural images (Fig. 2.1).

The distribution of natural images, i.e., the cloud of points, can be described by a probability density function, indicating the density of points at each position in the image space. The density at a position in the image space indicates how likely it is an image from that position would be chosen if it were sampled from the distribution of natural images. The density of the distribution is very uneven. In some positions of the image space, e.g., positions with image patches of forest scenes, the density is high, while in other positions, e.g., one with a random sample from a high-dimensional uniform distribution, the density of natural image distribution is virtually zero. Simply put, this is because, among natural images, image patches of a forest scene are relatively common, whereas image patches from a uniform distribution are less common. The concentration of the density function of natural image distribution around certain geometric regions is analogous to the distribution of mass in the universe. The density of mass is high at a star and low across nebulas, clouds of gas and dust in space. And the density is practically zero over large spans of space.

Now some notation should be introduced. Let $\mathbf{I}(x)$ be an image patch defined on a square or rectangular domain $\mathcal{D}$, where $x \in \mathcal{D}$ indexes the location within the image. The domain, or coordinate space, $\mathcal{D}$ for location x can be continuous or discrete. An example of a continuous domain is $\mathcal{D} = [0, 1] \times [0, 2]$, which denotes a rectangular grayscale image with continuous axes. The coordinates for location x are continuous, i.e., $x \in ([0, 1], [0, 2])$. An example of a discrete domain

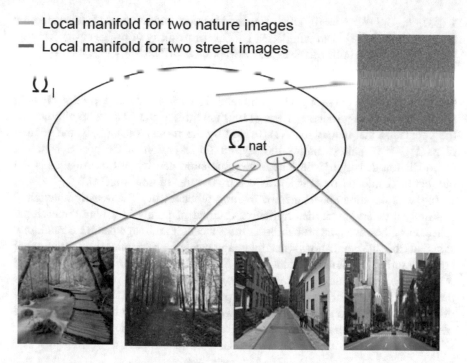

Fig. 2.1 The natural image population is a subset of the entire image space, i.e., $\Omega_{\text{nat}} \subset \Omega_{\text{I}}$. Lines are drawn from the four natural images below to their positions in Ω_{nat}, a high-dimensional space. Similar images lie on local manifolds of Ω_{nat}, separated by relatively small geodesic distances. The image of noise does not belong to the natural image space, Ω_{nat}

is $\mathcal{D} = \{1, \ldots, 10\} \times \{1, \ldots, 10\}$, which denotes a 10×10 pixel grayscale image with discrete axes. In this case, the image is composed of pixels. The coordinates for location x are discrete, i.e., $x \in (\{1, \ldots, 10\}, \{1, \ldots, 10\})$. The coordinate space $\mathcal{D}_{\text{color}}$ of a three-channel color image is simply the Cartesian product of the grayscale coordinate space $\mathcal{D}_{\text{gray}}$ and the set $\{1, 2, 3\}$ that indexes the color channel, i.e., $\mathcal{D}_{\text{color}} = \mathcal{D}_{\text{gray}} \times \{1, 2, 3\}$. $\mathbf{I}(x)$ may be treated as a function $\mathbf{I} : \mathcal{D} \to \mathcal{V}$, which gives the intensity values of each pixel in $\mathbf{I}$. For continuous pixel intensities between 0 and 1, pixel values $\mathcal{V} = [0, 1]$, and for discrete integer-valued pixel intensities between 0 and 255, pixel values $\mathcal{V} = \{0, \ldots, 255\}$.

When the coordinate space $\mathcal{D}$ is discrete and a finite set, i.e., for images with a finite number of pixels, $\mathbf{I}$ may be treated as a vector if an ordering for the pixels is fixed. As such, each $\mathbf{I}$ becomes a point in the image space $\Omega_{\mathbf{I}} = \mathcal{V}^{|\mathcal{D}|}$, in which $|\mathcal{D}|$ is the total number of pixels in the image. (For color images, each channel has its own separate pixels.) For example, every 10×10 pixel grayscale image patch $\mathbf{I}$, with pixel intensities between 0 and 1, is a point $\mathbf{I} \in \Omega_{\mathbf{I}} = [0, 1]^{100}$. (The notation $\Omega_{\mathbf{I}} = [0, 1]^{100}$ indicates that the sample space $\Omega_{\mathbf{I}}$ is made up of samples with 100 pixels that take a continuous value between 0 and 1.) Similarly, every 10×10 pixel color image, with pixel intensities for each channel between 0 and 1, is a point

$\mathbf{I} \in \Omega_{\mathbf{I}} = [0, 1]^{100 \times 3} = [0, 1]^{300}$. A more technical definition of $\Omega_{\mathbf{I}}$ is needed for the cases in which $\mathcal{D}$ is an infinite set, but for the majority of topics covered in this book, this discrete, finite, pixel-based description is sufficient.

Now consider the cloud of points $\Omega_{\text{nat}} \subset \Omega_{\mathbf{I}}$ made up of *natural* image patches. The distribution of Ω_{nat} can be described by a density function $f : \Omega_{\mathbf{I}} \to \mathbb{R}$, such that $f \geq 0$ and $\int_{\Omega_{\mathbf{I}}} f(\mathbf{I}) d\mathbf{I} = 1$. Intuitively, for each point $\mathbf{I} \in \Omega_{\mathbf{I}}$, let $\mathcal{N}(\mathbf{I})$ be a small neighborhood around $\mathbf{I}$. Then $f(\mathbf{I})$ is the limit of the proportion of points in the population that belong to $\mathcal{N}(\mathbf{I})$ divided by the volume of $\mathcal{N}(\mathbf{I})$, as the volume of $\mathcal{N}(\mathbf{I}) \to 0$. This is simply the asymptotic extension of the common notion of population density. For instance, the population density of Los Angeles is the number of people in Los Angeles divided by the area of Los Angeles.

In the explanation above, there is nothing probabilistic. The density function f is simply a deterministic description of a cloud of points in a high-dimensional space. Now, however, the probability is introduced. If an image patch $\mathbf{I}$ is randomly sampled from the population of natural image patches, i.e., a point from the cloud of points described by density function $f(\mathbf{I})$ is randomly sampled, then image patch $\mathbf{I}$ is said to be a random sample, for example, from $f(\mathbf{I})$. Accordingly, $f(\mathbf{I})$ becomes the density function for the random image patch $\mathbf{I}$, and $\mathbf{I} \sim f(\mathbf{I})$ is written. The goal of learning, then, is to gain some knowledge about the density function $f(\mathbf{I})$ based on a set of examples $\{\mathbf{I}_1, \ldots, \mathbf{I}_M\}$ that are sampled from $f(\mathbf{I})$ independently.

2.2   Information and Encoding

Recall that concepts (in the model space) are represented by, or correspond to, populations (in the image space). They can be defined by a distribution or density function f. A concept can be general, such as the concept of natural images, a certain texture, or white noise, or it can be more specific, such as the concept of the face of a tiger.

In high-dimensional image spaces, common to vision problems, a notable phenomenon called concentration of measure may occur. The density function $f(\mathbf{I})$ is close to zero outside of a typical set Ω_f. It is nearly uniform inside of the typical set Ω_f. So it can be roughly stated that $f(\mathbf{I}) = \frac{1}{|\Omega_f|}$ for $\mathbf{I} \in \Omega_f$ and $f(\mathbf{I}) = 0$ for $\mathbf{I} \notin \Omega_f$. The high-dimensional nature of the data in many vision problems makes it such that the image space $\Omega_{\mathbf{I}}$ is largely dominated by white noise. As a result, a nearly uniform density exists over $\Omega_{\mathbf{I}}$, as almost all of it is simply white noise. The goal of learning a concept, then, is to identify which set Ω_f, in the largely meaningless, noise-dominated set $\Omega_{\mathbf{I}}$, corresponds to the given concept, i.e., find a probability density function f defining it.

Now consider two models, an "unfocused" model $g(\mathbf{I})$ that has a uniform density within $\Omega_{\mathbf{I}}$ and a "focused" model $f(\mathbf{I})$ that has a uniform density within the subset of natural images Ω_{nat} and is approximately zero outside this set. The volume of the subset Ω_{nat} can be described using *entropy*. Entropy can be intuitively understood as a measure of disorder. Deterministic variables (with perfect order) possess zero entropy, while uniform random variables (with the perfect disorder) possess the maximum entropy of all possible distributions over the support of the uniform variable.

The fact that the volume of the subset Ω_{nat} can be described using entropy is an important point. It means the complexity of a concept f in the model space can ultimately be expressed by the entropy, i.e., the log volume, of its corresponding set Ω_f in the image space. The entropy of an image set, by giving the disorder of pixel information of that set, encodes the complexity of the corresponding concept. In computer vision, entropy and entropy rate are both measured in pixels.

The entropy H of a distribution $p(\mathbf{I})$ over a finite image space $\Omega_\mathbf{I}$ is

$$H(p) := - \sum_{\mathbf{I} \in \Omega_\mathbf{I}} p(\mathbf{I}) \log_2 p(\mathbf{I}). \tag{2.1}$$

The use of $\log_2$ gives a measurement of entropy in terms of bits, but, in general, any base can be used for the logarithm, as different bases only scale H differently. For a distribution q that is uniform over a set $\Omega_q \subseteq \Omega_\mathbf{I}$ and zero elsewhere, the entropy is

$$H(q) = - \sum_{\mathbf{I} \in \Omega_q} \frac{1}{|\Omega_q|} \log_2 \frac{1}{|\Omega_q|} = \log_2 |\Omega_q|, \tag{2.2}$$

in which the sum over $\Omega_\mathbf{I}$ reduces to a sum over Ω_q, as density is zero outside of Ω_q. Therefore, the entropy of a uniform distribution q is a measure of the log volume of the space Ω_q. In particular, for the "unfocused" model that is uniform over the entire image space $\Omega_\mathbf{I}$, $H(f) = \log_2 |\Omega_\mathbf{I}|$, and for the "focused" model that is uniform over only the natural image space Ω_{nat} and zero elsewhere, $H(g) = \log_2 |\Omega_{nat}|$. Later chapters describe a more detailed relationship between entropy and learning using information theory. For now, it is only important to note the connection between entropy and the volume of image space.

The entropy of a subset of the image space encodes the amount of information needed to represent members of the subset. Consider the space Ω_h of grayscale images that have the same fixed-width horizontal black bar against a white background. Each image can be described by a single number that represents the vertical position of the bar. Suppose that there are $|\Omega_h| = n_h$ possible positions for the horizontal bar. Now consider the space Ω_{hv} of grayscale images that have the same fixed-width horizontal black bar and the same fixed-width vertical black bar against a white background. All images from this set can be described by exactly two numbers representing the locations of the two bars. Suppose that there are n_h positions for the horizontal bar and n_v positions for the vertical bar, such that $|\Omega_{hv}| = n_h n_v$. The amount of encoding information needed in the space of vertical and horizontal bars is higher than that of horizontal bars alone, and this is reflected in the difference $\log_2 |\Omega_h| - \log_2 |\Omega_{hv}| = -\log_2 n_v < 0$.

In the case of images with $|\mathcal{D}|$ pixels, e.g., $|\mathcal{D}| = n^2$ for a grayscale image with n pixels for both its height and width, the *entropy rate* $\bar{H}$ scales entropy relative to the image size, lending a per-pixel entropy. For a density q uniformly distributed over Ω_q, the entropy is

$$\bar{H}(q) = \frac{H(q)}{|\mathcal{D}|} = \frac{1}{|\mathcal{D}|} \log_2 |\Omega_q|. \qquad (2.3)$$

For example, consider the space of $n \times n$ grayscale images where $\mathcal{V} = \{0, \ldots, 255\}$, indicating each pixel value is an integer from 0 to 255. In this case, there is the image space $\Omega_{\mathbf{I}} = \{0, \ldots, 255\}^{n^2}$ and the log volume of the image space $|\Omega_{\mathbf{I}}| = 256^{n^2}$. Therefore, the unfocused model $g(\mathbf{I})$ with uniform density in $\Omega_{\mathbf{I}}$ has entropy rate

$$\bar{H}(g) = \frac{1}{n^2} \log_2 |\Omega_{\mathbf{I}}| = \frac{n^2 \log_2 256}{n^2} = 8 \text{ bits/pixel}. \qquad (2.4)$$

This means that an image randomly sampled from $\Omega_{\mathbf{I}}$ can be encoded using an average of 8 bits per pixel.

Now, considering a natural image $I \in \Omega_{\text{nat}}$ sampled from f, the many regularities found in natural images allow for a drastic reduction in the number of bits needed for encoding the image. The empirical upper bound for the average bits per pixel, i.e., the entropy rate, needed to encode natural images is 0.3:

$$\bar{H}(f) = \frac{1}{|\mathcal{D}|} \log_2 |\Omega_{\text{nat}}| \leq 0.3 \text{ bits/pixel}. \qquad (2.5)$$

For the unfocused model $g(\mathbf{I})$ with uniform density in $\Omega_{\mathbf{I}}$, 8 bits per pixel are needed to encode any potential image that could be drawn from $\Omega_{\mathbf{I}}$; it is not possible to compress uniform random samples drawn from the entire image space $\Omega_{\mathbf{I}}$. However, for the unfocused model $f(\mathbf{I})$ with uniform density within Ω_{nat} and zero density elsewhere, the reduction in entropy rate from 8 bits/pixel to 0.3 bits/pixel allows for compression. This decreased entropy rate of 0.3 bits/pixel gives an intuitive measure for the amount of compression possible when the regular features found across previously observed natural images are used to encode new images.

Here the first statistical observation for natural images may be noted: redundancy. The redundancy in real-world images allows for continual improvements to image/video compression. Instead of needing 8 bits to encode a pixel, only an upper bound of 0.3 bits is needed. Using any more bits than this to encode a pixel would not take advantage of all the redundancies in natural images. Note that compression methods work well on natural images but poorly, for example, on a noisy image such as a QR code (a matrix barcode).

Now a rough estimate of the size of the observed natural image space Ω_{nat} may be given. Assume there are 10 billion people who have lived on the Earth, and each person lived or will live to be 100 years old, and each person observed 20 images per second. The volume of the natural image space, $|\Omega_{\text{nat}}|$, as seen by humans, may be estimated as

$$|\Omega_{\text{nat}}| \approx 10^7 \text{humans} \times 100 \text{years} \times 365 \text{days} \times 24 \text{hours} \times 3600 \text{sec} \times 20 \text{fps}$$

$$\approx 6.3 \times 10^{22} \text{images} \tag{2.6}$$

The size of the total image universe $|\Omega_I| = 256^{n^2}$ discussed previously is much larger than $|\Omega_{\text{nat}}|$, even for very small image spaces, e.g., image patches with $n = 10$ pixels for both height and width. Keep in mind that the models in this book are meant to represent samples from the natural image space Ω_{nat} and reflect the regularities observed therein.

2.3 Image Statistics and Power Law

The universe is composed of an ensemble of structures at many different scales. Using a coarse scale, a structure may be perceived as an atomic entity, while using a finer scale, the same structure may be perceived as a compound entity with many different parts. Structured ensembles that can be viewed as a singular entity are commonly referred to as objects. In natural images, objects can be identified by regularities (such as consistent color or texture or connected surfaces) that enable the perception of them as distinct concepts. The regularities observed in natural images reflect the hierarchically structured order of the universe. Noise images, on the other hand, do not obey such laws. As a result, natural images and noise can be distinguished based on numerical summaries of structural regularities. These statistical properties form the foundation for a more detailed understanding of images pursued in later chapters.

In natural images, nearby pixel intensities are typically similar in value, but occasional jumps or discontinuities can be observed as well. As two neighboring cities are not usually expected to have a large difference in elevation, two neighboring pixels are not usually expected to have a large difference in intensity, i.e., pixel values, or two neighboring patches to have a large difference in smoothness. A simple measure of similarity between the intensities of two pixels is covariance or correlation. For an image $\mathbf{I} \in \mathcal{V}^{|\mathcal{D}|} \subset \mathbb{R}^{|\mathcal{D}|}$ with $|\mathcal{D}|$ pixels, assume that the marginal mean is normalized to be 0 and the marginal variance to be 1, i.e., $\frac{1}{|\mathcal{D}|} \sum_x \mathbf{I}(x) = 0$ and $\frac{1}{|\mathcal{D}|} \sum_x \mathbf{I}(x)^2 = 1$. Then the covariance between pixel $x = (x_1, x_2)$ and $x + d = (x_1 + d_1, x_2 + d_2)$ can be calculated as $C(d) = \frac{1}{|\mathcal{D}|} \sum_x \mathbf{I}(x)\mathbf{I}(x + d)$. The covariance $C(d)$ is a function of the distance between pixels, d, and can be large when $|d| = \sqrt{d_1^2 + d_2^2}$ is small and decays to 0 as $|d|$ becomes large. An equivalent measure of the similarity between the intensities of two pixels is the power spectrum, or power spectral density, of the image in the Fourier domain. The power spectrum of the image, $A^2(f)$, is simply the Fourier transform of the covariance function

$C(d)$, so the covariance and the power spectrum are simply two ways of capturing the same statistical property: the second-order moment of the image distribution.

The power spectrum $A^2(f)$ is the square of the amplitude spectrum $A(f)$. The power spectrum provides a way to capture and represent these differences. It describes how relations between image pixels are distributed for different frequencies, i.e., how the image content fluctuates at different frequencies or scales. In general, the power spectrum describes the distribution of power over frequency components composing a given signal. As the relevant application is vision, the signal is an image and the frequency is spatial, not temporal. Spatial frequency is periodic across positions in space and a measure of how often sinusoidal components (as determined by the Fourier transform) of the structure repeat per unit of distance, in this case, pixels. According to the Fourier analysis, any physical signal can be composed of a number of discrete frequencies or a spectrum of frequencies over a continuous range. Accordingly, images may be composed of a number of discrete spatial frequencies. The statistical average of a signal as analyzed in terms of its frequency content is called its spectrum, and summation or integration of the spectral components yields the total power (for a physical process) or variance (for a statistical process).

As mentioned previously, it may be surprising to note that natural images, with their diverse patterns and orientations, share any consistent statistical features at all, but indeed they do. Natural images can easily be distinguished from non-natural images, such as images of random patterns, by their amplitude spectra or power spectra. In images of random patterns, such as white noise, the amplitude spectra are flat. Natural images, however, display the greatest amplitude at low frequencies (at the center of the plot in Fig. 2.2) and decreasing amplitude as the frequency increases, regardless of the orientation of the image, as depicted in Fig. 2.2.

The Fourier transform decomposes a function of time, i.e., a signal, in the time domain, into its constituent temporal frequencies in the frequency domain. In this case, it decomposes a function of space, i.e., an image, in the spatial domain, into its constituent spatial frequencies in the frequency domain. It is a linear transform that projects an image, in the form of a vector, onto an eigenvector space, transforming the image from a Euclidean basis to a Fourier basis. In many cases with discrete signals, the Fourier transform is invertible, and no data is lost if the inverse Fourier Transform is used to recover the original signal. Even when the original signal is real-valued, frequency signals contain both real and imaginary components, so the absolute value is used to represent the total power or energy of a signal. A Fast Fourier Transform (FFT) may be performed on an image $I \in R^{|\mathcal{D}_1| \times |\mathcal{D}_2|}$ using

$$\hat{\mathbf{I}}(\xi, \eta) = \sum_{(x,y) \in \mathcal{D}_1 \times \mathcal{D}_2} \mathbf{I}(x, y) e^{-i2\pi(\frac{x\xi}{H} + \frac{y\eta}{W})}, \qquad (2.7)$$

in which $|\mathcal{D}_1| = H$ is the number of pixel rows, $|\mathcal{D}_2| = W$ is the number of pixel columns, and (ξ, η) represents the horizontal and vertical frequencies. The absolute value of the resulting complex number

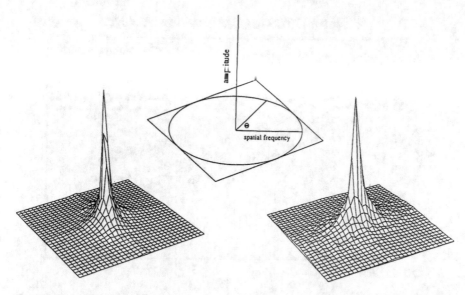

Fig. 2.2 From Field [54], two natural images display similar amplitude spectra. The center of the plots represents 0 spatial frequency, and frequency increases as a function of distance from the center. Orientation is represented by the angle from the horizontal. Note how amplitude decreases sharply with increasing frequency at all orientations. To be precise, the amplitude decreases by a factor of $1/f$, and power decreases by a factor of $1/f^2$, in which f is frequency [54]

$$A(\xi, \eta) = |\hat{\mathbf{I}}(\xi, \eta)| \tag{2.8}$$

is the Fourier amplitude, which gives the magnitude of the frequency (ξ, η) within the image signal $\mathbf{I}$. The Fourier power $A^2(\xi, \eta)$ is the square of the Fourier amplitude $A(\xi, \eta)$. Intuitively, the Fourier power, for given horizontal and vertical frequencies (ξ, η), encodes how much of the image vector is projected onto these frequencies at a given orientation.

An interesting empirical observation for natural images is that, for all frequency $f = \sqrt{\xi^2 + \eta^2}$,

$$A(f) \propto 1/f, \tag{2.9}$$

$$\log A(f) = \text{constant} - \log f. \tag{2.10}$$

Thus, the amplitude A of a Fourier coefficient is inversely related to the frequency f. The inverse relationship between amplitude and frequency is referred to as the inverse power law for natural images. Figure 2.3 plots the $\log A(f)$ over $\log f$ for the four natural images from Fig. 2.1, and the curves can be fitted by straight lines with a slope close to -1, showing the inverse relationship. Not all natural images

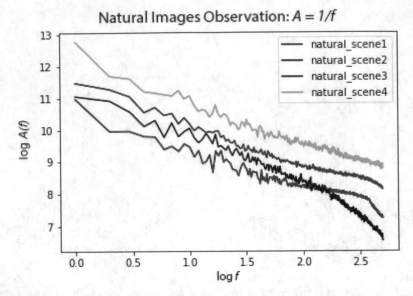

Fig. 2.3 Note the log of the amplitude, $\log A(f)$, for a given log of the frequency, $\log f$, for the four natural images from Fig. 2.1. In natural images, it is observed that the amplitude $A(f)$ is inversely related to the frequency f, i.e., $A(f) = 1/f$. Accordingly, the power $A^2(f) = 1/f^2$. This property for natural images is referred to as the inverse power law

exhibit the inverse power law, e.g., natural images of fields of grass or the night sky, but this can be expected for most natural images [54].

The inverse relationship between amplitude and frequency may be used to calculate the total Fourier power $A^2(f)$ in the frequency band, or octave, $[f, 2f]$:

$$\iint_{f^2 \leq \xi^2 + \eta^2 \leq (2f)^2} |\hat{\mathbf{I}}(\xi, \eta)|^2 d\xi d\eta = 2\pi \int_{f^2}^{4f^2} \frac{1}{f^2} df^2 = \text{constant}, \quad \forall f.$$

(2.11)

As depicted in Fig. 2.4, as frequency increases, and hence the total area within each frequency band increases, amplitude decreases. As a result, the power remains constant overall, i.e., the power for a given frequency band $[f, 2f]$ is equal to the power for frequency bands $[2f, 4f]$, $[4f, 8f]$, and so forth. This is what is communicated by Eq. (2.11). This observation that natural images contain equal power across frequency bands reveals that they are scale-invariant, lending the same power independent of frequency f, or scale, i.e., the viewing distance.

The second-order properties of the image distribution, captured by the covariance $C(d)$ and the power spectrum $A^2(f)$, can be fully reproduced by a multivariate Gaussian distribution with a variance–covariance matrix that agrees with $C(d)$ or $A^2(f)$. For instance, a Gaussian model that accounts for the inverse power law is surprisingly simple. It was shown by Mumford [175] that a Gaussian Markov Random Field (GMRF) model, as in Eq. (2.12), has exactly $1/f$-Fourier amplitude:

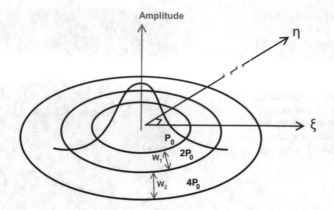

Fig. 2.4 Because amplitude decreases as the frequency (and hence the total area within each frequency band) increases, natural images effectively exhibit constant power within each band such as ω_2. η and ξ represent the vertical and horizontal frequencies, respectively. $2P_o$ indicates a doubling of the frequency of the original image or image distribution, P_o

$$p_{1/f}(\mathbf{I}; \beta) = \frac{1}{Z} \exp \left\{ -\sum_x \beta |\nabla \mathbf{I}(x)|^2 \right\}, \tag{2.12}$$

in which $|\nabla \mathbf{I}(x)|^2 = (\nabla_x \mathbf{I}(x))^2 + (\nabla_y \mathbf{I}(x))^2$, and

$$\nabla_x = \mathbf{I}(x_1 + 1, x_2) - \mathbf{I}(x_1, x_2) \quad \text{and} \quad \nabla_y = \mathbf{I}(x_1, x_2 + 1) - \mathbf{I}(x_1, x_2) \tag{2.13}$$

are discrete approximations of the gradients.

Since $p_{1/f}(\mathbf{I}; \beta)$ is a Gaussian model, one can easily draw a random sample $\mathbf{I} \sim p_{1/f}(\mathbf{I}; \beta)$. Figure 2.5 shows a typical sample image by Mumford [175]. This model matches the local regularity of natural image statistics and nothing else. It can be considered a natural image model with one constraint, making it only slightly better than a random noise model. Large values of the parameter β lead to samples with large regions of similar intensity, and small values of β lead to samples with more variation in nearby pixels. The limiting case $\beta \to \infty$ concentrates all mass on the zero image $\mathbf{I} = 0$. From this, it can be concluded that although certain statistics like the covariance and power spectrum found in natural images are also captured by this model, these statistics clearly do not contain sufficient complexity to capture all key features needed for natural image modeling and inference.

The power spectrum can also be pooled by Gabor filters centered at different frequencies. Gabor filters are sine and cosine waves multiplied by elongate Gaussian functions, and the filter response is a localized Fourier transform. Let F be a Gabor filter whose sine and cosine waves have frequency ω. Let $[F * \mathbf{I}](x)$ be the Gabor filter response at location x. Then, $[F * \mathbf{I}](x)$ measures the frequency content of $\mathbf{I}$ around frequency ω at location x. Due to the spatial localization of filter F, F

Fig. 2.5 Observe a random
sample from the Gaussian
Markov Random Field model.
Note that it has very little
structure, a large number of
edges, and minimal flatness in
its energy landscape. Clearly,
second-order properties such
as covariance and spectrum
do not sufficiently capture key
features of natural images

also responds to sine and cosine waves whose frequencies are close to frequency ω. Thus, if $\sum_x |F * \mathbf{I}|^2 / |\mathcal{D}|$ is pooled, it will be the average of the power spectrum within the band of frequencies to which filter F responds. By the inverse power law, $\sum_x |F * \mathbf{I}|^2 / |\mathcal{D}|$ remains scale-invariant, i.e., it is constant for different frequencies ω.

2.4   Kurtosis and Sparsity

As mentioned above, second-order properties such as the covariance, power spectrum, or average of the squared filter responses can be fully reproduced by a multivariate Gaussian distribution with a matching variance–covariance matrix. The Gaussian distribution can be thought of as a cloud of points, which forms an ellipsoid shape. The ellipsoid-shaped Gaussian distribution is rather dull. It cannot be expected to capture the presumably highly complex shape of the natural image distribution. For example, the distribution of natural images has many low-dimensional spikes, as illustrated in Fig. 2.6. As a result, researchers have put immense effort into finding patterns of deviation from the Gaussian distribution, attempting to pinpoint non-Gaussian features. The study of natural image statistics has leveraged many properties, from covariances to histograms of filter responses, e.g., Gabor filters. While covariances only measure second-order moments, histograms of filter responses include higher order information such as skewness and kurtosis.

To define skewness and kurtosis, let X denote a random variable with mean μ and variance σ^2, such that X can be normalized by the transform $(X - \mu)/\sigma$. Skewness is the third-order statistical moment $\mathrm{E}[((X - \mu)/\sigma)^3]$, and it measures the asymmetry of the distribution. Kurtosis is the fourth-order statistical moment

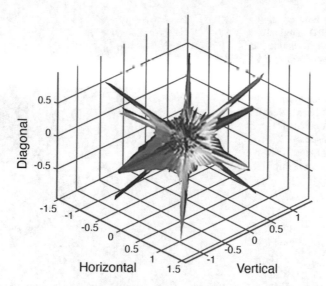

Fig. 2.6 The distribution of natural images has many low-dimensional spikes

$E[((X - \mu)/\sigma)^4]$, and a histogram with heavy tails demonstrates high kurtosis. In measuring skewness and kurtosis of a random variable, the first-order moment μ and second-order moment σ^2 only serve to normalize the random variable, making its mean equal to zero and variance equal to one.

With histograms, the idea of a 1D marginal projection is utilized in order to gain an understanding of a high-dimensional space such as Ω_{nat}. A histogram can be regarded as a 1D marginal projection because it contains statistics in one dimension, i.e., a line. Histograms allow peaking into a high-dimensional space Ω_f and learning about it by learning many 1D marginal densities of $f(\mathbf{I})$. By learning these densities, the idea is that a probability density could be constructed to represent Ω_f in terms of all of these learned 1D marginal densities of $f(\mathbf{I})$.

The use of histograms leads to another statistical observation about natural images: they exhibit high kurtosis. The histograms of Gabor filter responses to natural images, found by Eq. (2.14), are highly kurtotic, i.e., heavy-tailed [55]. Here, F is a Gabor filter and $f_n(\mathbf{I})$ is a "focused" model, which, recall, has zero density outside of the space of natural images Ω_{nat}. The fact that the histograms have tails heavier than a Gaussian distribution reveals that natural images have high-order, non-Gaussian structures. Gabor filters are primarily used to detect orientation, but other filters are also used such as gradient filters, Haar filters, and so forth.

The histogram of the filter response is calculated by the following equation:

$$h(a) = \int_{\Omega_{\mathbf{I}}} f(\mathbf{I})\delta(\langle F, \mathbf{I}\rangle - a)d\mathbf{I}, \tag{2.14}$$

Fig. 2.7 Note how the histograms or 1D marginal projections, which capture the filter responses of a filter F to a natural image, display high kurtosis. That is, they have a heavy-tailed structure, exhibiting great extremities of deviations or outliers. An important statistical observation of natural images is that they have a highly kurtotic structure

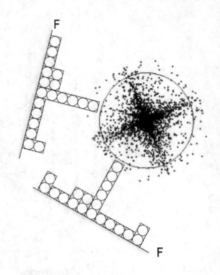

where $\delta(r)$ is the Dirac δ function. The histograms defined by Eq. (2.14) can also be visualized, as shown in Fig. 2.7. These histograms represent 1D statistical properties for an image **I**, located in $\Omega_{\text{nat}} \subset \Omega_{\text{I}}$, a high-dimensional space. F is a local filter, one that has different responses to different regions of image I that is represented by the points; in linear algebra, the term "base" is often used instead of "filter." The inner product $\langle F, \mathbf{I} \rangle$ is zero when $F \perp \mathbf{I}$. The two filter responses illustrate the binning operation performed in Eq. (2.14). High kurtosis is observed due to the vast amount of orthogonality between filter F and image **I**. Remember that kurtosis is associated with the tails of the distribution and not its peak. Accordingly, it is the greater extremity of deviations, or outliers, displayed in the histograms in Fig. 2.7, not the high peaks of these histograms, that portray the high kurtosis of the natural image. Any filter F response would show the same highly kurtotic structure (Fig. 2.8).

The high kurtosis in natural images is only marginal evidence for hidden structures in natural scenes. A direct way to discover structures and reduce image redundancy is to transform an image into a superposition of image components. This can be done, for example, with Fourier transforms, wavelet transforms [166], various image pyramids [218] for generic images, and principal component analysis (PCA) for particular ensembles of images, such as face images. The transformation from image pixels to a linear basis, such as a Fourier basis, wavelets basis, or PCA basis, achieves two desirable properties. First, the transformation induces variable decoupling; coefficients of these bases are less correlated or become independent in an ideal case. Second, the transformation induces dimension reduction; the number of basis vectors needed to approximately reconstruct an image is often much smaller than the number of pixels.

If one treats an image $\mathbf{I}(x)$ as a function defined on the domain $\mathcal{D}$, then one may also use harmonic analysis to perform image dimension reduction. Harmonic

Fig. 2.8 A natural image scaled through 1×, 2×, 4×, and 8× downsampling can be observed. Although the fixed-size filter kernel (shown in red) captures more information for each downsampled scale, the image statistics remain invariant

analysis decomposes various classes of functions (i.e., mathematical spaces) or signals into a superposition of basic waves or basis systems. But the population of natural images cannot be fully captured by such functional classes and so a better solution is needed for image decomposition. From this, an inspiring idea came about sparse coding with an over-complete basis, or dictionary, introduced by Olshausen and Field in 1996 [190]. Sparse coding algorithms learn useful sparse representations of data. Given a certain number of dimensions, they learn an over-complete basis to represent the data. An over-complete basis signifies that there is redundancy in the basis, as basis vectors "compete" to represent data more efficiently. This means not all dimensions are needed to represent a data point; some may be set to 0. With an over-complete basis, an image may be reconstructed by a small, i.e., sparse, number of basis vectors in the dictionary. Olshausen and Field then learned an over-complete dictionary from many natural images. This often leads to 10 to 100 folds of potential dimension reduction. For example, an image of 200 × 200 pixels can be approximately reconstructed by roughly 100–500 base images.

The idea that natural images may be represented with sparse coding reflects the nature of the point cloud formed by the population of natural images, which has many low-dimensional spikes, as illustrated by Fig. 2.6. Indeed, as we will see in the following, high kurtosis motivates sparse representations.

2.5   Scale Invariance

In addition to the heavy-tailed histograms of Gabor filter responses, another interesting observation reported by Ruderman (94) [210] and Zhu and Mumford (97) [279] is that marginal histograms of gradient-filtered images are consistent over a range of scales (see Fig. 2.9). This is yet another statistical observation for natural images: scale invariance. Specifically, for a natural image $\mathbf{I}$, a pyramid with a number of n scales can be built, such that $\mathbf{I} = \mathbf{I}^{(0)}, \mathbf{I}^{(1)}, \ldots, \mathbf{I}^{(n)}$, and $\mathbf{I}^{(s+1)}$ is obtained by averaging every block of 2×2 pixels in $\mathbf{I}^{(s)}$. The histograms of gradients $\nabla_1 \mathbf{I}^{(s)}(x)$ for natural images (first two plots) and for Gaussian noise (last plot) are shown in Fig. 2.9, for three scales $s = 1, 2, 3$ shown in Fig. 2.8. Note that, for natural images, despite scaling down the image, roughly the same amount of kurtosis is observed in the histograms for three different scales (Fig. 2.9).

To better illustrate the scale invariance property of natural images, here a toy example is presented: a 2D invariant world consisting of only 1D line segments. Figure 2.10a shows an image of the simulated world. In the image, each line segment is determined uniquely by its center location (x_i, y_i), orientation θ_i, and length r_i. The lines are independently distributed in terms of geometric features. The center of a line is selected from the image plane uniformly. Its orientation, measured as the angle formed with the horizontal line, is also uniformly distributed over $[0, \pi]$. The length of a line follows a cubic probability density function, i.e., $p(r) \propto \frac{1}{r^3}$. In reality, r_i can be sampled from $p(r)$ using inverse transform sampling. While the lines are i.i.d. (independent and identically distributed) in geometric features, the overall density of the segments is controlled by a Poisson distribution. That is, in each unit area, the number of line segments has a constant mean. The toy world can be constructed by the above rules, and it can be observed that it has a scale-invariant property similar to that of natural images. To be more specific, Fig. 2.10a is of size 1024×1024 pixels. Figure 2.10b and c is obtained by down-sampling the original image to 512×512 pixels and 256×256 pixels, respectively. Notice that in down-sampling, the long lines are truncated and the lines shorter than a pixel are discarded. Then, 128×128 pixel patches can be cropped from these three images. Indeed, the

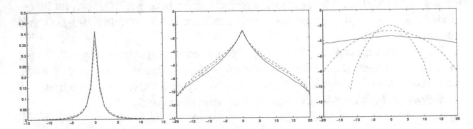

Fig. 2.9 First two plots: histograms of gradient-filtered natural image from Fig. 2.8 at three different scales. Last plot: log of the histograms of gradient-filtered Gaussian noise image at the same three scales. From this, it can be observed that high kurtosis is a property of natural images and not, e.g., Gaussian noise, and the level of kurtosis persists over different scales

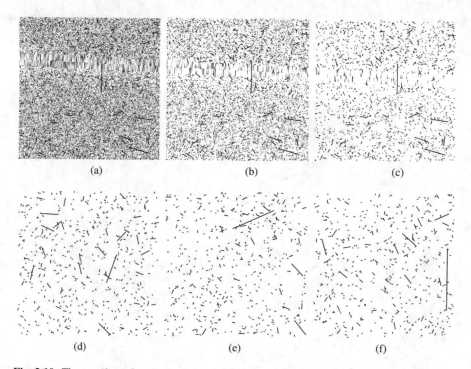

Fig. 2.10 The top three figures represent the 2D toy world with different pixel sizes, and the bottom ones are 128 × 128 pixel crops from the corresponding figures. (**a**) 1024 × 1024. (**b**) 512 × 512. (**c**) 256 × 256. (**d**) Crop from 1024 × 1024 world. (**e**) Crop from 512 × 512 world. (**f**) Crop from 256 × 256 world

three crops are identical to one another and hard to tell apart. This shows that the 2D toy world is scale-invariant in the sense that the image features are identical across scales. In fact, parallels can be drawn between the geometric features of the line segments and the gradient histograms of natural images. These statistics remain the same no matter at which scales the images are viewed.

Chapter 3
Textures

Texture is an important characteristic of the appearance of objects in natural scenes and is a crucial cue in visual perception. It plays an important role in computer vision, graphics, and image encoding. Understanding texture is an essential part of understanding human vision.

Texture analysis and synthesis has been an active research area, and a large number of methods have been proposed, with different objectives or assumptions about the underlying texture formation processes. In computer vision and psychology, instead of modeling specific texture formation processes, the goal is to search for a general model which should be able to describe a wide variety of textures in a common framework and which should also be consistent with the psychophysical and physiological understanding of human texture perception.

3.1 Julesz Quest

Imagine a scenario in which you walk into a store to purchase marble tiles for your new home. You search for the most consistent pattern, so your new floor looks uniform and the pieces are indistinguishable from one another. Suddenly, you notice a piece of marble, as shown in Fig. 3.1a,b, whose patterns you appreciate. However, as you continue inspecting this piece of marble, you suddenly catch attention of an area such as depicted in Fig. 3.1c. You ask yourself the question, "Is the texture in this marble patch consistent with the rest of the marble?" "What features of a texture make it distinguishable from another texture?" These questions may seem ostensibly simple to answer, but it is fundamentally difficult to exactly define the distinguishing features that set textures apart from one another.

Differentiating differences in texture patterns seems an intuitive and easy task for humans, but why are we able to differentiate the textures so easily? "The defects in the texture," one may answer, but how are the raw visual signals converted so that human brains can easily distinguish the defective patterns? How are texture

© Springer Nature Switzerland AG 2023
S.-C. Zhu, Y. N. Wu, *Computer Vision*, https://doi.org/10.1007/978-3-030-96530-3_3

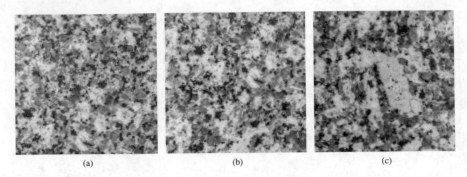

(a)   (b)   (c)

Fig. 3.1 (**a**) Presents a section of the marble you observe. (**b**) Presents another section of the same marble piece. (**c**) Is a defective area of the marble

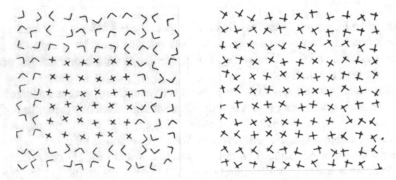

Fig. 3.2 Julesz texture pairs: the texture pattern formed by the + signs pops out from the surrounding pattern in the left panel, but it merges with the surrounding pattern in the right panel

patterns represented in human brains in general? This is a question that troubled psychologists for many years until Julesz initiated formal research on texture modeling.

In his seminal paper in 1962 [123], Julesz conducted research on textures and asked the following fundamental question: *What features and statistics are characteristic of a texture pattern, so that texture pairs that share the same features and statistics cannot be told apart by pre-attentive human visual perception?*

As illustrated in Fig. 3.2, the texture pattern formed by the + signs pops out from the surrounding pattern at first glance (so-called pre-attentive vision) in the left panel, but it merges with the surrounding pattern in the right panel. We have to examine the right panel attentively in order to tell the two texture patterns apart.

The first general texture model was proposed by Julesz in the 1960s. Julesz suggested that texture perception might be explained by extracting the so-called "k-th order" statistics, i.e., the co-occurrence statistics for intensities at k-tuples of pixels [123]. A key drawback of this model is that, on the one hand, the amount of data contained in the k-th order statistics is big and thus very hard to handle when

$k > 2$. On the other hand, psychophysical experiments show that the human visual system does extract at least some statistics of order higher than 2 [41].

In mathematical terms, a set of all texture images with the same features and statistics can be written as

$$\Omega_\mathbf{I} = \{\mathbf{I} : \mathbf{H}_i(\mathbf{I}) = \mathbf{h}_i, i = 1, \ldots, K\}, \tag{3.1}$$

where $\mathbf{I}$ is a texture image, each $\mathbf{H}_i$ is a chosen feature statistics, and K is the number of features chosen to be extracted such that texture patterns become indistinguishable when all features are extracted and matched. The search for the $\mathbf{H}_i$ has gone a long way beyond Julesz's statistical conjecture. Methods employed include co-occurrence matrices, run-length statistics, sizes and orientations of various textons, cliques in Markov Random Fields, and dozens of other measures. All these features have rather limited expressive power.

Here, it is worth noting that the modeling of texture not only concerns the data and distributions in the image space but also depends on human perception influenced by task or purpose. In Julesz's quest, the criterion for judging whether two texture images belong to the same category depends on the human visual system, which is trained for various tasks. Julesz carefully reduces task dependence by testing in a pre-attentive stage with no specific purpose; nevertheless, the human visual system has been trained ahead of the experiments. Later experiments show that human vision can learn and adapt to tell apart texture images after long training. This task dependence sets apart models in vision from those in physics.

In the subsequent chapters, let us illuminate the quest for features and statistics describing texture patterns.

3.2   Markov Random Fields

One approach for pursuing texture features is statistical modeling, which characterizes texture images as arising from probability distributions or random fields [30]. These modeling approaches involve only a small number of parameters and thus provide concise representation for textures. More importantly, texture analysis can be posed as a well-defined statistical inference problem. These statistical approaches enable us to not only infer the parameters of underlying probability models of texture images but also synthesize texture images by sampling from these models. Checking whether the synthesized images have similar visual appearances to the textures being modeled provides a way to test the model.

An issue, however, is that many of these statistical models, such as Markov random fields and clique-based Gibbs models, are too simple and thus suffer from a lack of expressive power to capture the fidelity of natural images.

Markov Random Field (MRF)

Markov Random Field (MRF) models were popularized by Besag in 1974 [19] for modeling spatial interactions on lattice systems and were used by Cross and Jain in 1983 [30] for texture modeling. An important characteristic of MRF modeling is that global patterns are formed via stochastic propagation of local interactions, which is particularly appropriate for modeling textures as they are characterized by global but not predictable repetitions of similar local structures.

Concretely, a Markov random field is a probability distribution over random variables $X_1, \ldots, X_n$ defined by an undirected graph $G = (V, E)$, where the nodes in V correspond to the variables X_i. Before illustrating the special properties of MRF models, we first introduce some notation to formalize graphs. In an undirected graph G, two nodes s and t are called neighbors if there is an edge between them. The neighborhood of a site s is defined as the set of all its neighbors, i.e., $\mathcal{N}_s = \{t : (s, t) \in E, t \in V\}$. Moreover, the neighborhood system of G is the set of all neighborhoods in the graph. We denote it as $\mathcal{N} = \{\mathcal{N}_s : s \in V\}$. For any patch A in graph G, define X_A as a window of observation. The boundary of patch A is thus $\mathcal{N}_A = \{t : (s, t) \in E, s \in A, t \notin A\}$. Also, we call $C \subset V$ a clique of G if for any two sites $s, t \in C$, we have $(s, t) \in E$. The depiction on the right-hand side of Fig. 3.3 shows a simple general MRF, in which, for example, the neighbors of node B are C, D, and E, and nodes B, D, and E form a three-node clique. A distinct feature of MRF models is that the random variables defined on the nodes satisfy the local Markov properties. That is, each variable X_i is conditionally independent of all other variables given its neighbors. Therefore, MRF can be used to model a set of local characteristics.

Based on the qualities of MRF models, we can consider a texture as a realization from a random field $\mathbf{I}$ defined over a spatial configuration $\mathcal{D}$. For example, $\mathcal{D}$ can be an array or a lattice. We denote $\mathbf{I}_s$ as the random variable at a location $s \in \mathcal{D}$, and let $\mathcal{N} = \{\mathcal{N}_s, s \in \mathcal{D}\}$ be the neighborhood system of $\mathcal{D}$ satisfying: (1) $s \notin \mathcal{N}_s$ and (2) $s \in \mathcal{N}_t \Longleftrightarrow t \in \mathcal{N}_s$. The pixels in $\mathcal{N}_s$ are the neighbors of s. A subset C of $\mathcal{D}$

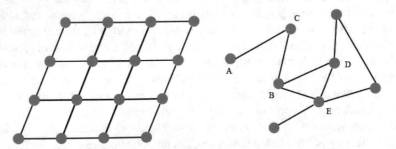

Fig. 3.3 On the left is an example lattice structure of an MRF. On the right is a toy example of a general MRF in which A and C are neighbors of each other, thus forming a two-node clique, and B, D, and E form a three-node clique

is a clique if every pair of distinct pixels in C is neighbors of each other; $\mathcal{C}$ denotes the set of all cliques.

Formally, the following gives a definition of MRF distributions on images.

Definition 1 $p(\mathbf{I})$ is an *MRF distribution* with respect to $\mathcal{N}$ if $p(\mathbf{I}_s \mid \mathbf{I}_{-s}) = p(\mathbf{I}_s \mid \mathbf{I}_{\mathcal{N}_s})$, in which $\mathbf{I}_{-s}$ denotes the values of all pixels other than s, and for $A \subset \mathcal{D}$, $\mathbf{I}_A$ denotes the values of all pixels in A.

In addition, an MRF distribution is closely related to a Gibbs distribution, which is defined below. In statistical mechanics, a Gibbs distribution gives the probability that a system will be in a certain state as a function of the state's local properties, such as the energy and the temperature of the system. Consequently, the Gibbs model can be viewed as a set of potentials defined on cliques.

Definition 2 $p(\mathbf{I})$ is a *Gibbs distribution* with respect to $\mathcal{N}$ if

$$p(\mathbf{I}) = \frac{1}{Z} \exp\left\{ -\sum_{C \in \mathcal{C}} \lambda_C(\mathbf{I}(C)) \right\}, \tag{3.2}$$

where Z is the normalizing constant (or partition function), and $\lambda_C(\cdot)$ is a function of intensities of pixels in clique C (called potential of C). Some constraints can be imposed on λ_C for them to be uniquely determined.

In fact, an MRF distribution is equivalent to a Gibbs distribution. The Hammersley–Clifford theorem establishes their equivalence [19]:

Theorem 1 *For a given $\mathcal{N}$, $p(\mathbf{I})$ is an MRF distribution $\Longleftrightarrow$ $p(\mathbf{I})$ is a Gibbs distribution.*

This equivalence provides a general method for specifying an MRF on $\mathcal{D}$, i.e., first choose an $\mathcal{N}$, and then specify λ_C. The MRF is *stationary* if for every $C \in \mathcal{C}$, λ_C depends only on the relative positions of its pixels. This is often pre-assumed in texture modeling.

Here the $\lambda_C(\cdot)$ function, which extracts statistics from an input image, can be seen as the feature representation $\mathbf{H_i}$ of the image as introduced in Sect. 3.1.

Often enough, a texture is considered as an MRF on a lattice system with each pixel represented by a node, as shown on the left of Fig. 3.3. The neighboring pixels form a clique and pixels farther away have less effect on the pixel in question. This paired system leads us to consider auto-models [19], which are MRF models with pair potentials.

Ising and Potts Models

Two important instantiations of MRF models with pair potentials have emerged throughout history, the Ising and the Potts model.

In general, auto-models with pair potentials have characteristics of $\lambda_C \equiv 0$ if $|C| > 2$, where $p(\mathbf{I})$ has the following form:

$$p(\mathbf{I}) = \frac{1}{Z} \exp \left\{ \sum_s \alpha_s \mathbf{I}_s + \sum_{t,s} \beta_{t-s} \mathbf{I}_t \mathbf{I}_s \right\}, \tag{3.3}$$

in which $\beta_{-t} = \beta_t$ and $\beta_{t-s} \equiv 0$ unless t and s are neighbors. An MRF model with pair potentials, as defined above, is commonly specified through conditional distributions:

$$p(\mathbf{I}_s \mid \mathbf{I}_{-s}) \propto \exp \left\{ \alpha_s \mathbf{I}_s + \sum_t \beta_{s-t} \mathbf{I}_t \mathbf{I}_s \right\}, \tag{3.4}$$

in which the neighborhood is usually of the order less than or equal to three pixels.

One of the classic MRF models with pair potentials is the Ising Model. The Ising Model was first proposed by Ernst Ising to study ferromagnetism. Similar to texture models, the Ising Model also considers lattice systems, in which each node $\mathbf{I}_s \in \{+1, -1\}$. To attach physical meaning to this construction, each node, or site, can be viewed as an electron having a particular "spin." With $\mathbf{I}_s = -1$, electron s points down, and with $\mathbf{I}_s = +1$, the electron points up. An Ising Model with all positive spins is shown in Fig. 3.4.

How do the electrons in the lattice interact with each other? What is the energy of the system? To answer these questions, the Ising Model considers two types of interactions that affect the energy of the system: the *external field* and *interaction* between neighboring electrons. Together, they form an energy function called a *Hamiltonian*, written as

$$H(\mathbf{I}) = - \sum_{s \sim t} \beta \mathbf{I}_s \mathbf{I}_t - \sum_t \alpha \mathbf{I}_t, \tag{3.5}$$

where $s \sim t$ means that s and t are neighbors, β represents the strength of magnetization and dictates electron interactions, and α represents the strength of the external field on each electron.

Fig. 3.4 Graphical view of the Ising Model with all positive spins

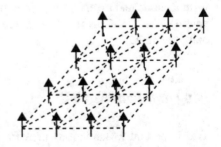

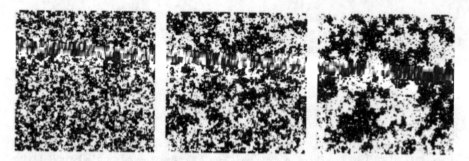

Fig. 3.5 Example sampling of Ising Model with different mean and variance. From left to right, β is 0.35, 0.40, and 0.43, respectively

The entire configuration of the lattice system is, similar to a general MRF, given by

$$p(\mathbf{I}) = \frac{1}{Z} \exp\left\{-\frac{H(\mathbf{I})}{kT}\right\}, \tag{3.6}$$

in which k is the Boltzmann constant and T is the temperature. The higher the temperature, the more random the particles become, and the more uniform the probability distribution becomes. An example sampling of the Ising Model using different β values is shown in Fig. 3.5.

The Potts Model is a generalization of the Ising Model on a lattice system. Instead of binary states, each particle has n spin angles, i.e., $\mathbf{I}_s \in \{\theta_0, \theta_1, \ldots, \theta_{n-1}\}$. Therefore, the energy Hamiltonian is given by

$$H(\mathbf{I}) = -\sum_{s \sim t} \beta \delta(\mathbf{I}_s, \mathbf{I}_t) - \sum_t \alpha \mathbf{I}_t, \tag{3.7}$$

in which $\delta(\cdot, \cdot)$ is the Kronecker delta function. Its distribution is in the same form as that of the Ising Model.

The advantage of pair-potential models such as the Ising and Potts models is that the parameters in the models can be easily inferred by auto-regression. However, these models are severely limited in the following two ways: (1) the cliques are too small to capture features of texture, and (2) the statistics on the cliques specify only first-order and second-order moments, i.e., mean and covariance, respectively. Yet many textures have local structures much larger than three or four pixels, and the covariance information, or equivalently the spectrum, cannot adequately characterize textures, as suggested by the existence of distinguishable texture pairs with identical second-order or even third-order moments, as well as indistinguishable texture pairs with different second-order moments [41]. Moreover, many textures are strongly non-Gaussian, regardless of neighborhood size.

The underlying reason for these limitations is that Eq. (3.2) involves too many parameters if we increase the neighborhood size or the order of the statistics, even for the simplest auto-models. This suggests that we need carefully designed functional forms for $\lambda_C(\cdot)$ to efficiently characterize local interactions as well as the statistics on local interactions.

Gaussian Markov Random Field (GMRF)

An early statistical model for texture patterns was the Gaussian Markov Random Field (GMRF). This family of models satisfies the properties of an MRF model with the additional restriction that the joint distribution of all nodes is multivariate normal. Important statistical properties of natural images can be observed in the GMRF model, suggesting that certain basic features of natural images are represented by this well-defined parametric family. The idea of matching fundamental features of observed images to synthesized new images with the same appearance provides inspiration for more complex models such as the FRAME model (see Sect. 3.4).

Formally, a GMRF is a graph $\mathcal{G} = (\mathcal{V}, \mathcal{E})$ where each node corresponds to a single dimension of a multivariate Gaussian $x \sim N(\mu, Q^{-1})$ with a non-singular precision matrix Q. Edges of $\mathcal{G}$ are determined by the relation

$$Q_{k,\ell} \neq 0 \Leftrightarrow \{k, \ell\} \in \mathcal{E}. \tag{3.8}$$

Conditional independence satisfies $x_k \perp x_\ell \,|\, x_{-k\ell} \Leftrightarrow Q_{k,\ell} = 0$, as can be seen from direct inspection of the multivariate Gaussian density. Therefore, $x_k \,|\, x_{-k} \sim x_k \,|\, \mathcal{N}(x_k)$, in which $\mathcal{N}(x_k) = \{x_\ell : Q_{k,\ell} \neq 0, k \neq \ell\}$ is the neighborhood of x_k, i.e., the distribution of a single node x_k given the rest of the nodes depends only on a subset of nodes which are connected to x_k in the GMRF graph. Note that the edges often represent nodes that are spatially related, such as nearby pixels within an image.

Consider an $N \times M$ image $\mathbf{I} \sim N(0, \beta^{-1} Q^{-1})$, in which $\mathbf{I}(x, y)$ denotes pixel at location (x, y) (note that this is different from the previous graph structure notation, in which $\mathbf{I}_s$ denotes node s in $\mathbf{I}$), $\beta > 0$ is the coupling strength between pixels, and

$$Q_{(x_1, y_1), (x_2, y_2)} = \begin{cases} 1 & (x_1, y_1) = (x_2, y_2) \\ -\frac{1}{4} & (x_2, y_2) \in \mathcal{N}(\mathbf{I}(x_1, y_1)) \\ 0 & \text{else} \end{cases} \tag{3.9}$$

with neighborhood structure $\mathcal{N}(\mathbf{I}(x, y)) = \{\mathbf{I}(x + 1, y), \mathbf{I}(x - 1, y), \mathbf{I}(x, y + 1), \mathbf{I}(x, y - 1)\}$. Suppose that the torus boundary condition is used, so that $\mathbf{I}(x + N, y) = \mathbf{I}(x, y)$ and $\mathbf{I}(x, y + M) = \mathbf{I}(x, y)$. This model has the density

$$p(\mathbf{I}) = \frac{1}{Z}\exp\left\{-\beta\sum_x\sum_y(\mathbf{I}(x+1,y)-\mathbf{I}(x,y))^2 + (\mathbf{I}(x,y+1)-\mathbf{I}(x,y))^2\right\},$$

$$(3.10)$$

and the conditional expectation for each pixel

$$E[\mathbf{I}(x,y)|\mathcal{N}(\mathbf{I}(x,y))] = \frac{1}{4}(\mathbf{I}(x+1,y)+\mathbf{I}(x-1,y)+\mathbf{I}(x,y+1)+\mathbf{I}(x,y-1)) \quad (3.11)$$

is the average of the neighboring pixels in the image. In the limiting case of infinitesimally small pixels, the discrete GMRF model converges to the continuous density

$$p(\mathbf{I}) = \frac{1}{Z}\exp\left\{-\beta\int_x\int_y(\nabla_x\mathbf{I}(x,y))^2 + (\nabla_y\mathbf{I}(x,y))^2 dy\, dx\right\}. \quad (3.12)$$

Several important observations follow from the analysis of the continuous analogue of the discrete GMRF density. First, the power law phenomenon observed in natural images can be explicitly derived for the density in Eq. (3.12). Recall that the Fourier transform $\hat{\mathbf{I}}(\xi,\eta)$ of an image $\mathbf{I}(x,y)$ is given by

$$F(\mathbf{I}) = \hat{\mathbf{I}}(\xi,\eta) = \int_x\int_y\mathbf{I}(x,y)e^{-i2\pi(x\xi+y\eta)}\,dy\,dx. \quad (3.13)$$

Algebraic manipulation shows that $F(\nabla_x\mathbf{I}) = 2\pi i\xi\hat{\mathbf{I}}$ and $F(\nabla_y\mathbf{I}) = 2\pi i\eta\hat{\mathbf{I}}$. The well-known Plancherel Theorem states that a function g and its Fourier transform $G = F(g)$ satisfy the relation

$$\int|g(t)|^2\,dt = \int|G(\eta)|^2\,d\eta, \quad (3.14)$$

meaning that the L_2 functional norm is preserved by the Fourier transform. Bringing all this together shows

$$\beta\int_x\int_y(\nabla_x\mathbf{I}(x,y))^2 + (\nabla_y\mathbf{I}(x,y))^2 dy\, dx = 4\pi^2\beta\int_\xi\int_\eta(\xi^2+\eta^2)|\hat{\mathbf{I}}(\xi,\eta)|^2\,d\eta d\xi, \quad (3.15)$$

so that the potential function in Eq. (3.12) can be rewritten in terms of $\hat{\mathbf{I}}$. Moreover, the separable form of the right-hand side of the above equation shows

$$p(\hat{\mathbf{I}}(\xi,\eta)) \propto \exp\left\{-4\pi^2\beta(\xi^2+\eta^2)|\hat{\mathbf{I}}(\xi,\eta)|^2\right\}. \quad (3.16)$$

Hence, $\hat{\mathbf{I}}(\xi,\eta)$ is independent of other states of $\hat{\mathbf{I}}$, and $\hat{\mathbf{I}}(\xi,\eta)$ is a Gaussian with parameters

$$E[\hat{\mathbf{I}}(\xi, \eta)] = 0 \quad \text{and} \quad \text{Var}[\hat{\mathbf{I}}(\xi, \eta)] = \frac{1}{8\pi^2\beta(\xi^2 + \eta^2)}. \tag{3.17}$$

Therefore, $E[|\hat{\mathbf{I}}(f)|^2]^{1/2} \propto 1/f$ so that the GMRF model $p(\mathbf{I})$ satisfies the scale invariance observed in natural images, as discussed in Sect. 2.3.

A second important property of the GMRF model is its connection to the heat equation. Considering again the potential function of a GMRF model

$$H(\mathbf{I}(x, y)) = \beta \int_x \int_y (\nabla_x \mathbf{I}(x, y))^2 + (\nabla_y \mathbf{I}(x, y))^2 dy\, dx, \tag{3.18}$$

we will show that generating an image according to the dynamics

$$\frac{d\mathbf{I}(x, y, t)}{dt} = -\frac{\delta H(\mathbf{I}(x, y, t))}{\delta \mathbf{I}}, \quad \forall (x, y), \tag{3.19}$$

is equivalent to the heat diffusion equation

$$\frac{d\mathbf{I}(x, y, t)}{dt} = \Delta \mathbf{I}(x, y), \tag{3.20}$$

in which $\Delta = \frac{\partial^2}{\partial x^2} + \frac{\partial^2}{\partial y^2}$ is the Laplacian operator.

Using an Euler–Lagrangian equation for two variables,

$$\frac{\delta J}{\delta f} = \frac{\partial L}{\partial f} - \frac{d}{dx}\left(\frac{\partial L}{\partial f_x}\right) - \frac{d}{dy}\left(\frac{\partial L}{\partial f_y}\right), \tag{3.21}$$

in which $J(f) = \int_y \int_x L(x, y, f, f_x, f_y) dx dy$ and f is a function of x and y.

Setting $f = \mathbf{I}$, $L = (\nabla_x \mathbf{I})^2 + (\nabla_y \mathbf{I})^2$, and $J = H$, we obtain

$$\frac{d\mathbf{I}(x, y, t)}{dt} = -\frac{\delta H(\mathbf{I}(x, y, t))}{\delta \mathbf{I}}$$

$$= -\left[\frac{\partial[(\nabla_x\mathbf{I})^2 + (\nabla_y\mathbf{I})^2]}{\partial \mathbf{I}} - \frac{d}{dx}\left[\frac{\partial[(\nabla_x\mathbf{I})^2 + (\nabla_y\mathbf{I})^2]}{\partial(\nabla_x\mathbf{I})}\right] - \frac{d}{dy}\left[\frac{\partial[(\nabla_x\mathbf{I})^2 + (\nabla_y\mathbf{I})^2]}{\partial(\nabla_y\mathbf{I})}\right]\right]$$

$$= 0 + \frac{d}{dx}[2(\nabla_x\mathbf{I})] + \frac{d}{dy}[2(\nabla_y\mathbf{I})]$$

$$= 2\left(\frac{\partial^2\mathbf{I}}{\partial x^2} + \frac{\partial^2\mathbf{I}}{\partial y^2}\right)$$

$$= 2\Delta\mathbf{I}. \tag{3.22}$$

Ignoring the constant factor, we have shown that in fact, the learning dynamics of an image modeled by a GMRF potential function is equivalent to the dynamics modeled by the heat diffusion equation.

Despite interesting connections to statistical properties of natural images and the heat diffusion equation, the GMRF model is still quite restricted. Interesting local structures exist but the joint pixel density is a unimodal Gaussian. The GMRF model can capture aspects of the local regularity found in natural images but nothing else. Therefore, the GMRF model from Eq. (3.10) is not capable of serving as a probability density for even simple image patterns. The GMRF model can represent certain properties of an image (specifically, the tendency of nearby pixels to have similar intensity), but it cannot account for the formation of visual patterns. In the following sections, we will see how the GMRF potential from Eq. (3.10) can be adapted as one of many filters (in particular, a gradient filter) whose joint features can be used to synthesize realistic images.

Advanced Models: Hierarchical MRF and Mumford–Shah Model

In reality, we often encounter images with multiple types of textures, each occupying a set of regions in the image domain. In these situations, we need more powerful models to deal with the interaction of distinct texture regions.

When several texture patterns are present in an image, the boundaries of different regions need to be specified. Often, the simple MRF models, including the GMRF model, are not expressive enough in encoding this kind of knowledge in images. For example, we might expect long, straight edges to be penalized in some images, but the pixel intensities alone do not accurately reflect the existence of edge elements. Geman and Geman [74] introduce the hierarchical MRF model for the maximum a posteriori (MAP) image restoration problem and describe the images by both pixel intensities and edge continuity. Specifically, an image $\mathbf{I} = (\mathbf{F}, \mathbf{L})$ is modeled as an MRF of two processes. The intensity process $\mathbf{F}$ is a simple MRF of observable pixel intensities as discussed in Sect. 3.2). The line process $\mathbf{L}$ is another MRF of unobservable edge elements. We define a line site d as placed midway between each vertical or horizontal pair of pixels. The set $\mathcal{D}$ of all line sites is thus all possible locations of the edge elements. While $\mathbf{F}(x, y)$ measures the intensity at a pixel, $\mathbf{L}(x, y)$ represents a lack or presence (and orientation) of an edge at this specific site. The line process $\mathbf{L}$ can affect a pixel's neighborhood. For instance, if an edge is present at a line site, the potential over the pair-clique consisting of the pixels separated by this line site is zero. Therefore, these two pixels will not influence the other's intensity, and the bonding between them is broken. The Gibbs distribution used to define the hierarchical MRF is

$$p(\mathbf{F} = f, \mathbf{L} = l) = \frac{1}{Z} \exp\left\{ -\sum_{C \in \mathcal{C}} \lambda_C(f, l) \right\}. \tag{3.23}$$

With the hierarchical MRF, we can model more complicated image structures and define new types of tasks in computer vision, as it allows us to analyze the discontinuity in images from a new perspective.

Another technique that facilitates the modeling of multiple textures is the Mumford–Shah model. This model aims to segment an image into a few simple regions, keeping the color of each region as smooth as possible. It utilizes the Mumford–Shah energy functional to complete three tasks simultaneously. First, the functional measures how well the model approximates the observed image. Second, it asks that the model varies slowly except at boundaries. Third, it requires the set of contours to be as short, and hence as simple and straight, as possible.

Selecting Filters and Learning Potential Functions

The models mentioned thus far have equipped us with the tools needed to describe and generate a variety of textures, as long as we select appropriate image filters and learn optimal potential functions and parameters for the learned probability distributions. The questions then become, given one or a set of observed texture images, how do we select filters that will most completely capture the image statistics? And how do we learn the optimal potential functions and parameters? These questions will be answered as we explain the FRAME model in Sect. 3.4. First, though, it will be helpful to review some of the more well-known filters in vision.

3.3 Filters for Early Vision

Due to the limitations of clique-based models, the researchers have also explored feature extraction from the perspective of image filtering. In the various stages along the visual pathway, from the retina to V1, to the extra-striate cortex, cells with increasing sophistication and abstraction have been discovered: center-surround isotropic retinal ganglion cells, frequency and orientation selective simple cells, and complex cells that perform nonlinear operations. Here we focus on filtering theory inspired by the multi-channel filtering mechanism discovered and generally accepted in neurophysiology, which proposes that the visual system decomposes the retinal image into a set of sub-bands. These sub-bands are computed by convolving the image with a bank of linear filters, followed by some nonlinear procedures [18]. Considering again the definition $\Omega(\mathbf{h}) = \{\mathbf{I} : \mathbf{H}_i(\mathbf{I}) = \mathbf{h}_i, i = 1, \ldots, K\}$, it is now natural, after convolving the image with a bank of linear filters, to use a marginal statistic (e.g., a histogram) to represent a feature $\mathbf{H}_i$ of an image.

Filtering is a process that changes pixel values of a given image, in an effort to extract valuable information such as clusters and edges. Figure 3.6 provides an example of typical image filtering.

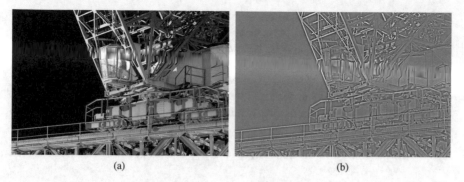

(a)   (b)

Fig. 3.6 (**a**) Presents an image and (**b**) presents a filtered version of the image

An integral part of image processing, filters provide biologically plausible ways of extracting visual information from raw input signals. In the coming section, we will explore different types of filters that extract different feature information from images.

Correlation and Convolution

Before we introduce specific types of filters, it is important to understand how filters are applied to images. One method of linear filtering is correlation, which uses filters to obtain a weighted combination of all pixels in a small neighborhood. Suppose an image $\mathbf{I}$ is defined on an $N \times N$ lattice D, and $\mathbf{I}(x, y) \in \mathcal{L}$ is the intensity value of an image $\mathbf{I}$ at location (x, y) and $\mathcal{L}$ is an interval on the real line or a set of integers representing pixel intensity. Suppose also that filter F is defined on an $M \times M$ lattice much smaller than the lattice of $\mathbf{I}$. Correlation is computed as

$$\mathbf{I}(x, y) \otimes F = \sum_{(k,l)} \mathbf{I}(x + k, y + l) F(k, l) \tag{3.24}$$

in which $\otimes$ denotes correlation and each $F(k, l)$ is a filter coefficient. To maintain the image size, we pad the boundary of the original image with enough zeros so that the filtered output image is the same size.

Correlation measures similarity between a filtered region and the filter itself, but to extract meaningful information, a more commonly used method is convolution. Different from correlation, convolution is defined as

$$\mathbf{I}(x, y) * F = \sum_{(k,l)} \mathbf{I}(x - k, y - l) F(k, l), \tag{3.25}$$

where $*$ denotes the convolution operation. It appears that convolution is correlation with both axes flipped. The motivation for using convolution instead of correlation is that it is associative, in addition to being commutative and distributive. This means that if an image needs to be convolved with multiple filters, the filters can be convolved with each other first before being applied to the image. A detailed proof of the mathematical properties of convolution is not of our interest for this book, but to intuit why convolution is associative, notice that the Fourier transform of convolution is a product in the frequency domain, which is clearly associative.

Knowing some basic operations for image processing, let us introduce some classical filters.

Edge Detection Filters

An edge in an image consists of a sudden change of pixel intensity, and to detect them, a natural decision is to detect the change of pixel values throughout the image. If an image exists on a continuous domain, we can represent the change in pixel intensities by derivatives

$$\nabla \mathbf{I} = \left[\frac{\partial \mathbf{I}}{\partial x}, \frac{\partial \mathbf{I}}{\partial y} \right]. \tag{3.26}$$

To only account for horizontal edges or vertical edges, we can represent changes in pixel intensities by $[\frac{\partial \mathbf{I}}{\partial x}, 0]$ or $[0, \frac{\partial \mathbf{I}}{\partial y}]$, respectively. The gradient direction θ, as in calculus, is

$$\theta = \tan^{-1} \left(\frac{\frac{\partial \mathbf{I}}{\partial y}}{\frac{\partial \mathbf{I}}{\partial x}} \right), \tag{3.27}$$

and the gradient strength is

$$\|\nabla \mathbf{I}\| = \sqrt{\left(\frac{\partial \mathbf{I}}{\partial x} \right)^2 + \left(\frac{\partial \mathbf{I}}{\partial y} \right)^2}. \tag{3.28}$$

In practice, however, images consist of discrete pixels, so we need a discrete approximation of image derivatives. Note that the definition of the derivative with respect to the x-coordinate is

$$\frac{\partial \mathbf{I}(x, y)}{\partial x} = \lim_{h \to 0} \frac{\mathbf{I}(x + h, y) - \mathbf{I}(x, y)}{h}, \tag{3.29}$$

and its discrete counterpart is

$$\frac{\partial \mathbf{I}(x, y)}{\partial x} \approx \frac{\mathbf{I}(x + 1, y) - \mathbf{I}(x, y)}{1}. \tag{3.30}$$

To implement this derivative using a convolution, we can use a simple filter

$$F = \begin{bmatrix} -1 & 1 \end{bmatrix}. \tag{3.31}$$

However, a more commonly used 1D filter for edge detection is

$$F = \begin{bmatrix} -1 & 0 & 1 \end{bmatrix}, \tag{3.32}$$

which corresponds to, in the continuous domain, a derivative definition that extends by a small amount in both directions:

$$\frac{\partial \mathbf{I}(x, y)}{\partial x} = \lim_{h \to 0} \frac{\mathbf{I}(x + h, y) - \mathbf{I}(x - h, y)}{2h}. \tag{3.33}$$

Note that the constant factor 2 is neglected in the filter since applying the unnormalized filter consistently will not affect the relative intensity of the resulting image.

In the 2D case, three well-known edge detection filters are the Prewitt, Sobel, and Roberts filters, which are displayed in the same order below:

$$\begin{bmatrix} -1 & 0 & 1 \\ -1 & 0 & 1 \\ -1 & 0 & 1 \end{bmatrix} \begin{bmatrix} -1 & 0 & 1 \\ -2 & 0 & 2 \\ -1 & 0 & 1 \end{bmatrix} \begin{bmatrix} 0 & 1 \\ -1 & 0 \end{bmatrix}. \tag{3.34}$$

Both the Prewitt and Sobel filters detect edges in the x-direction, and detection in the y-direction is similar but with the filter weights transposed. However, the Sobel filter is more commonly used between these two because it provides smoothing in the direction perpendicular to the direction of the edge detection. Details on smoothing will be provided in the next section. Images filtered by the Prewitt filter may also suffer from any noise in the original image. The Roberts filter detects edges in the diagonal direction instead. Figure 3.7 shows an image filtered by each filter.

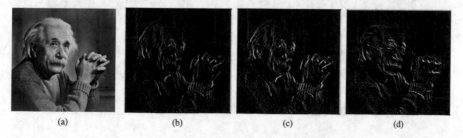

(a)   (b)   (c)   (d)

Fig. 3.7 An image (**a**) filtered by Prewitt (**b**), Sobel (**c**), and Roberts (**d**) filters, respectively

Gaussian Filters

A Gaussian filter is used for smoothing and blurring. As indicated by its name, the filter resembles a Gaussian density function. Note that a multivariate Gaussian function can be written as

$$G(\vec{x} \mid \vec{x}_0, \Sigma) = \frac{1}{\sqrt{(2\pi)^n |\Sigma|}} \exp\left\{ -\frac{1}{2}(\vec{x} - \vec{x}_0)^\mathsf{T} \Sigma^{-1}(\vec{x} - \vec{x}_0) \right\} \qquad (3.35)$$

in which G denotes a Gaussian function, $\vec{x}$ is an n-dimensional variable centered at $\vec{x}_0$, and Σ is the covariance matrix. However, for the purpose of image smoothing, a Gaussian filter is typically assumed to be independent among different coordinates. That is, in two dimensions, the filter can be written as

$$G(x, y \mid x_0, y_0, \sigma_x, \sigma_y) = \frac{1}{2\pi \sigma_x \sigma_y} e^{-((x-x_0)^2/2\sigma_x^2 + (y-y_0)^2/2\sigma_y^2)} \qquad (3.36)$$

where by convention it is assumed that $\sigma_x = \sigma_y = \sigma$. Its discrete counterpart can be represented by filter

$$\frac{1}{16} \begin{bmatrix} 1 & 2 & 1 \\ 2 & 4 & 2 \\ 1 & 2 & 1 \end{bmatrix}. \qquad (3.37)$$

Varying σ allows one to vary the width of the Gaussian distribution, controlling the degree of smoothing. A larger σ corresponds to a larger filter size with each convolution accounting for more neighboring pixels. Note that Gaussian filtering is a form of a weighted sum with the highest weight at the center, effectively smoothing out noises in a close vicinity. A Gaussian-smoothed image is shown in Fig. 3.8.

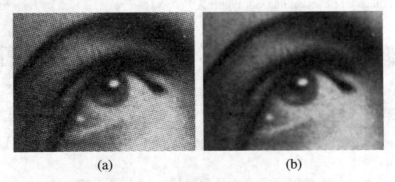

(a)   (b)

Fig. 3.8 An image (**a**) smoothed by a Gaussian filter (**b**)

Derivative of Gaussian and Laplacian of Gaussian Filters

In performing an edge detection, a major problem for Sobel and Prewitt filters is that they are very sensitive to noise due to a hard subtraction of neighboring values. It is therefore common to first conduct Gaussian smoothing before applying image gradients. Since convolution is associative, we can directly apply gradient operations to a Gaussian filter before applying them to the image itself.

This suggests that we need to calculate the first derivative of the Gaussian, and Fig. 3.9 shows the first derivatives, with respect to both the x- and y-coordinates, and their corresponding filters. Convolving a derivative of the Gaussian filter with an image allows for both smoothing and edge detection, and it is less sensitive to noise than Prewitt or Sobel filters.

For edge detection purposes, first derivatives of the Gaussian are commonly used, but they are still somewhat undesirable due to the resulting thickness of detected edges, induced by σ in the Gaussian function. Second derivatives, on the other hand, represent edges with zero intensity, as the first derivative transforms edges to max/min intensity after filtering. This infinitely thin line of zero-crossing is more desirable for edge detection. To calculate the second derivative, the Laplacian $\Delta = \frac{\partial^2}{\partial x^2} + \frac{\partial^2}{\partial y^2}$ is most commonly used.

Similar to the derivation of image gradients, the second derivative in the x-direction can be written as

$$
\frac{\partial^2 \mathbf{I}(x, y)}{\partial x^2} = \lim_{h \to 0} \frac{\frac{\partial \mathbf{I}(x+h, y)}{\partial x} - \frac{\partial \mathbf{I}(x-h, y)}{\partial x}}{2h}
$$

$$
= \lim_{h \to 0} \frac{\lim_{t \to 0} \frac{\mathbf{I}(x+h, y) - \mathbf{I}(x+h-t, y)}{t} - \lim_{t \to 0} \frac{\mathbf{I}(x-h+t, y) - \mathbf{I}(x-h, y)}{t}}{2h}
$$

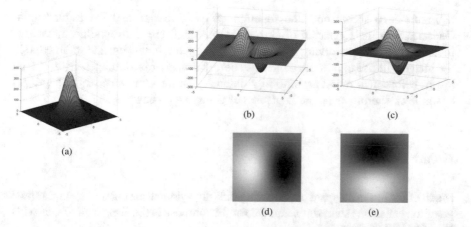

Fig. 3.9 (**a**) Gaussian density. (**b**) Function of x-derivative of Gaussian. (**c**) Function of y-derivative of Gaussian. (**d**) Filter visualization of x-derivative of Gaussian. (**e**) Filter visualization of y-derivative of Gaussian

$$\approx \frac{\frac{\mathbf{I}(x+1,y)-\mathbf{I}(x,y)}{1} - \frac{\mathbf{I}(x,y)-\mathbf{I}(x-1,y)}{1}}{1}$$

$$= \mathbf{I}(x+1,y) - 2\mathbf{I}(x,y) + \mathbf{I}(x-1,y). \qquad (3.38)$$

This corresponds to an (unnormalized) discrete filter

$$\begin{bmatrix} 1 & -2 & 1 \end{bmatrix}. \qquad (3.39)$$

In 2D, it is not difficult to see that a simple Laplacian filter becomes

$$\begin{bmatrix} 0 & 1 & 0 \\ 1 & -4 & 1 \\ 0 & 1 & 0 \end{bmatrix}. \qquad (3.40)$$

However, similar to image gradients, the Laplacian operator is extremely sensitive to noise in the input. Therefore, Gaussian smoothing is usually applied before a Laplacian operation. As the Laplacian can also be represented by a convolution as shown above, it can be applied to a Gaussian filter before being applied to the image. The resulting filter is named the Laplacian of Gaussian (LoG or LG). This gives a continuous formulation

$$LoG(x, y \mid x_0, y_0, \sigma_x, \sigma_y) =$$

$$- \left\{ \frac{1}{2\pi\sigma_x^3\sigma_y} \left[1 - \frac{(x-x_0)^2}{\sigma_x^2} \right] + \frac{1}{2\pi\sigma_x\sigma_y^3} \left[1 - \frac{(y-y_0)^2}{\sigma_y^2} \right] \right\} \exp\left\{ -\left(\frac{(x-x_0)^2}{2\sigma_x^2} + \frac{(y-y_0)^2}{2\sigma_y^2} \right) \right\}. \qquad (3.41)$$

Figure 3.10 shows both the function of the Gaussian and the Laplacian of Gaussian. For the Laplacian of Gaussian, the uphill area surrounding a "valley" indicates that, by convolving it with an image, the low-intensity side of an edge in the image will rise to have high intensity before dropping to zero intensity and even lower, before then rising again to high intensity on the other side of the edge. An example of filtering using the Laplacian of Gaussian is shown in Fig. 3.11.

Gabor Filters

Much effort has been spent on modeling radially symmetric center-surround retinal ganglion cells. One such simple model for this purpose is the Laplacian of Gaussian introduced above with $\sigma_x = \sigma_y = \sigma$.

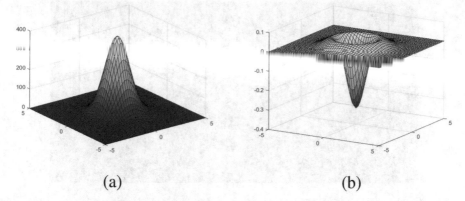

Fig. 3.10 (a) Density plot of the Gaussian, and (b) the second derivative of the Gaussian

(a)   (b)

Fig. 3.11 An example of filtering using the Laplacian of Gaussian

Yet, as retinal ganglion cells respond to specific frequency bands, Gabor filters also provide a good way to extract texture patterns from an image under specific frequencies [35]. In addition to being frequency-selective, Gabor filters are also directional-selective. Mathematically, a Gabor filter is a pair of cosine and sine waves with angular frequency ω and η, respectively, and amplitude modulated by the Gaussian function. Its general form is

$$F_\omega(x, y) = G(x, y \mid x_0, y_0, \sigma_x, \sigma_y) \cdot S(x, y \mid \omega, \eta), \qquad (3.42)$$

in which $G(x, y \mid x_0, y_0, \sigma_x, \sigma_y)$ is a Gaussian function, $S(x, y \mid \omega, \eta) = \exp\{-i(\omega x + \eta y)\}$ is a wave function, and $\phi = \arctan(\eta/\omega)$ is the direction of the wave.

Here F_ω defines a Gabor filter that matches image regions with frequency ω in the x-direction and frequency η in the y-direction. High-frequency filters will match high-frequency patterns whereas low-frequency filters will match low-frequency patterns.

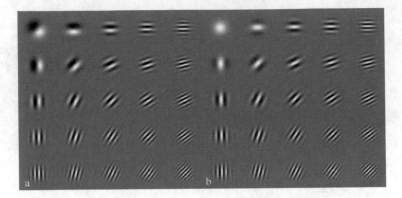

Fig. 3.12 Gabor wavelets in different orientations and frequencies

Gabor filters are closely related to Gabor wavelets, which serve as basis functions for a Fourier Transform of an image. Some examples of Gabor wavelets are shown in Fig. 3.12.

The filters mentioned above are linear. Some functions are further applied to these linear filters to model the nonlinear functions of the complex cells. One way to model the complex cells is to use the power of each pair of Gabor filters, $|(F * I)(x, y)|^2$. In fact, $|(F * I)(x, y)|^2$ is the local spectrum of I at (x, y) smoothed by a Gaussian function. Thus, it serves as a local spectrum analyzer.

Although these filters are very efficient in capturing local spatial features, some problems are not well understood. First, given a bank of filters, how can we choose the best set of filters, especially when some are linear while others are nonlinear, or the filters are highly correlated with each other? Second, after selecting the filters, how can we encapsulate the features they capture into a single texture model? These questions will be answered in the remainder of the chapter.

3.4   FRAME Model

Markov Random Fields provide us with physical inspiration for modeling texture patterns despite some notable limitations. Filters, on the other hand, afford a powerful and biologically plausible way of extracting features. Is there a way to combine the two classes of texture models? In this coming section, we propose a modeling methodology that is built on and directly combines the above two important themes for texture modeling. It is called the FRAME (Filters, Random field, and Maximum Entropy) model [282]. Before continuing, however, you are encouraged to go over Chap. 9 and understand Maximum Entropy, Minimum Entropy, and Minimax Entropy Principles, as they will be employed in this chapter in derivations.

Intuition and the Big Picture

Let image **I** be defined on a domain $\mathcal{D}$ where each $\mathbf{I}_s \in \mathcal{L}$ is the intensity value of image **I** at pixel s, and $\mathcal{L}$ is an interval on the real line or a set of integers to represent pixel intensity. Without loss of generality, we denote a feature as $\phi \in S = \{\phi^{(\alpha)}, \alpha = 1, 2, \ldots, K\}$, where K is the number of features in question. In the FRAME model, each feature is typically a vector of histograms of activations (normalized or unnormalized) that result from image **I** being convolved with a selected filter. As a reminder, a filter can be seen as a low-dimensional space that pierces through the higher dimensional image space. Accordingly, convolution of the image **I** with the filter produces the projection of the image onto this low-dimensional space. The set of projected histograms is a set of marginal distributions for the image; our goal is to match "enough" marginal distributions that we can accurately model the observed image. An intuitive drawing is presented as Fig. 3.13.

Now given a set of observed images $\{\mathbf{I}_i^{obs}, i = 1, 2, \ldots, M\}$ from a distribution $f(\mathbf{I})$, we define feature statistics of the observation as

$$\mu_{obs}^{(\alpha)} = \frac{1}{M} \sum_{i=1}^{M} \phi^{(\alpha)}(\mathbf{I}_i^{obs}), \quad \text{for } \alpha = 1, \ldots, K, \tag{3.43}$$

and the set of images that match the statistics of these features as

$$\Omega_{\mathbf{I}} = \left\{ \mathbf{I} : \phi^{(\alpha)}(\mathbf{I}) = \mu_{obs}^{(\alpha)}, \ \alpha = 1, \ldots, K \right\}. \tag{3.44}$$

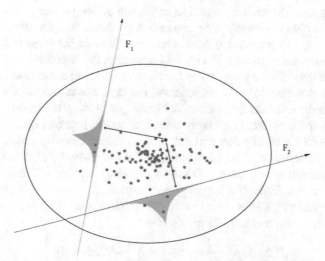

Fig. 3.13 Suppose each point is an image and the red dot represents the selected image. Two filters are seen as axes piercing through the image space. The features described by these two filters are represented by the image's projection (i.e., convolution with a filter) against the two axes

Similarly, we can define the set of distributions that match the statistics as

$$\Omega_p = \left\{ p(\mathbf{I}) \; : \; E_p[\phi^{(\alpha)}(\mathbf{I})] = \mu_{obs}^{(\alpha)}, \; \alpha = 1, \dots, K \right\}. \tag{3.45}$$

We now have defined our feature statistics. The next task is to pursue a suitable distribution $p \in \Omega_p$ that matches the observed distributions.

This pursuit process is iterative and twofold. First, we select K suitable filters from a given set of filters to completely describe the most distinctive features of a texture. Second, we find the best-parameterized distribution to explain the current statistics given a fixed set of filters. This parameterized distribution can be written as $p(\mathbf{I}; \Lambda; S)$, where S denotes the set of selected filters, and Λ denotes the distribution parameters. Filter selection and distribution matching is an iterative process, reminiscent of coordinate descent, minimizing differences between the true distribution and the model distribution. Note that as we try to model pre-attentive processes (i.e., the subconscious accumulation and processing of environmental information) in human brains when detecting textures, we generally neglect the training of filters and simply use pre-trained filters.

We make use of the Maximum Entropy Principle to pursue the best parameters Λ when the set of features (parameterized by a set of filters) to be matched is fixed. The parameterized distribution is constrained to match histograms of the filters' activations generated from texture images. However, we do not want to over-constrain our distribution since we should avoid adding extra information that may change the feature statistics. That is, we only match statistics given by the selected features and no more, meaning that we maximize the entropy of our distribution while matching all the given features. More intuitively, we want to learn the distribution of the texture, but not to learn each training texture image precisely, so that we are able to generate new texture images that follow the same distribution as texture images in our training dataset but not reproduce the same images that are already in our training dataset. Figure 3.14 illustrates this intuition.

To match the statistics provided by the observed data, an extreme solution would be a distribution that only concentrates on the data points themselves. However, this solution over-shrinks the constrained image set, adding too many unnecessary constraints, and it results in a highly overfitted model that can hardly generalize. Instead, we seek a distribution that exhibits maximum capacity under the current constraints. Remember in the earlier chapters, we discussed that the capacity of a space is measured by the entropy of the distribution defined for it. Thus, maximizing the entropy means finding the largest possible space.

As explained by the Maximum Entropy Principle, the solution to this constrained problem is in the form of a Gibbs distribution:

$$p(\mathbf{I}; \Lambda, S) = \frac{1}{Z(\Lambda)} \exp\left\{ -\sum_{\alpha=1}^{K} \langle \lambda^{(\alpha)}, \phi^{(\alpha)}(\mathbf{I}) \rangle \right\}, \tag{3.46}$$

where $\lambda^{(\alpha)}$ is the Lagrange multiplier associated with feature $\phi^{(\alpha)}$ controlling the relative strength of activation, $\langle \cdot \; , \; \cdot \rangle$ denotes inner product, and $\Lambda = (\lambda^{(\alpha)}, \alpha = 1, \dots, K)$.

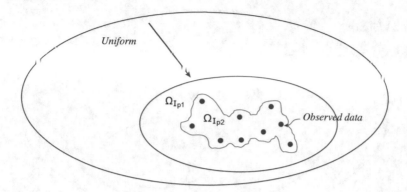

Fig. 3.14 The Maximum Entropy Principle gives us the most general distribution that matches the feature statistics we choose. Here $\Omega_{\mathbf{I}_{p_1}}$ and $\Omega_{\mathbf{I}_{p_2}}$ are two possible distributions we learned to try to match the feature statistics of the observed data points. $\Omega_{\mathbf{I}_{p_2}}$ concentrates too much on the observed data points, which is representative of overfitting and poor generalization, while $\Omega_{\mathbf{I}_{p_1}}$ is the more general one by applying the Maximum Entropy Principle

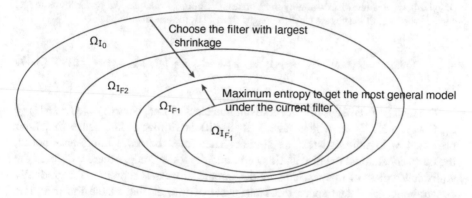

Fig. 3.15 We can observe the min-max entropy process for training the FRAME model. Suppose we are at $\Omega_{\mathbf{I}_0}$ and we have two filters we could select: $\mathbf{F}_1$ and $\mathbf{F}_2$. If either of them is selected, we update our model based on the Maximum Entropy principle, which lends the largest capacity based on the current constraint (e.g., preferring $\Omega_{\mathbf{I}_{F_1}}$ to $\Omega_{\mathbf{I}_{F_1'}}$). In choosing between $\mathbf{F}_1$ and $\mathbf{F}_2$, we want the one that lends the largest reduction in the entropy of the parameterized distribution with respect to the feature space

The Lagrange multipliers are unique. After the current distribution is found, we proceed to find a new pre-trained filter to better model the texture pattern. Intuitively, we select the best filter that minimizes the Kullback–Leibler (KL) divergence from $f(\mathbf{I})$ to $p(\mathbf{I}; \Lambda, S)$. As derived in the Minimum Entropy Principle section, minimizing this KL divergence is equivalent to minimizing the entropy of the parameterized distribution with respect to the feature space (Fig. 3.15).

This pursuit process can be visualized as below:

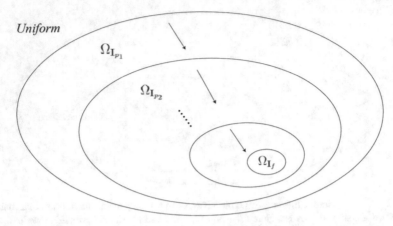

Fig. 3.16 Reduction of constrained set of distributions from those sampled from uniform distribution, which corresponds to the entire pixel space, to an equivalence class of images that match
projected statistics of the true samples. Here $\Omega_{I_{p_1}}$ represents the set of images drawn from p_1,
which follows a Gibbs distribution by the Maximum Entropy Principle. We perform this model
pursuit process until images are close to images from $f(\mathbf{I})$, the true distribution

$$\text{Uniform} = p(\mathbf{I}; \Lambda_0, S_0) \; \rightarrow \; p(\mathbf{I}; \Lambda_1, S_1) \; \rightarrow \; \ldots \; \rightarrow \; p(\mathbf{I}; \Lambda_K, S_K) \sim f(\mathbf{I}). \quad (3.47)$$

The matched statistics give an equivalence class of images $\Omega(\mathbf{h}) = \{\mathbf{I} : \mathbf{H_i}(\mathbf{I}) =$
$\mathbf{h_i}, i = 1, \ldots, K\}$ in which each $\mathbf{I} \in \Omega_{\mathbf{I}}(\mathbf{h})$ is similar in the sense that their
projected histogram statistics are similar. Therefore, the more filters are chosen,
the more constrained the distributions are, and the fewer images there will be in the
equivalence class of images. This model pursuit process is a process for gradually
constraining the image space to obtain a set of images that match the true data
distribution. The pursuit process is also shown in Fig. 3.16.

The figure can be interpreted as follows. We start from a random uniform distribution and start selecting a filter according to Minimum Entropy Principle. Then we
use Maximum Entropy Principle to best match the selected feature statistics. This
gives $\Omega_{\mathbf{I}_{p_1}}$ described by distribution $p(\mathbf{I}; \Lambda_1, S_1)$. The set of equivalence class of
images also decreases, as indicated by the shrinking ellipse. This iterative process
continues until our distribution is constrained enough so that it is indistinguishable
from the observed distribution $f(\mathbf{I})$.

An interesting analogy to this pursuit process is shepherding. At first, we can
imagine that sheep wander in an infinitely large area. At a certain time step,
shepherds start guiding the sheep into a newly constructed fenced area, which
defines a constrained space. Not wanting to be constrained further, the sheep push
back against the fence, trying to escape. Shepherds then construct a new, smaller
fenced area inside the last one and guide the sheep into this smaller area, with
the sheep again pushing back. This process is iterated until the shepherds stop
constructing fences.

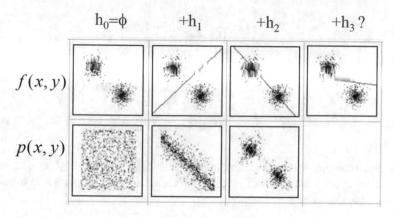

Fig. 3.17 A toy example pursuit process as described by FRAME

Figure 3.17 shows a toy example (images with 2 pixels) of this pursuit process. As shown, $f(x, y)$ is the true data distribution we try to model, and our distribution $p(x, y)$ starts from random uniform and a feature h_1 is selected. Maximizing entropy of the constrained distribution gives us the second distribution. We further select the next best feature h_2 and match our distribution against it. We can see that now the $p(x, y)$ is already very close to $f(x, y)$ and the process is stopped.

In the coming section, we derive the FRAME model and how it can be learned.

Deriving the FRAME Model

With the bigger picture in mind, we now proceed to derive the FRAME model. We first fix a set of filters and try to learn Λ.

To re-iterate our definition, let image $\mathbf{I}$ be defined on a discrete domain $\mathcal{D}$, in which $\mathcal{D}$ can be an $N \times N$ lattice. For each pixel $s \in \mathcal{D}$, $\mathbf{I}_s \in \mathcal{L}$, and $\mathcal{L}$ is an interval of $\mathcal{R}$ or $\mathcal{L} \subset \mathcal{Z}$. For each texture, we assume that there exists a "true" joint probability density $f(\mathbf{I})$ over the image space $\mathcal{L}^{|\mathcal{D}|}$, and $f(\mathbf{I})$ should concentrate on a subspace of $\mathcal{L}^{|\mathcal{D}|}$ (projection of $\mathbf{I}$ onto the subspace), which corresponds to texture images that have perceptually similar texture appearances.

Learning Potential Functions

Given an image $\mathbf{I}$ and a filter $F^{(\alpha)}$ with $\alpha = 1, 2, \ldots, K$ for the indices of the filter, we let $\mathbf{I}_s^{(\alpha)} = F^{(\alpha)} * \mathbf{I}_s$ be the filter response at location s and $\mathbf{I}^{(\alpha)}$ the filtered image. The marginal empirical distribution (histogram) of $\mathbf{I}^{(\alpha)}$ is

$$H^{(\alpha)}(z) = \frac{1}{|\mathcal{D}|} \sum_{s \in \mathcal{D}} \delta(z - \mathbf{I}_s^{(\alpha)}), \tag{3.48}$$

where z is a specific filter response and $\delta(\cdot)$ is the Dirac delta function. The marginal distribution of $f(\mathbf{I})$ with respect to $F^{(\alpha)}$ at location s is denoted by

$$f_s^{(\alpha)}(z) = \int_{\mathbf{I}_s^{(\alpha)}=z} f(\mathbf{I})d\mathbf{I} = \mathrm{E}_f[\delta(z - \mathbf{I}_s^{(\alpha)})]. \tag{3.49}$$

At first thought, it seems an intractable problem to estimate $f(\mathbf{I})$ due to the overwhelming dimensionality of image $\mathbf{I}$. To reduce dimensions, we first introduce the following theorem.

Theorem 1 *Let $f(\mathbf{I})$ be the $|\mathcal{D}|$-dimensional continuous probability distribution of a texture, then $f(\mathbf{I})$ is a linear combination of $f^{(\xi)}$, the latter are the marginal distributions on the linear filter response $F^{(\xi)} * \mathbf{I}$.*

Proof By the inverse Fourier transform, we have

$$f(\mathbf{I}) = \frac{1}{(2\pi)^{|\mathcal{D}|}} \int e^{2\pi i \langle \mathbf{I}, \xi \rangle} \hat{f}(\xi)d\xi, \tag{3.50}$$

where $\hat{f}(\xi)$ is the characteristic function of $f(\mathbf{I})$ and

$$\begin{aligned}
\hat{f}(\xi) &= \int e^{-2\pi i \langle \xi, \mathbf{I} \rangle} f(\mathbf{I})d\mathbf{I} \\
&= \int e^{-2\pi i z} \int_{\langle \xi, \mathbf{I} \rangle = z} f(\mathbf{I})dz d\mathbf{I} \\
&= \int e^{-2\pi i z} \int \delta(z - \langle \xi, \mathbf{I} \rangle) f(\mathbf{I})dz d\mathbf{I} \\
&= \int e^{-2\pi i z} f^{(\xi)}(z)dz, \tag{3.51}
\end{aligned}$$

where $\langle \cdot, \cdot \rangle$ is the inner product, by definition $f^{(\xi)}(z) = \int \delta(z - \langle \xi, \mathbf{I} \rangle) f(\mathbf{I})d\mathbf{I}$ is the marginal distribution of $F^{(\xi)} * \mathbf{I}$, and we define $F^{(\xi)}(s) = \xi(s)$ as a linear filter. $\qquad \square$

Theorem 1 transforms $f(\mathbf{I})$ into a linear combination of its one-dimensional marginal distributions. Thus, it motivates a new method for inferring $f(\mathbf{I})$; namely, construct a distribution $p(\mathbf{I})$ so that $p(\mathbf{I})$ has the same marginal distributions $f^{(\xi)}$. If $p(\mathbf{I})$ matches all marginal distributions of $f(\mathbf{I})$, then $p(\mathbf{I}) = f(\mathbf{I})$. But this method will involve an uncountable number of filters, and each filter $F^{(\xi)}$ is as large as image $\mathbf{I}$.

Our second motivation comes from psychophysical research on human texture perception, which suggests that two homogeneous textures are often difficult to discriminate when they produce similar *marginal distributions* for responses from *a bank of filters* [18]. This means that it is plausible to ignore some statistical properties of $f(\mathbf{I})$ which are not important for human texture discrimination.

To make texture modeling a tractable problem, we make the following assumptions to limit the number of filters and the window size of each filter for computational reasons, though these assumptions are not necessary conditions for our theory to hold true. First, we limit our model to homogeneous textures; thus, $f(\mathbf{I})$ is stationary with respect to location s. Second, for a given texture, all features which concern texture perception can be captured by "locally" supported filters. In other words, the sizes of filters should be smaller than the size of the image. For example, the size of the image is 256×256 pixels, and the sizes of filters we used are limited to less than 33×33 pixels. These filters can be linear or nonlinear as we discussed in Sect. 3.3. Third, only a finite set of filters are used to estimate $f(\mathbf{I})$.

The first and second assumptions are made because we often have access to only one observed (training) texture image. For a given observed image $\mathbf{I}^{obs}$ and a filter $F^{(\alpha)}$, we let $\mathbf{I}^{obs(\alpha)}$ denote the filtered image and $H^{obs(\alpha)}(z)$ the histogram of $\mathbf{I}^{obs(\alpha)}$. According to the first assumption, $f_s^{(\alpha)}(z) = f^{(\alpha)}(z)$ is independent of s. By ergodicity, $H^{obs(\alpha)}(z)$ makes a consistent estimator of $f^{(\alpha)}(z)$. The second assumption ensures that the image size is larger relative to the support of filters so that ergodicity takes effect for $H^{obs(\alpha)}(z)$ to be an accurate estimate of $f^{(\alpha)}(z)$.

Now for a specific texture, let $S_K = \{F^{(\alpha)}, \alpha = 1, \ldots, K\}$ be a finite set of well-selected filters and $f^{(\alpha)}(z), \alpha = 1, \ldots, K$, be the corresponding marginal distributions of $f(\mathbf{I})$. We denote the probability distribution $p(\mathbf{I})$ that matches these marginal distributions as a set

$$\Omega_{pK} = \left\{ p(\mathbf{I}) \mid E_p[\delta(z - \mathbf{I}_s^{(\alpha)})] = f^{(\alpha)}(z) \quad \forall z \in R, \ \forall \alpha = 1, \ldots, K, \ \forall s \in \mathcal{D} \right\},$$
(3.52)

where $E_p[\delta(z - \mathbf{I}_s^{(\alpha)})]$ is the marginal distribution of $p(\mathbf{I})$ with respect to filter $F^{(\alpha)}$ at location s. Thus according to the third assumption, any $p(\mathbf{I}) \in \Omega_{pK}$ is perceptually an adequate model for the texture, provided that we have enough well-selected filters. Then we choose from Ω_{pK} a distribution $p(\mathbf{I})$ which has the maximum entropy:

$$p(\mathbf{I}) = \arg\max \left\{ -\int p(\mathbf{I}) \log p(\mathbf{I}) d\mathbf{I} \right\},$$
(3.53)

subject to $E_p[\delta(z - \mathbf{I}_s^{(\alpha)})] = f^{(\alpha)}(z), \quad \forall z \in R, \ \forall \alpha = 1, \ldots, K, \ \forall s \in \mathcal{D},$

and $\int p(\mathbf{I}) d\mathbf{I} = 1.$
(3.54)

We can achieve the maximum entropy (ME) distribution by making $p(\mathbf{I})$, while it satisfies constraints in some dimensions, as random as possible in the other unconstrained dimensions. This makes it such that $p(\mathbf{I})$ is the most universal set it could be given its constraints; it does not capture, or represent, any additional constraints that are not explicitly defined by $p(\mathbf{I})$. In this case, by not representing more information than we have available, we say the ME distribution exhibits the purest fusion of the extracted features.

Using Lagrange multipliers, solving the constrained optimization problem in Eq. (3.53) is equal to solving the underlying unconstrained problem:

$$
p(\mathbf{I}) = \arg\max \left\{ -\int p(\mathbf{I}) \log p(\mathbf{I}) d\mathbf{I} \right.
$$

$$
\left. - \sum_s \sum_{a=1}^{K} \int \lambda^{(\alpha)}(z)(E_p[\delta(z - \mathbf{I}_s^{(\alpha)})] - f^{(\alpha)}(z)) dz - \lambda^{(\beta)} \left(\int p(\mathbf{I}) d\mathbf{I} - 1 \right) \right\} \quad (3.55)
$$

$$
= \arg\max \left\{ -\int p(\mathbf{I}) \log p(\mathbf{I}) d\mathbf{I} \right.
$$

$$
\left. - \sum_s \sum_{a=1}^{K} \int \lambda^{(\alpha)}(z) \left(\int \delta(z - \mathbf{I}_s^{(\alpha)}) p(\mathbf{I}) d\mathbf{I} - f^{(\alpha)}(z) \right) dz - \lambda^{(\beta)} \left(\int p(\mathbf{I}) d\mathbf{I} - 1 \right) \right\},
$$

$$(3.56)$$

where $\lambda^{(\alpha)}(z)\alpha = 1, 2, \ldots, K$ and $\lambda^{(\beta)}$ are Lagrange multipliers we introduce. Note that because the constraints on Eq. (3.53) differ from the ones given in Sect. 3.4 in that z takes continuous real values and there is an uncountable number of constraints. Therefore, the Lagrange parameter λ takes the form of a function of z. Also, since the constraints are the same for all locations $s \in \mathcal{D}$, λ should be independent of s. According to the Euler–Lagrange equation, if we want to find the stationary point (maximum or minimum) for the functional

$$
S(f(x)) = \int_a^b L(x, f(x), \dot{f}(x)) dx, \quad (3.57)
$$

then the function $f(x)$ should satisfy

$$
\frac{\partial L}{\partial f} - \frac{d}{dx} \frac{\partial L}{\partial \dot{f}} = 0. \quad (3.58)
$$

Applying this to Eq. (3.56), we can derive that the optimal $p(\mathbf{I})$ should satisfy

$$
-\log p(\mathbf{I}) - 1 - \sum_s \sum_{\alpha=1}^{K} \int \lambda^{(\alpha)}(z)\delta(z - \mathbf{I}_s^{\alpha}) dz - \lambda^{(\beta)} = 0,
$$

$$
p(\mathbf{I}) = \exp \left\{ -1 - \lambda^{(\beta)} - \sum_s \sum_{\alpha=1}^{K} \int \lambda^{(\alpha)}(z)\delta(z - \mathbf{I}_s^{\alpha}) dz \right\}. \quad (3.59)
$$

By reorganizing Eq. (3.59), we can get the following ME distribution:

$$p(\mathbf{I}; \Lambda_K, S_K) = \frac{1}{Z(\Lambda_K)} \exp\left\{ -\sum_s \sum_{\alpha=1}^{K} \int \lambda^{(\alpha)}(z)\delta(z - \mathbf{I}_s^{\alpha})dz \right\} \quad (3.60)$$

$$= \frac{1}{Z(\Lambda_K)} \exp\left\{ -\sum_s \sum_{\alpha=1}^{K} \lambda^{(\alpha)}(\mathbf{I}_s^{(\alpha)}) \right\}, \quad (3.61)$$

where $S_K = \{F^{(1)}, F^{(2)}, \ldots, F^{(K)}\}$ is a set of selected filters and $\Lambda_K = (\lambda^{(1)}(\cdot), \lambda^{(2)}(\cdot), \ldots, \lambda^{(K)}(\cdot))$ is the Lagrange parameter. $Z(\Lambda_K)$ is the normalizing constant that contains the term $\exp\{-1-\lambda^{(\beta)}\}$. Note that in Eq. (3.61), for each filter $F^{(\alpha)}$, $\lambda^{(\alpha)}(\cdot)$ takes the form as a continuous function of the filter response $\mathbf{I}_s^{(\alpha)}$.

To proceed further, let us derive a discrete form of Eq. (3.61). Assume that the filter response $\mathbf{I}_s^{(\alpha)}$ is discretized into L discrete gray levels, and as such z takes values from $\{z_1^{(\alpha)}, z_2^{(\alpha)}, \ldots, z_L^{(\alpha)}\}$. In general, the width of these bins does not have to be equal, and the number of gray levels L for each filter response may vary. As a result, the marginal distributions and histograms are approximated by piecewise constant functions of L bins, and we denote these piecewise functions as vectors. $H^{(\alpha)} = (H_1^{(\alpha)}, H_2^{(\alpha)}, \ldots, H_L^{(\alpha)})$ is the histogram of $\mathbf{I}^{(\alpha)}$, $H^{obs(\alpha)}$ denotes the histogram of $\mathbf{I}^{obs(\alpha)}$, and the potential function $\lambda^{(\alpha)}(\cdot)$ is approximated by vector $\lambda^{(\alpha)} = (\lambda_1^{(\alpha)}, \lambda_2^{(\alpha)}, \ldots, \lambda_L^{(\alpha)})$.

So Eq. (3.60) is rewritten as

$$p(\mathbf{I}; \Lambda_K, S_K) = \frac{1}{Z(\Lambda_K)} \exp\left\{ -\sum_s \sum_{\alpha=1}^{K} \sum_{i=1}^{L} \lambda_i^{(\alpha)} \delta(z_i^{(\alpha)} - \mathbf{I}_s^{(\alpha)}) \right\}, \quad (3.62)$$

and by changing the order of summations:

$$p(\mathbf{I}; \Lambda_K, S_K) = \frac{1}{Z(\Lambda_K)} \exp\left\{ -\sum_{\alpha=1}^{K} \sum_{i=1}^{L} \lambda_i^{(\alpha)} H_i^{(\alpha)} \right\}$$

$$= \frac{1}{Z(\Lambda_K)} \exp\left\{ -\sum_{\alpha=1}^{K} \langle \lambda^{(\alpha)}, H^{(\alpha)} \rangle \right\}. \quad (3.63)$$

The virtue of Eq. (3.63) is that it provides us with a simple parametric model for the probability distribution on $\mathbf{I}$, and this model has the following properties:

- $p(\mathbf{I}; \Lambda_K, S_K)$ is specified by $\Lambda_K = (\lambda^{(1)}, \lambda^{(2)}, \ldots, \lambda^{(K)})$ and S_K.
- Given an image $\mathbf{I}$, its histograms $H^{(1)}, H^{(2)}, \ldots, H^{(K)}$ are sufficient statistics, i.e., $p(\mathbf{I}; \Lambda_K, S_K)$ is a function of $(H^{(1)}, H^{(2)}, \ldots, H^{(K)})$.

We plug Eq. (3.63) into the constraints of the ME distribution and solve for $\lambda^{(\alpha)}$, $\alpha = 1, 2, \ldots, K$ iteratively by the following equations:

$$\frac{d\lambda^{(\alpha)}}{dt} = E_{p(\mathbf{I};\Lambda_K,S_K)}\left[H^{(\alpha)}\right] - H^{obs(\alpha)}. \tag{3.64}$$

In Eq. (3.64), we have substituted $H^{obs(\alpha)}$ for $f^{(\alpha)}$, and $E_{p(\mathbf{I};\Lambda_K,S_K)}(H^{(\alpha)})$ is the expected histogram of the filtered image $\mathbf{I}^{(\alpha)}$ where $\mathbf{I}$ follows $p(\mathbf{I}; \Lambda_K, S_K)$ with the current Λ_K. We converge to the unique solution at $\Lambda_K = \hat{\Lambda}_K$, and $\hat{\Lambda}_K$ is called the ME-estimator.

It is worth mentioning that this ME-estimator is equivalent to the maximum likelihood estimator (MLE):

$$\hat{\Lambda}_K = \underset{\Lambda_K}{\arg\max} \log p(\mathbf{I}^{obs}; \Lambda_K, S_K)$$

$$= \underset{\Lambda_K}{\arg\max} \left[-\log Z(\Lambda_K) - \sum_{\alpha=1}^{K} \langle \lambda^{(\alpha)}, H^{obs(\alpha)} \rangle \right]. \tag{3.65}$$

By gradient ascent, maximizing the log-likelihood lends Eq. (3.64), following the property of the partition function $Z(\Lambda_K)$. In Eq. (3.64), at each step, given Λ_K and hence $p(\mathbf{I}; \Lambda_K, S_K)$, the analytic form of $E_{p(\mathbf{I};\Lambda_K,S_K)}(H^{(\alpha)})$ is not available; instead, we propose the following method to estimate it, as we did for $f^{(\alpha)}$ before. We draw a typical sample from $p(\mathbf{I}; \Lambda_K, S_K)$ and thus synthesize a texture image $\mathbf{I}^{syn}$. Then we use the histogram $H^{syn(\alpha)}$ of $\mathbf{I}^{syn(\alpha)}$ to approximate $E_{p(\mathbf{I};\Lambda_K,S_K)}(H^{(\alpha)})$. This requires that the size of $\mathbf{I}^{syn}$ that we are synthesizing should be large enough.

To draw a typical sample image from $p(\mathbf{I}; \Lambda_K, S_K)$, we use the Gibbs sampler which simulates a Markov chain in the image space $\mathcal{L}^{|\mathcal{D}|}$. The Markov chain starts from any random image, e.g., a white noise image, and it converges to a stationary process with distribution $p(\mathbf{I}; \Lambda_K, S_K)$. Thus, when the Gibbs sampler converges, the images synthesized follow distribution $p(\mathbf{I}; \Lambda_K, S_K)$.

In summary, we propose Algorithm 1 for inferring the underlying probability model $p(\mathbf{I}; \Lambda_K, S_K)$ and for synthesizing the texture according to $p(\mathbf{I}; \Lambda_K, S_K)$. The algorithm stops when the sub-band histograms of the synthesized texture closely match the corresponding histograms of the observed images (Fig. 3.18).[1]

In Algorithm 2, to compute $p(\mathbf{I}_s = val \mid \mathbf{I}_{-s})$, we set $\mathbf{I}_s$ to val, and due to the Markov property we only need to compute the changes of $\mathbf{I}^{(\alpha)}$ in the neighborhood of s. The size of the neighborhood is determined by the size of filter $F^{(\alpha)}$. With the updated $\mathbf{I}^{(\alpha)}$, we calculate $H^{(\alpha)}$, and the probability is normalized such that $\sum_{val} p(\mathbf{I}_s = val \mid \mathbf{I}_{-s}) = 1$.

---

[1] We assume the histogram of each sub-band $\mathbf{I}^{(\alpha)}$ is normalized such that $\sum_i H_i^{(\alpha)} = 1$, and therefore all the $\{\lambda_i^{(\alpha)}, i = 1, \ldots, L\}$ computed in this algorithm have one extra degree of freedom for each α, i.e., we can increase $\{\lambda_i^{(\alpha)}, i = 1, \ldots, L\}$ by a constant without changing $p(\mathbf{I}; \Lambda_K, S_K)$. This constant will be absorbed by the partition function $Z(\Lambda_K)$.

Input a texture image $\mathbf{I}^{obs}$.
Select a group of K filters $S_K = \{F^{(1)}, F^{(2)}, \ldots, F^{(K)}\}$.
Compute $\{H^{obs(\alpha)}, \quad \alpha = 1, \ldots, K\}$.
Initialize $\lambda_i^{(\alpha)}$, 0, $i = 1, 2, \ldots, L$, or $= 1, 2, \ldots, K$
Initialize $\mathbf{I}^{syn}$ as a uniform white noise texture.[2]
Repeat
     Calculate $H^{syn(\alpha)}$ $\alpha = 1, 2, \ldots, K$ from $\mathbf{I}^{syn}$, use it for
$\mathbb{E}_{p(\mathbf{I}; \Lambda_K, S_K)}(H^{(\alpha)})$.
     Update $\lambda^{(\alpha)}$ $\alpha = 1, 2, \ldots, K$ by Equation 3.64), so $p(\mathbf{I}; \Lambda_K, S_K)$ is
updated.
     Apply Gibbs sampler to flip $\mathbf{I}^{syn}$ for w sweeps under $p(\mathbf{I}; \Lambda_K, S_K)$
Until $\frac{1}{2} \sum_i^L | H_i^{obs(\alpha)} - H_i^{syn(\alpha)} | \leq \epsilon$ for $\alpha = 1, 2, \ldots, K$.

Algorithm 1: The learning algorithm

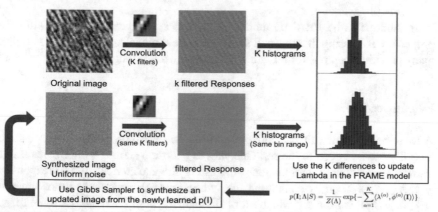

Stop until the 2 histograms look the same for each K filters. If a model can synthesize an image that has K similar filtered responses as those of original image, then the synthesized image is similar to the original image.

Fig. 3.18 Algorithm 1: Given a texture image and K filters (after filter selection), perform convolution to extract K histograms from the filtered response, denoting each as $h^{obs(\alpha)}$. Synthesize an image (initialized as white noise) that is of the same size as the input image, and perform convolution to extract another K histograms from the filtered response, denoting each as $h^{syn(\alpha)}$. Update the coefficient of the FRAME model λ by $\lambda_{t+1}^{\alpha} = \lambda_t^{\alpha} + \eta(h^{syn(\alpha)} - h^{obs(\alpha)})$. With this new model, we can again use the Gibbs sampler for W sweeps to synthesize another image for the next iteration

In the Gibbs sampler, flipping a pixel is a step of the Markov chain, and we define flipping $|\mathcal{D}|$ pixels as a sweep, where $|\mathcal{D}|$ is the size of the synthesized image. Then the overall iterative process becomes an inhomogeneous Markov chain. At the beginning of the process, $p(\mathbf{I}; \Lambda_K, S_K)$ is a "hot" uniform distribution. By updating

---

[2] The white noise image with uniform distribution is the samples from $p(\mathbf{I}; \Lambda_K, S_K)$ with $\lambda_i^{(\alpha)} = 0$.

the parameters, the process gets closer and closer to the target distribution, which is much colder. So the algorithm is very much like a simulated annealing algorithm, which is helpful for getting around local modes of the target distribution.

> Given image $\mathbf{I}_s$, flip_counter$\leftarrow 0$
> Repeat
>   Randomly pick a location s under the uniform distribution.
>   For $val = 0, \ldots, G - 1$ with G being the number of grey levels of $\mathbf{I}$
>   Calculate $p(\mathbf{I}_s = val \mid \mathbf{I}_{-s})$ by $p(\mathbf{I}; \Lambda_K, S_K)$.
>   Randomly flip $\mathbf{I}_s \leftarrow val$ under $p(val \mid \mathbf{I}_{-s})$.
>   flip_counter $\leftarrow$ flip_counter $+ 1$
> Until flip_counter $= w \times M \times N$.

Algorithm 2: The Gibbs Sampler for w sweeps

In summary, the FRAME model incorporates and generalizes the attractive properties of filtering theories and the random field models. Moreover, it interprets many previous methods for texture modeling with a unifying perspective.

Filter Selection

After Λ is learned with the above algorithm, we can now proceed to select the next filter to be added to the set. One way to choose S_K from a filter bank $\mathcal{B}$ is to search for all possible combinations of K filters in $\mathcal{B}$ and compute the corresponding $p(\mathbf{I}; \Lambda_K, S_K)$. Then by comparing the synthesized texture images following each $p(\mathbf{I}; \Lambda_K, S_K)$, we can see which set of filters is the best. Such a brute force search is computationally infeasible, and for a specific texture, we often do not know what K is. Instead, we propose a stepwise greedy strategy. We start from $S_0 = \emptyset$ (hence $p(\mathbf{I}; \Lambda_0, S_0)$ is a uniform distribution) and then sequentially introduce one filter at a time. Namely, we want to find a filter that reduces KL$(f \| p(\mathbf{I}; \Lambda))$ the most (by the minimum entropy principle).

Suppose that at the k-th step we have chosen $S_k = \{F^{(1)}, F^{(2)}, \ldots, F^{(k)}\}$ and obtained a maximum entropy distribution

$$p(\mathbf{I}; \Lambda_k, S_k) = \frac{1}{Z(\Lambda_k)} \exp \left\{ -\sum_{\alpha=1}^{k} \langle \lambda^{(\alpha)}, H^{(\alpha)} \rangle \right\}, \tag{3.66}$$

so that $E_{p(\mathbf{I}; \Lambda_k, S_k)}[H^{(\alpha)}] = f^{(\alpha)}$ for $\alpha = 1, 2, \ldots, k$. Then at the $(k + 1)$-th step, for each filter $F^{(\beta)} \in \mathcal{B}/S_k$, we denote by $d(\beta) = D\left(E_{p(\mathbf{I}; \Lambda_k, S_k)}[H^{(\beta)}], f^{(\beta)}\right)$ the distance between $E_{p(\mathbf{I}; \Lambda_k, S_k)}[H^{(\beta)}]$ and $f^{(\beta)}$, which are, respectively, the marginal distributions of $p(\mathbf{I}; \Lambda_k, S_k)$ and $f(\mathbf{I})$ with respect to filter $F^{(\beta)}$. Intuitively, the bigger $d(\beta)$ is, the more information $F^{(\beta)}$ carries, since it reports a big difference between $p(\mathbf{I}; \Lambda_k, S_k)$ and $f(\mathbf{I})$. Therefore we should choose the filter which has the maximal distance, i.e.,

$$F^{k+1} = \underset{F^{(\beta)} \in \mathcal{B}/S_k}{\arg\max} \ D\left(\mathrm{E}_{p(\mathbf{I}; \Lambda_k, S_k)}[H^{(\beta)}], f^{(\beta)}\right). \qquad (3.67)$$

More formally, define the Kullback–Leibler divergence between the data distribution f and a model distribution p to be

$$\mathrm{KL}(f \,\|\, p) = \mathrm{E}_f\left[\log \frac{f(\mathbf{I})}{p(\mathbf{I})}\right], \qquad (3.68)$$

which measures the "distance" between f and p.

We can choose the next filter to be the one that maximizes reduction in $\mathrm{KL}(f \,\|\, p(\mathbf{I}; \Lambda))$, where $p(\mathbf{I}; \Lambda)$ is the maximum entropy distribution. This indicates that we want to maximize

$$D\left(\mathrm{E}_{p(\mathbf{I}; \Lambda_{k+1}, S_{k+1})}[H^{(\beta)}], f^{(\beta)}\right) = \mathrm{KL}(f \,\|\, p(\mathbf{I}; \Lambda_k, S_k)) - \mathrm{KL}(f \,\|\, p(\mathbf{I}; \Lambda_{k+1}, S_{k+1})). \qquad (3.69)$$

A key observation is

$$\mathrm{E}_f\left[\log p(\mathbf{I}; \Lambda_k, S_k)\right] = \mathrm{E}_f\left[-\sum_{\alpha=1}^{k} \langle \lambda^{(\alpha)}, H^{(\alpha)} \rangle - \log Z(\Lambda_k)\right] \qquad (3.70)$$

$$= \mathrm{E}_{p(\mathbf{I}; \Lambda_k, S_k)}\left[-\sum_{\alpha=1}^{k} \langle \lambda^{(\alpha)}, H^{(\alpha)} \rangle - \log Z(\Lambda_k)\right] \quad (3.71)$$

$$= \mathrm{E}_{p(\mathbf{I}; \Lambda_k, S_k)}\left[\log p(\mathbf{I}; \Lambda_k, S_k)\right], \qquad (3.72)$$

because $\mathrm{E}_f\left[H^{(\alpha)}\right] = \mathrm{E}_{p(\mathbf{I}; \Lambda_k, S_k)}\left[H^{(\alpha)}\right]$.
For the same reason,

$$\mathrm{E}_f\left[\log p(\mathbf{I}; \Lambda_{k+1}, S_{k+1})\right] = \mathrm{E}_{p(\mathbf{I}; \Lambda_{k+1}, S_{k+1})}\left[\log p(\mathbf{I}; \Lambda_{k+1}, S_{k+1})\right], \quad (3.73)$$

and

$$\mathrm{E}_{p(\mathbf{I}; \Lambda_{k+1}, S_{k+1})}\left[\log p(\mathbf{I}; \Lambda_k, S_k)\right] = \mathrm{E}_{p(\mathbf{I}; \Lambda_k, S_k)}\left[\log p(\mathbf{I}; \Lambda_k, S_k)\right], \quad (3.74)$$

again due to the matching of marginal histograms.

Thus we have the following:

Proposition 1

$$KL(f \| p(\mathbf{I}; \Lambda_k, S_k)) - KL(f \| p(\mathbf{I}; \Lambda_{k+1}, S_{k+1}))) = KL(p(\mathbf{I}; \Lambda_{k+1}, S_{k+1}) \| p(\mathbf{I}; \Lambda_k, S_k))$$

$$(3.75)$$

$$= \text{entropy}(p(\mathbf{I}; \Lambda_k, S_k) - \text{entropy}(p(\mathbf{I}; \Lambda_{k+1}, S_{k+1})). \quad (3.76)$$

Proof To simplify notation, we denote $p(\mathbf{I}; \Lambda_k, S_k)$ as p and $p(\mathbf{I}; \Lambda_{k+1}, S_{k+1}))$ as p_+:

$$
\begin{aligned}
KL(f \| p) - KL(f \| p_+) &= E_f[\log f(\mathbf{I}) - \log p(\mathbf{I})] - E_f[\log f(\mathbf{I}) - \log p_+(\mathbf{I})] \\
&= E_f[-\log p(\mathbf{I})] - E_f[-\log p_+(\mathbf{I})] \\
&= E_{p_+}[-\log p(\mathbf{I})] - E_{p_+}[-\log p_+(\mathbf{I})] \\
&= KL(p_+ \| p) \\
&= E_p[-\log p(\mathbf{I})] - E_{p_+}[-\log p_+(\mathbf{I})] \\
&= \text{entropy}(p) - \text{entropy}(p_+). \quad (3.77)
\end{aligned}
$$

$\square$

Thus we want to select the next filter to maximally reduce the entropy of the maximum entropy model.

Simplifying the distance measure as $D(h)$, where h is the evaluated histogram feature using the newly added filter, L_p-norm is a reasonable approximation to $D(h)$. Specifically, denote p as the distribution modeled prior to adding the new filter, and then let h_0 be the feature evaluated using the new filter on samples from p and h^{obs} be the feature evaluated using the new filter on samples from true distribution f. Using Taylor expansion,

$$D(h^{obs}) \approx D(h)\Big|_{h=h_0} + \frac{\partial D}{\partial h}\Big|_{h=h_0}(h^{obs} - h_0) + \frac{1}{2}(h^{obs} - h_0)^{\top}\frac{\partial^2 D}{\partial h^2}\Big|_{h=h_0}(h^{obs} - h_0),$$

$$(3.78)$$

where $\frac{\partial D}{\partial h} = -\lambda^{k+1}$ and $\frac{\partial^2 D}{\partial h^2}$ is the covariance matrix $\Sigma_{h^{obs}}$ of each component of h^{obs} evaluated using the new filter. As λ^{k+1} is typically initialized to 0, we also have $D(h)\Big|_{h=h_0} = 0$ and $\frac{\partial D}{\partial h}\Big|_{h=h_0}(h^{obs} - h_0) = 0$.

Furthermore, since we assume each component is independent of each other, we can approximate the second order term by L_2 distance. In general, L_p distance metric is a good surrogate for $D(h)$.

As L_p-norm is directly computable and an approximation to information gain, we typically choose it to measure the distance $d(\beta)$, which is also an approximation of the reduction in bits, i.e.,

$$F^{k+1} = \arg\max_{F^{(\beta)} \in \mathcal{B}/S_k} \frac{1}{2} \left\| f^{(\beta)} - E_{p(\mathbf{I}; \Lambda_k, S_k)} \left[H^{(\beta)} \right] \right\|_p . \tag{3.79}$$

For the following, we choose $p = 1$. To estimate $f^{(\eta)}$ and $E_{p(\mathbf{I}; \Lambda_k, S_k)}[H^{(\eta)}]$, we applied $F^{(\beta)}$ to the observed image $\mathbf{I}^{obs}$ and the synthesized image $\mathbf{I}^{syn}$ sampled from the $p(\mathbf{I}; \Lambda_k, S_k)$ and substitute the histograms of the filtered images for $f^{(\beta)}$ and $E_{p(\mathbf{I}; \Lambda_k, S_k)}[H^{(\beta)}]$, i.e.,

$$F^{k+1} = \arg\max_{F^{(\beta)} \in \mathcal{B}/S_k} \frac{1}{2} \left| H^{obs(\beta)} - H^{syn(\beta)} \right| . \tag{3.80}$$

The filter selection procedure stops when the $d(\beta)$ for all filters $F^{(\beta)}$ in the filter bank is smaller than a constant ϵ. Due to fluctuation, various patches of the same observed texture image often have a certain amount of histogram variance, and we use such a variance for ϵ.

Finally, we propose Algorithm 3 for filter selection.

Let $\mathcal{B}$ be a bank of filters, S the set of selected filters, $\mathbf{I}^{obs}$ the observed texture image,
and $\mathbf{I}^{syn}$ the synthesized texture image.
    Initialize $k = 0$, $S \leftarrow \emptyset$, $p(\mathbf{I}) \leftarrow$ uniform dist. $\mathbf{I}^{syn} \leftarrow$ uniform noise.
    For $\alpha = 1, \ldots, |\mathcal{B}|$ do
        Compute $\mathbf{I}^{obs(\alpha)}$ by applying $F^{(\alpha)}$ to $\mathbf{I}^{obs}$.
        Compute histogram $H^{obs(\alpha)}$ of $\mathbf{I}^{obs(\alpha)}$.
    Repeat
        For each $F^{(\beta)} \in \mathcal{B}/S$ do
            Compute $\mathbf{I}^{syn(\beta)}$ by applying $F^{(\beta)}$ to $\mathbf{I}^{syn}$
            Compute histogram $H^{syn(\beta)}$ of $\mathbf{I}^{syn(\beta)}$
            $d(\beta) = \frac{1}{2} | H^{obs(\beta)} - H^{syn(\beta)} |$,[3]
        Choose F^{k+1} so that $d(k + 1) = \max\{d(\beta) : \forall F^{(\beta)} \in \mathcal{B}/S\}$
        $S \leftarrow S \cup \{F^{k+1}\}$, $k \leftarrow k + 1$.
        Starting from $p(\mathbf{I})$ and $\mathbf{I}^{syn}$, run algorithm 1 to compute new $p^*(\mathbf{I})$ and $\mathbf{I}^{syn}$.
        $p(\mathbf{I}) \leftarrow p^*(\mathbf{I})$ and $\mathbf{I}^{syn} \leftarrow \mathbf{I}^{syn}$.
    Until $d^{(\beta)} < \epsilon$.

Algorithm 3: Filter selection

---

[3] Since both histograms are normalized to have $sum = 1$, then $error \in [0, 1]$. We note this measure is robust with respect to the choice of the bin number L (e.g. we can take $L = 16, 32, 64$), as well as the normalization of the filters.

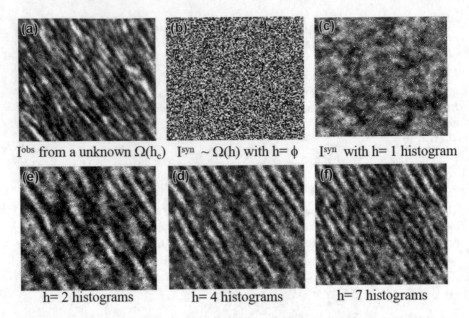

Fig. 3.19 Synthesis of a fur texture: (**a**) is the observed texture; (**b**),(**c**),(**d**),(**e**),(**f**) are the synthesized textures using $K = 0, 1, 2, 4, 7$ filters, respectively

Figure 3.19a is the observed image of animal fur. We start from the uniform noise image in Fig. 3.19b. The first filter picked by the algorithm is a Laplacian of Gaussian filter $LG(1.0)$ and its window size is 5×5. It has the largest error $d(\beta) = 0.611$ among all the filters in the filters bank. Then we synthesize texture as shown in Fig. 3.19c, which has almost the same histogram at the sub-band of this filter (the error $d(\beta)$ drops to 0.035).

To provide a final insight into the pursuit process from a distribution point of view, we refer to Fig. 3.20. Note that this is only a 2D view of the very high-dimensional distribution space. First, selecting a filter gives a (high-dimensional) constrained set of probability distributions satisfying a given feature statistic. Then we try to look for a distribution in this constrained set that is the closest to the random uniform distribution (by the Maximum Entropy Principle), as represented by the right angle. In fact, we are minimizing $\mathrm{KL}(p \| p_0)$, where $p \in \Omega_{p_1}$ and p_0 is the starting distribution. This objective is equivalent to maximizing entropy if and only if p_0 is random uniform. We leave this proof as an exercise for the reader. Second, we find the new filter that gives the largest reduction in $\mathrm{KL}(f \| p)$ (by the Minimum Entropy Principle), constraining the equivalence class of distributions further to a lower dimensional set. It is important to note that Ω_{p_2} has a lower dimension than Ω_{p_1} and every set after Ω_{p_2} has a lower dimension due to additional constraints. The true distribution f again also remains in each set, as it satisfies each additional feature statistic. In each new constrained set, we start again from the random uniform distribution and find a new closest distribution. The algorithm

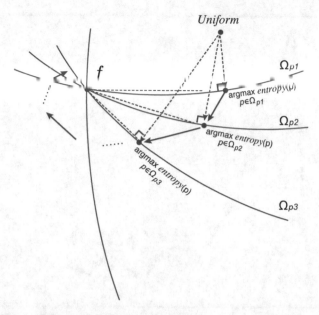

Fig. 3.20 A graphical view of the FRAME learning algorithm in distribution space

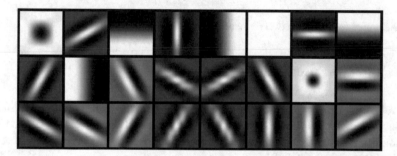

Fig. 3.21 Note the selected filter after each filter selection. Each time, we add the filter that reduces $\mathrm{KL}(f \| p(\mathbf{I}; \Lambda))$ the most

continues and we see from the solid line that our projected distribution "spirals" into the true data distribution as more constraints are put onto the distribution. Further details on information projection are presented in Chap. 9.

Figure 3.21 is the filter we selected and added into the filter set, which we use it the model. As we can see in the figure, most of the filters we selected are Gabor filters and we also need some other filters to help detect the edges and corners.

Figure 3.22 shows the curve of the weighted error per bin over the number of filters used for synthesis. As it is shown in the plot, after about 10 iterations, the weighted error is stable enough and can generate very reasonable images, which on the other hand perfectly match the result that given in Fig. 3.19.

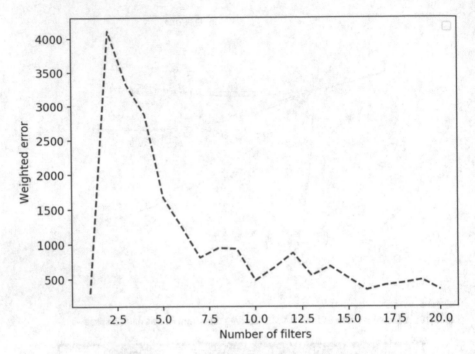

Fig. 3.22 The weighted error over the number of filters used

3.5   Texture Ensemble

The following section presents theoretical work on texture theories and establishes connections between FRAME, Statistical Physics, and the Julesz Ensemble. This section is highly technical and it is encouraged for interested readers to go through the original works [258, 259].

Ensembles in Statistical Physics

We have developed the FRAME model based on the maximum entropy principle, and in the following, we will show that there is a simpler and deeper way to derive or justify the FRAME model. This insight again originates from Statistical Physics, specifically the equivalence of ensembles.

Statistical physics is the subject of studying the macroscopic properties of a system involving a large number of elements (see Chandler [24]). Figure 3.23 illustrates three types of physical systems which are interesting to us.

1. Micro-canonical ensembles. Figure 3.23.a is an insulated system of N elements. The elements could be atoms, molecules, or electrons in systems such as gas, ferromagnetic material, fluid, etc. N is big, say $N = 10^{23}$, and is considered infinity.

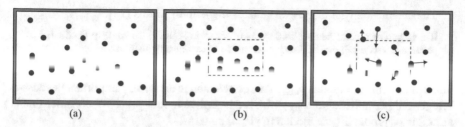

Fig. 3.23 Three types of ensembles in statistical physics. (**a**) Micro-canonical ensembles. (**b**) Canonical ensembles. (**c**) Grand-Canonical ensembles

The system is decided by a configuration or state $s = (\mathbf{x}^N, \mathbf{m}^N)$, where $\mathbf{x}^N$ describes the coordinates of the N elements and $\mathbf{m}^N$ their momenta [24]. It is impractical to study the exact motions of $6N$ vector s, and in fact, these microscopic states are less relevant, and people are more interested in the macroscopic properties of the system as a whole, say the number of elements N, the total energy $E(s)$, and total volume V. Other derivative properties are temperature, pressure, etc.

If we denote by $\mathbf{h}(s) = (N, E, V)$ the macroscopic properties, at thermodynamic equilibrium all microscopic states that satisfy this property make up a *micro-canonical ensemble*:

$$\Omega_{mce}(\mathbf{h}_o) = \{s = (\mathbf{x}^N, \mathbf{m}^N) : \mathbf{h}(s) = \mathbf{h}_o = (N, V, E)\}. \tag{3.81}$$

s is an *instance* and $\mathbf{h}(s)$ is *summary* of the system state for practical purposes. Obviously Ω_{mce} is a deterministic set or an equivalence class for all states that satisfy descriptive constraints $\mathbf{h}(s) = \mathbf{h}_o$.

An essential assumption in statistical physics is, as a first principle,

all microscopic states are equally likely at thermodynamic equilibrium.

This is simply a maximum entropy assumption. Let $\Omega \ni s$ be the space of all possible states, then $\Omega_{mce} \subset \Omega$ is associated with a uniform probability,

$$p_{\text{unif}}(s; \mathbf{h}_o) = \begin{cases} 1/|\Omega_{mce}(\mathbf{h}_o)| & \text{for } s \in \Omega_{mce}(\mathbf{h}_o), \\ 0 & \text{for } s \in \Omega/\Omega_{mce}(\mathbf{h}_o). \end{cases} \tag{3.82}$$

2. Canonical ensembles. The canonical ensemble refers to a small system (with fixed volume V_1 and elements N_1) embedded in a micro-canonical ensemble (see Fig. 3.23b). The canonical ensemble can exchange energy with the rest of the system (called the heat bath or reservoir). The system is relatively small, e.g., $N_1 = 10^{10}$, so the bath can be considered a micro-canonical ensemble itself.

At thermodynamic equilibrium, the microscopic state s_1 for the small system follows a Gibbs distribution,

$$p_{\text{Gib}}(s_1; \beta) = \frac{1}{Z} \exp\{-\beta E(s_1)\}. \tag{3.83}$$

The conclusion was stated as a general theorem by Gibbs [78],

If a system of a great number of degrees of freedom is micro-canonically distributed in phase, any very small part of it may be regarded as canonically distributed.

In accordance with this theorem, the Gibbs model p_{Gib} is a conditional probability of the uniform model p_{unif}. This conclusion is extremely important because it bridges a deterministic set Ω_{mce} with a descriptive model p_{Gib}. We consider this is an origin of probability in modeling visual patterns.

3. *Grand-Canonical ensembles.* When the small system with fixed volume V_1 can also exchange elements with the bath as in liquid and gas materials, then it is called a grand-canonical ensemble (see Fig. 3.23c). The grand-canonical ensemble follows a distribution,

$$p_{\text{gce}}(s_1; \beta_o, \beta) = \frac{1}{Z} \exp\{-\beta_o N_1 - \beta E(s_1)\}, \tag{3.84}$$

where an extra parameter β_o controls the number of elements N_1 in the ensemble. If we replace the energy of the physical system with the feature statistics of the texture image, then we will arrive at a mathematical model for textures.

Texture Ensemble

To study real-world textures, one needs to characterize the dependency between pixels by extracting spatial features and calculating some statistics averaged over the image. One main theme of texture research is to seek the essential ingredients in terms of features and statistics $\mathbf{h}(\mathbf{I})$, which are the bases for human texture perception. From now on, we use the bold font $\mathbf{h}$ to denote statistics of image features.

In the literature, the search for $\mathbf{h}$ has converged to marginal histograms of Gabor filter responses. We believe that some bins of joint statistics may also be important as long as they can be reliably estimated from finite observations.

Given K Gabor filters as feature detectors $\{F^{(1)}, \ldots, F^{(K)}\}$, we convolve the filters with image $\mathbf{I}$ to obtain the sub-band filtered images $\{\mathbf{I}^{(1)}, \ldots, \mathbf{I}^{(K)}\}$, where $\mathbf{I}^{(k)} = F^{(k)} * \mathbf{I}$. Let $h^{(k)}$ be the normalized intensity histogram of $\mathbf{I}^{(k)}$; then the feature statistics $\mathbf{h}$ collects the normalized histograms of these K sub-band images,

$$\mathbf{h}(\mathbf{I}) = (h^{(1)}(\mathbf{I}), \ldots, h^{(K)}(\mathbf{I})). \tag{3.85}$$

We use $\mathbf{H}(\mathbf{I}) = (H^{(1)}(\mathbf{I}), \ldots, H^{(K)}(\mathbf{I}))$ to denote the unnormalized histograms. We assume that the boundary conditions of the images are properly handled (e.g., periodic boundary condition). It should be noted that the conclusions here hold as long as $\mathbf{h}(\mathbf{I})$ can be expressed as spatial averages of local image features. The marginal histograms of Gabor filter responses are only special cases.

Given statistics $\mathbf{h}(\mathbf{I})$, one can partition the image space Ω_D into equivalence classes $\Omega_D(\mathbf{h}) = \{\mathbf{I} : \mathbf{h}(\mathbf{I}) = \mathbf{h}\}$. For finite D, the exact constraint $\mathbf{h}(\mathbf{I}) = \mathbf{h}$ may not be satisfied, so we relax this constraint and replace $\Omega_D(\mathbf{h})$ by

$$\Omega_D(\mathcal{H}) = \{\mathbf{I} : \mathbf{h}(\mathbf{I}) \in \mathcal{H}\}, \tag{3.86}$$

with $\mathcal{H}$ being a small set around $\mathbf{h}$. Then we can define the uniform counting measure or the uniform probability distribution on $\Omega_D(\mathcal{H})$ as

$$q(\mathbf{I}; \mathcal{H}) = \begin{cases} 1/|\Omega_D(\mathcal{H})|, & \text{if } \mathbf{I} \in \Omega_D(\mathcal{H}), \\ 0, & \text{otherwise,} \end{cases} \tag{3.87}$$

where $|\Omega_D(\mathcal{H})|$ is the volume of or the number of images in $\Omega_D(\mathcal{H})$. Now we can define the Julesz ensemble as follows.

Definition 3 Given a set of feature statistics $\mathbf{h}(\mathbf{I}) = (h^{(1)}(\mathbf{I}), \ldots, h^{(K)}(\mathbf{I}))$, a Julesz ensemble of type $\mathbf{h}$ is a limit of $q(\mathbf{I}; \mathcal{H})$ as $D \to Z^2$ and $\mathcal{H} \to \mathbf{h}$ with some boundary condition.[4]

The Julesz ensemble is defined mathematically as the limit of a uniform counting measure. It is always helpful to imagine the Julesz ensemble of type $\mathbf{h}$ as the image set $\Omega_D(\mathbf{h})$ on a large D. Also, in the later calculation, we shall often ignore the minor complication that constraint $\mathbf{h}(\mathbf{I}) = \mathbf{h}$ may not be exactly satisfied.

Then, we are ready to give a mathematical definition for textures.

Definition 4 A texture pattern is a Julesz ensemble defined by a type $\mathbf{h}$ of the feature statistics $\mathbf{h}(\mathbf{I})$ (Figs. 3.24 and 3.25).

Type Theory and Entropy Rate Functions

A Simple Independent Model

In this section, we introduce basic concepts, such as type, ensemble, entropy function, typical images, and equivalence of ensembles, using a simple image model where the pixel intensities are independently and identically distributed (i.i.d.).

Let $\mathbf{I}$ be an image defined on a finite lattice $D \subset Z^2$, and the intensity at pixel $v \in D$ is denoted by $\mathbf{I}(v) \in \mathcal{G} = \{1, 2, \ldots, g\}$. Thus $\Omega_D = \mathcal{G}^{|D|}$ is the space of images on D, with $|D|$ being the number of pixels in D.

(1) The FRAME model for i.i.d. images. We consider a simple image model where pixel intensities are independently and identically distributed according to

---

[4] We assume $D \to Z^2$ in the sense of van Hove, i.e., the ratio between the boundary and the size of D goes to 0.

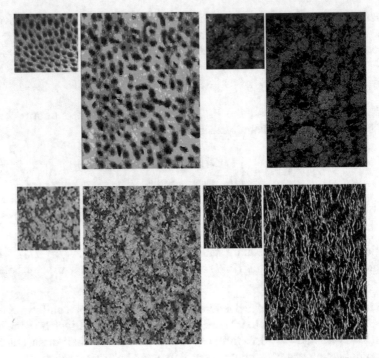

Fig. 3.24 For each pair of texture images, the image on the left is the observed image, and the image on the right is the image randomly sampled from the Julesz texture ensemble

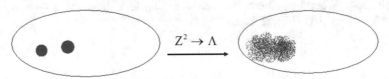

Fig. 3.25 Julesz ensembles of different textures are disjoint in the space of infinite images. They will overlap in the space of image patches of finite size

a probability $p = (p_1, \ldots, p_g)$ where $\sum_i p_i = 1$. The distribution of $\mathbf{I}$ can be written as a FRAME model

$$p(\mathbf{I}; \beta) = \prod_{v \in D} p_{\mathbf{I}(v)} = \prod_{i=1}^{g} p_i^{H_i(\mathbf{I})} = \exp\{\langle \log p, H(\mathbf{I}) \rangle\} = \exp\{\langle \beta, H(\mathbf{I}) \rangle\},$$

(3.88)

where $H(\mathbf{I}) = (H_1(\mathbf{I}), \ldots, H_g(\mathbf{I}))$ is the unnormalized intensity histogram of $\mathbf{I}$, i.e., H_i is the number of pixels whose intensities are equal to i. $\beta = (\log p_1, \ldots, \log p_g)$ is the parameter of $p(\mathbf{I}; \beta)$—a special case of the FRAME model.

(2) Type. Let $h(\mathbf{I}) = H(\mathbf{I})/|D|$ be the normalized intensity histogram. We call $h(\mathbf{I})$ the *type* of image $\mathbf{I}$.

Fig. 3.26 The partition of
the image space into
equivalence classes, and each
class corresponds to an h on
the probability simplex

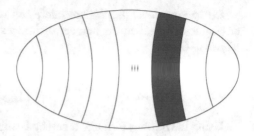

(3) Equivalent class. Let $\Omega_D(h)$ be the set of images with $h(\mathbf{I}) = h$,[5] i.e., $\Omega_D(h) = \{\mathbf{I} : h(\mathbf{I}) = h\}$. Then the image space is partitioned into equivalence classes

$$\Omega_D = \cup_h \Omega_D(h). \tag{3.89}$$

As shown in Fig. 3.26, each equivalence class $\Omega_D(h)$ is mapped into one type h on a *simplex*—a plane defined by $h_1 + \cdots + h_g = 1$ and $h_i \geq 0$, $\forall i$ in a g-dimensional space.

(4) The Julesz ensemble for i.i.d. images. The hard constraint in defining the equivalence class $\Omega_D(h)$ makes sense only in the limit as $D \to Z^2$, where statistical fluctuations vanish. Therefore, we may attempt to define the Julesz ensemble as the limit of $\Omega_D(h)$ as $D \to Z^2$, or even more directly, as the set of images $\mathbf{I}$ defined on Z^2 with $h(\mathbf{I}) = h$.

Unfortunately, the above "definitions" are not mathematically well defined. Instead, we need to define the Julesz ensemble in a slightly indirect way. First, we associate with each equivalence class $\Omega_D(h)$ a probability distribution $q(\mathbf{I}; h)$, which is uniform over $\Omega_D(h)$ and vanishes outside. Then, the *Julesz ensemble* of type h is defined to be the limit of $q(\mathbf{I}; h)$ as $D \to Z^2$.

For finite D, the equivalence class $\Omega_D(h)$ may be empty because $|D|h$ may not be integers. Thus, to be more rigorous, we should replace h by a small set $\mathcal{H}$ around h, and let $\mathcal{H}$ go to h as $D \to \infty$. For simplicity, however, we shall neglect this minor complication and simply treat $|D|h$ as integers.

The uniform distribution $q(\mathbf{I}; h)$ only serves as a counting measure of the equivalence class $\Omega_D(h)$, i.e., all the images in $\Omega_D(h)$ are counted equally. Therefore, any probability statement under the uniform distribution $q(\mathbf{I}; h)$ is equivalent to a frequency statement of images in $\Omega_D(h)$. For example, the probability that image $\mathbf{I}$ has a certain property under $q(\mathbf{I}; h)$ is actually the frequency or the proportion of images in $\Omega_D(h)$ that have this property. The limit of $q(\mathbf{I}; h)$ thus essentially defines a counting measure of the set of infinitely large images (defined on Z^2) with histogram h. With a little abuse of language, we sometimes also call the equivalence class $\Omega_D(h)$ defined on a large lattice D a Julesz ensemble, and it is always helpful to imagine a Julesz ensemble as such an equivalence class if the reader finds the limit of probability measures too abstract.

---

[5] We hope that the notation $h(\mathbf{I}) = h$ will not confuse the reader. The h on the left is a function of $\mathbf{I}$ for extracting statistics, while the h on the right is a specific value of the statistics.

(5) Entropy function. We are interested in computing the volume of the Julesz ensemble $\Omega_D(h)$, i.e., the number of images in $\Omega_D(h)$. We denote this volume by $|\Omega_D(h)|$. Clearly

$$|\Omega_D(h)| = \frac{|D|!}{\prod_{i=1}^{g}(h_i|D|)!}. \tag{3.90}$$

Using the Stirling formula, it can be easily shown that

$$\lim_{D \to Z^2} \frac{1}{|D|} \log |\Omega_D(h)| = \lim_{D \to Z^2} \frac{1}{|D|} \log \frac{|D|!}{\prod_{i=1}^{g}(h_i|D|)!} \tag{3.91}$$

$$= -\sum_{i=1}^{g} h_i \log h_i = \text{entropy}(h). \tag{3.92}$$

Thus for a large enough lattice, the volume of $\Omega_D(h)$ is said to be in the order of entropy(h), i.e.,

$$|\Omega_D(h)| \sim \exp^{|D|\text{entropy}(h)}. \tag{3.93}$$

For notational simplicity, we denote the entropy function by $s(h) = \text{entropy}(h)$.

(6) The probability rate function. Now we are ready to compute the total probability mass that $p(\mathbf{I}; \beta)$ assigns to an equivalence class $\Omega_D(h)$. We denote this probability by $p(\Omega_D(h); \beta)$. Because images in $\Omega_D(h)$ all receive equal probabilities, it can be shown that

$$\lim_{D \to Z^2} \frac{1}{|D|} \log p(\Omega_D(h); \beta) = \lim_{D \to Z^2} \frac{1}{|D|} \log \left\{ |\Omega_D(h)| \prod_{i=1}^{g} p_i^{|D|h_i} \right\} \tag{3.94}$$

$$= -\sum_{i=1}^{g} h_i \log \frac{h_i}{p_i} = -\text{KL}(h\|p), \tag{3.95}$$

where $\text{KL}(h\|p)$ denotes the Kullback–Leibler distance from h to p. $\text{KL}(h\|p) \geq 0$ for all h and p, with equality holding when $h = p$.

Thus, on a large enough lattice, the total probability mass of an equivalence class $\Omega_D(h)$ is said to be on the order of $-\text{KL}(h\|p)$,

$$p(\Omega_D(h); \beta) \sim \exp^{-|D|\text{KL}(h\|p)}, \tag{3.96}$$

where $-\text{KL}(h\|p)$ is the probability rate function and is denoted by $s_\beta(h) = -\text{KL}(h\|p)$.

(7) Typical vs. most likely images. Suppose among $p_1, \ldots, p_g$, p_m is the largest probability. Consider one extreme type h, with $h_m = 1$, and $h_i = 0, \forall i \neq m$. Then the image in this $\Omega_D(h)$ is the most likely image under model $p(\mathbf{I}; \beta)$, i.e.,

it receives the highest probability. However, $\Omega_D(h)$ has only one constant image, and the probability that $p(\mathbf{I}; \beta)$ assigns to this $\Omega_D(h)$ is essentially zero for large lattice. Now consider the equivalence class $\Omega_D(h = p)$. Each image in $\Omega_D(h = p)$ receives less probability than the most likely image, but overall, the total probability received by the whole $\Omega_D(h = p)$ is essentially 1 for large lattice. That is, if we sample from the FRAME model $p(\mathbf{I}; \beta)$, we will almost always get an image $\mathbf{I}$ from $\Omega_D(h = p)$. Such images are called the *typical images* of the FRAME model.

Having introduced the basic concepts, we now explain the basic ideas of ensemble equivalence in the next two subsections by going in both directions from one to the other.

From FRAME Model to Julesz Ensemble on Infinite Lattice

We study the limit of the FRAME model $p(\mathbf{I}; \beta)$ as $D \to Z^2$. A simple fact will be repeatedly used in this section. To see this fact, let us consider the following example. Suppose we have two terms, one is e^{5n} and the other is e^{3n}. Consider their sum $e^{5n} + e^{3n}$. As $n \to \infty$, the sum $e^{5n} + e^{3n}$ is dominated by e^{5n}, and the order of this sum is still 5, i.e., $\lim_{n \to \infty} \frac{1}{n} \log(e^{5n} + e^{3n}) = 5$. This means that for the sum of many terms, the term with the largest exponential order dominates the sum and the order of the sum is the largest order among the individual terms.

Suppose $\mathcal{H}$ is a set of types and

$$\Omega_D(\mathcal{H}) = \{\mathbf{I}; h(\mathbf{I}) \in \mathcal{H}\} \tag{3.97}$$

is the set of all images $\mathbf{I}$ whose type is within $\mathcal{H}$. Then from Eq. (3.96), we have

$$p(\mathbf{I} \in \Omega_D(\mathcal{H}); \beta) = \int_{\mathcal{H}} p(\Omega_D(h); \beta)dh \tag{3.98}$$

$$\sim \int_{\mathcal{H}} \exp^{-|D|\mathrm{KL}(h\|p)} dh \sim \exp^{-|D|\mathrm{KL}(h^*\|p)}, \tag{3.99}$$

where h^* is the type in $\mathcal{H}$ which has the minimum Kullback–Leibler divergence to p,

$$h^* = \arg \min_{h \in \mathcal{H}} \mathrm{KL}(h\|p). \tag{3.100}$$

That is, the total probability for $\Omega_D(\mathcal{H})$ has an exponential order $\mathrm{KL}(h^*\|p)$ and is dominated by the heaviest type h^*. In a special case $\Omega_D(\mathcal{H}) = \Omega_D$, i.e., the whole image space, we have $h^* = p$ and $\mathrm{KL}(h^*\|p) = 0$.

Clearly, as $|D| \to \infty$, the FRAME model quickly concentrates its probability mass on $\Omega_D(h = p)$ and assigns equal probabilities to images in $\Omega_D(h = p)$. For other $h \neq p$, the probabilities will decrease to 0 at an exponential rate. Thus, the FRAME model becomes a Julesz ensemble.

From Julesz Ensemble to FRAME Model on Finite Lattice

In this section, we tighten up the notation a little bit. We use $\mathbf{I}_D$ to denote the image defined on lattice D, and we use $\mathbf{I}_{D_0}$ to denote the image patch defined on $D_0 \subset D$. For a fixed type h of feature statistics, consider the uniform distribution $q(\mathbf{I}; h)$ on $\Omega_D(h)$. Under $q(\mathbf{I}; h)$, the distribution of $\mathbf{I}_{D_0}$, denoted by $q(\mathbf{I}_{D_0}; h)$, is well defined.[6] We shall show that if we fix D_0 and let $D \rightarrow Z^2$, then $q(\mathbf{I}_{D_0}; h)$ goes to the FRAME model (see Eq. (3.88)) with $p = h$.

First, the number of images in $\Omega_D(h)$ is

$$|\Omega_D(h)| = \frac{|D|!}{\prod_{i=1}^{m}(h_i|D|)!}. \tag{3.101}$$

Then, let us fix $\mathbf{I}_{D_0}$ and calculate the number of images in $\Omega_D(h)$ whose image value (i.e., intensities) on D_0 is $\mathbf{I}_{D_0}$. Clearly, for every such image, its image value on the rest of the lattice D/D_0, i.e., $\mathbf{I}_{D/D_0}$ must satisfy

$$H(\mathbf{I}_{D/D_0}) = h|D| - H(\mathbf{I}_{D_0}), \tag{3.102}$$

where $H(\mathbf{I}_{D_0}) = |D_0|h(\mathbf{I}_{D_0})$ is the unnormalized histogram of $\mathbf{I}_{D_0}$. Therefore

$$\mathbf{I}_{D/D_0} \in \Omega_{D/D_0}\left(\frac{h|D| - H(\mathbf{I}_{D_0})}{|D/D_0|}\right). \tag{3.103}$$

So the number of such images is $|\Omega_{D/D_0}((h|D| - H(\mathbf{I}_{D_0}))/|D/D_0|)|$. Thus,

$$
\begin{aligned}
q(\mathbf{I}_{D_0}; h) &= \frac{|\Omega_{D/D_0}(\frac{h|D| - H(\mathbf{I}_{D_0})}{|D/D_0|})|}{|\Omega_D(h)|} \\[2mm]
&= \frac{(|D| - |D_0|)! / \prod_{i=1}^{g}(h_i|D| - H_i(\mathbf{I}_{D_0}))!}{|D|! / \prod_{i=1}^{g}(h_i|D|)!} \\[2mm]
&= \frac{\prod_{i=1}^{g}(h_i|D|)(h_i|D| - 1)\ldots(h_i|D| - H_i(\mathbf{I}_{D_0}) + 1)}{|D|(|D| - 1)\ldots(|D| - |D_0| + 1)} \\[2mm]
&= \frac{\prod_{i=1}^{g} h_i(h_i - 1/|D|)\ldots(h_i - (H_i(\mathbf{I}_{D_0}) - 1)/|D|)}{(1 - 1/|D|)\ldots(1 - (|D_0| - 1)/|D|)} \\[2mm]
&\rightarrow \prod_{i=1}^{g} h_i^{H_i(\mathbf{I}_{D_0})} \quad \text{as } |D| \rightarrow \infty. \tag{3.104}
\end{aligned}
$$

---

[6] In the i.i.d. case, $q(\mathbf{I}_{D_0}; h)$ is both the marginal distribution and the conditional distribution of $q(\mathbf{I}; h)$, while in random fields, we only consider the conditional distribution.

Therefore, the distribution of $\mathbf{I}_{D_0}$ is the FRAME model (see Eq. (3.88)) with $p = h$ under the Julesz ensemble defined by h.

The above calculation can be easily interpreted in a non-probabilistic way. That is, $q(\mathbf{I}_{D_0}; h)$ is the frequency or the proportion of images in $\Omega_D(h)$ (on large D) whose patches on D_0 are $\mathbf{I}_{D_0}$. In other words, if we look at all the images in the Julesz ensemble through D_0, then we will find a population of images on D_0, and the distribution of this population is described by the FRAME model. The reason that the FRAME model assigns probabilities to all images on D_0 is also quite clear. Under the hard constraint on $h(\mathbf{I}_D)$, $h(\mathbf{I}_{D_0})$ can still take any possible values.

Equivalence of FRAME and Julesz Ensemble

In this section, we show the equivalence between the Julesz ensembles and the FRAME models, using the fundamental principle of equivalence of ensembles in statistical mechanics.

From Julesz Ensemble to FRAME Model

In this subsection, we derive the local Markov property of the Julesz ensemble, which is globally defined by type h. This derivation is adapted from a traditional argument in statistical physics (e.g., [141]). It is not as rigorous as modern treatments, but it is much more revealing.

Suppose the feature statistics are $h(\mathbf{I})$ where $\mathbf{I}$ is defined on D. For a fixed value of feature statistics h, consider the image set $\Omega_D(h) = \{\mathbf{I} : h(\mathbf{I}) = h\}$, and the associated uniform distribution $q(\mathbf{I}; h)$. First, we fix $D_1 \subset D$ and then fix $D_0 \subset D_1$. We are interested in the conditional distribution of the local patch $\mathbf{I}_{D_0}$ given its local environment $\mathbf{I}_{D_1/D_0}$ under the model $q(\mathbf{I}; h)$ as $D \to Z^2$. We assume that D_0 is sufficiently smaller than D_1 so that the neighborhood of D_0, ∂D_0, is contained in D_1.

Let $\mathbf{H}_0 = \mathbf{H}(\mathbf{I}_{D_0}|\mathbf{I}_{\partial D_0})$ be the unnormalized statistics computed for $\mathbf{I}_{D_0}$ where filtering takes place within $D_0 \cup \partial D_0$. Let $\mathbf{H}_{01}$ be the statistics computed by filtering inside the fixed environment D_1/D_0. Let $D_{-1} = D/D_1$ be the big patch outside of D_1. Then the statistics computed for D_{-1} is $h|D| - \mathbf{H}_0 - \mathbf{H}_{01}$. Let $h_- = (h|D| - \mathbf{H}_{01})/|D_{-1}|$, then the normalized statistics for D_{-1} is $h_- - \mathbf{H}_0/|D_{-1}|$.

For a certain $\mathbf{I}_{D_0}$, the number of images in $\Omega_D(h)$ with such a patch $\mathbf{I}_{D_0}$ and its local environment $\mathbf{I}_{D_1/D_0}$ is $|\Omega_{D_{-1}}(h_- - \mathbf{H}_0/|D_{-1}|)|$. Therefore the conditional probability, as a function of $\mathbf{I}_{D_0}$, is

$$q(\mathbf{I}_{D_0} \mid \mathbf{I}_{D_1/D_0}, h) \propto \left| \Omega_{D_{-1}}(h_- - \frac{\mathbf{H}_0}{|D_{-1}|}) \right|. \qquad (3.105)$$

Unlike the simple i.i.d. case, however, the above volume cannot be computed analytically. However, the volume $|\Omega_D(\mathbf{h})|$ still shares the same asymptotic behavior as that of the simple i.i.d. example, namely,

$$\lim_{D \to \mathbf{Z}^2} \frac{1}{|D|} \log |\Omega_D(\mathbf{h})| \to s(\mathbf{h}), \tag{3.106}$$

where $s(\mathbf{h})$ is a concave entropy function of $\mathbf{h}$.

Like the simple i.i.d. case, in the above derivation, we ignore the minor technical complication that $\Omega_D(\mathbf{h})$ may be empty because the exact constraint may not be satisfied on a finite lattice. A more careful treatment is to replace $\mathbf{h}$ by a small set $\mathcal{H}$ around $\mathbf{h}$, and let $\mathcal{H} \to \mathbf{h}$ as $D \to \mathbf{Z}^2$. Let $\Omega_D(\mathcal{H}) = \{\mathbf{I} : \mathbf{h}(\mathbf{I}) \in \mathcal{H}\}$, then we have the following:

Proposition 2 *The limit*

$$\lim_{D \to \mathbf{Z}^2} \frac{1}{|D|} \log \Omega_D(\mathcal{H}) = s(\mathcal{H}) \tag{3.107}$$

exists. Let $s(\mathbf{h}) = \lim_{\mathcal{H} \to \mathbf{h}} s(\mathcal{H})$, *then* $s(\mathbf{h})$ *is concave, and* $s(\mathcal{H}) = \sup_{\mathbf{h} \in \mathcal{H}} s(\mathbf{h})$.

See Lanford [142] for a detailed analysis of the above result. The $s(\mathbf{h})$ is a measure of the volume of the Julesz ensemble of type $\mathbf{h}$. It defines the randomness of the texture appearance of type $\mathbf{h}$.

With such an estimate, we are ready to compute the conditional probability. Note that the conditional distribution, $q(\mathbf{I}_{D_0} \mid \mathbf{I}_{D_1/D_0}, \mathbf{h})$, as a function of $\mathbf{I}_{D_0}$, is decided only by $\mathbf{H}_0$, which is the sufficient statistics. Therefore, we only need to trace $\mathbf{H}_0$ while leaving other terms as constants. For large D, a Taylor expansion at $\mathbf{h}_-$ gives

$$\log q(\mathbf{I}_{D_0} \mid \mathbf{I}_{D_1/D_0}, \mathbf{h}) = \text{constant} + \log \left| \Omega_{D_{-1}} \left(\mathbf{h}_- - \frac{\mathbf{H}_0}{|D_{-1}|} \right) \right|$$

$$= \text{constant} + |D_{-1}| s \left(\mathbf{h}_- - \frac{\mathbf{H}_0}{|D_{-1}|} \right)$$

$$= \text{constant} - \langle s'(\mathbf{h}_-), \mathbf{H}_0 \rangle + o\left(\frac{1}{|D|} \right). \tag{3.108}$$

Assuming the entropy function s has continuous derivative at $\mathbf{h}$, and let $\boldsymbol{\beta} = s'(\mathbf{h})$, then, as $D \to \mathbf{Z}^2$, $\mathbf{h}_- \to \mathbf{h}$, and $s'(\mathbf{h}_-) \to \boldsymbol{\beta}$. Therefore,

$$\log q(\mathbf{I}_{D_0} \mid \mathbf{I}_{D_1/D_0}, \mathbf{h}) \to \text{const} - \langle s'(\mathbf{h}), \mathbf{H}_0 \rangle$$

$$= \text{const} - \langle \boldsymbol{\beta}, \mathbf{H}_0 \rangle, \tag{3.109}$$

so

$$q(\mathbf{I}_{D_0} \mid \mathbf{I}_{D_1/D_0}, \mathbf{h}) \rightarrow \frac{1}{Z_{D_0}(\boldsymbol{\beta})} \exp\left\{-\langle \boldsymbol{\beta}, \mathbf{H}(\mathbf{I}_{D_0} \mid \mathbf{I}_{\partial D_0})\rangle\right\}, \qquad (3.110)$$

which is exactly the Markov property imposed by the FRAME model. This derivation shows that local computation using the FRAME model is justified under the Julesz ensemble. It also reveals an important relationship, i.e., the parameter $\boldsymbol{\beta}$ can be identified as the derivative of the entropy function $s(\mathbf{h})$, $\boldsymbol{\beta} = s'(\mathbf{h})$.

As in the simple i.i.d. case, this result can be interpreted in a non-probabilistic way in terms of frequencies.

From FRAME Model to Julesz Ensemble

In this subsection, we study the statistical properties of the FRAME models as $D \rightarrow \mathbf{Z}^2$.

Consider the FRAME model

$$p(\mathbf{I}; \boldsymbol{\beta}) = \frac{1}{Z(\boldsymbol{\beta})} \exp\left\{-|D|\langle \boldsymbol{\beta}, \mathbf{h}(\mathbf{I})\rangle\right\}, \qquad (3.111)$$

which assigns equal probabilities to the $|\Omega_D(\mathbf{h})|$ images in $\Omega_D(\mathbf{h})$. The probability that $p(\mathbf{I}; \boldsymbol{\beta})$ assigns to $\Omega_D(\mathbf{h})$ is

$$p(\Omega_D(\mathbf{h}); \boldsymbol{\beta}) = \frac{1}{Z_D(\boldsymbol{\beta})} \exp\left\{-|D|\langle \boldsymbol{\beta}, \mathbf{h}\rangle\right\} |\Omega_D(\mathbf{h})|. \qquad (3.112)$$

The asymptotic behavior of this probability is

$$s_{\boldsymbol{\beta}}(\mathbf{h}) = \lim_{D \rightarrow \mathbf{Z}^2} \frac{1}{|D|} \log p(\Omega_D(\mathbf{h}); \boldsymbol{\beta}) \qquad (3.113)$$

$$= -\langle \boldsymbol{\beta}, \mathbf{h}\rangle + s(\mathbf{h}) - \lim_{D \rightarrow \mathbf{Z}^2} \frac{1}{|D|} \log Z_D(\boldsymbol{\beta}). \qquad (3.114)$$

For the last term, we have the following:

Proposition 3 *The limit*

$$\rho(\boldsymbol{\beta}) = \lim_{D \rightarrow \mathbf{Z}^2} \frac{1}{|D|} \log Z_D(\boldsymbol{\beta}) \qquad (3.115)$$

exists and is independent of the boundary condition. $\rho(\boldsymbol{\beta})$ is convex.

See Griffiths and Ruelle [86] for a proof. Therefore, we have the following:

Proposition 4 *The probability rate function $s_{\boldsymbol{\beta}}(\mathbf{h})$ of the FRAME model $p(\mathbf{I}; \boldsymbol{\beta})$ is*

$$s_\beta(\mathbf{h}) = \lim_{D \to \mathbf{Z}^2} \frac{1}{|D|} \log p(\Omega_D(\mathbf{h}); \boldsymbol{\beta}) = s(\mathbf{h}) - \langle \boldsymbol{\beta}, \mathbf{h} \rangle - \rho(\boldsymbol{\beta}). \qquad (3.116)$$

Therefore, the probability that the FRAME model puts on $\Omega_D(\mathbf{h})$ behaves like $\exp\{|D|s_\beta(\mathbf{h})\}$, and clearly $s_\beta(\mathbf{h}) \leq 0$ for any $\mathbf{h}$ and any $\boldsymbol{\beta}$ (otherwise, the probability will go unbounded). This $s_\beta(\mathbf{h})$ can be identified with $-\mathrm{KL}(h\|p)$ in the simple i.i.d. case.

The probability rate function $s_\beta(\mathbf{h})$ tells us that $p(\mathbf{I}; \boldsymbol{\beta})$ will eventually concentrate on the $\mathbf{h}_*$ that maximizes $s_\beta(\mathbf{h}) = s(\mathbf{h}) - \langle \boldsymbol{\beta}, \mathbf{h} \rangle - \rho(\boldsymbol{\beta})$, or $s(\mathbf{h}) - \langle \boldsymbol{\beta}, \mathbf{h} \rangle$. So we have the following:

Theorem 2 *If there is a unique* $\mathbf{h}_*$ *where* $s_\beta(\mathbf{h})$ *achieves its maximum 0, then* $p(\mathbf{I}; \boldsymbol{\beta})$ *eventually concentrates on* $\mathbf{h}_*$ *as* $D \to \mathbf{Z}^2$. *Therefore the FRAME model* $p(\mathbf{I}; \boldsymbol{\beta})$ *goes to a Julesz ensemble defined by* h_*. *If* $s(\mathbf{h})$ *is differentiable at* $\mathbf{h}_*$, *then* $s'(\mathbf{h}_*) = \boldsymbol{\beta}$.

3.6 Reaction and Diffusion Equations

In the previous sections, we have motivated feature extraction using filters and matching marginal distributions of observed data. This design of features has led to FRAME, a unified view of both clique-based and filter-based methods in texture modeling. Using Gibbs distribution as the probability distribution for textures, we further learn its potential function with dynamics

$$\frac{d\lambda^{(\alpha)}}{dt} = \mathrm{E}_{p(\mathbf{I}; \Lambda_K, S_K)}\left[H^{(\alpha)}\right] - H^{obs(\alpha)}, \qquad (3.117)$$

as in Eq. (3.64).

Computing $\mathrm{E}_{p(\mathbf{I}; \Lambda_K, S_K)}[H^{(\alpha)}]$ is difficult and it involves sampling synthetic images using the current model distribution. A class of methods for image synthesis involves nonlinear Partial Differential Equations (PDEs) in the form of

$$\frac{d\mathbf{I}}{dt} = T(\mathbf{I}), \qquad (3.118)$$

where T is a function of current $\mathbf{I}$. In this section, we connect Gibbs distribution to PDE paradigms for texture formation and derive a common framework under which many previous PDE methods can be similarly derived.

We first introduce historical methods adopting PDEs for image processing inspired by physics and chemistry, and then we introduce our discovery of reaction-diffusion functions as Gibbs potential functions, leading to Gibbs Reaction and Diffusion Equation (GRADE) as a family of PDEs for texture patterns formation.

Turing Diffusion-Reaction

A set of nonlinear PDEs was first studied in [237] for modeling the formation of animal coat patterns by the diffusion and reaction of chemicals, which Turing called the "morphogens." These equations were further explored by Murray in theoretical biology [178]. For example, let $A(x, y, t)$ and $B(x, y, t)$ be the concentrations of two morphogens at location (x, y) and time t , the typical reaction-diffusion equations are

$$\begin{cases} \frac{\partial A}{\partial t} = D_a \Delta A + R_a(A, B) \\ \frac{\partial B}{\partial t} = D_b \Delta B + R_b(A, B), \end{cases} \tag{3.119}$$

where D_a and D_b are constraints, $\Delta = \frac{\partial^2}{\partial x^2} + \frac{\partial^2}{\partial y^2}$ is the Laplace operator, and $R_a(A, B) and R_b(A, B)$ are nonlinear functions for the reaction of chemicals, e.g., $R_a(A, B) = A * B - A - 12$ and $R_b(A, B) = 16 - A * B$.

The morphogen theory provides a way for synthesizing some texture patterns. In the texture synthesis experiments, chemical concentrations are replaced by various colors, and the equations run for a finite number of steps with free boundary condition starting with some initial patterns. In some cases, both the initial patterns and the running processes have to be controlled manually in order to generate realistic textures. Two canonical textures synthesized by the Turing reaction-diffusion equation are the leopard blobs and zebra stripes.

In the reaction-diffusion equation above, the reaction terms are responsible for pattern formation, however, they also make the equations unstable or unbounded. Even for a small system, the existence and stability problems for these PDEs are intractable [87]. In fact, we believe that running any nonlinear PDEs for a finite number of steps will render some patterns, but it is unknown how to design a set of PDEs for a given texture pattern.

Heat Diffusion

As introduced in Sect. 3.2, generating an image according to the heat diffusion equation

$$\frac{d\mathbf{I}}{dt} = \Delta \mathbf{I}(x, y) \tag{3.120}$$

is equivalent to minimizing the potential energy of GMRF and

$$\frac{d\mathbf{I}}{dt} = -\frac{\delta U(\mathbf{I}(x, y, t))}{\delta \mathbf{I}}, \tag{3.121}$$

where $U(\mathbf{I}(x, y)) = \beta \int_x \int_y (\nabla_x \mathbf{I}(x, y))^2 + (\nabla_y \mathbf{I}(x, y))^2 dy\, dx$.

In the following, we show that anisotropic diffusion can also be written in a similar form of minimizing a potential energy.

Anisotropic Diffusion

Perona and Malik introduced a family of anisotropic diffusion equations for generating image scale space $\mathbf{I}(x, y, t)$ [195]. Similar to the heat diffusion equation, their equation also simulates the "heat" diffusion process,

$$\frac{d\mathbf{I}}{dt} = \text{div}(c(x, y, t)\nabla\mathbf{I}), \quad \mathbf{I}(x, y, 0) = \mathbf{I}_0, \qquad (3.122)$$

where $\nabla\mathbf{I} = (\frac{\partial \mathbf{I}}{\partial x}, \frac{\partial \mathbf{I}}{\partial y})$ is the intensity gradient and div is the divergence operator, $\text{div}(\vec{V}) = \nabla_x P + \nabla_y Q$, for any vector $\vec{V} = (P, Q)$. In practice, the heat conductivity $c(x, y, t)$ is defined as a function of location gradients. For example, Perona and Malik [195] chose

$$\frac{d\mathbf{I}}{dt} = \nabla_x \left(\frac{1}{1 + (\nabla_x \mathbf{I}/b)^2} \nabla_x \mathbf{I} \right) + \nabla_y \left(\frac{1}{1 + (\nabla_y \mathbf{I}/b)^2} \nabla_y \mathbf{I} \right), \qquad (3.123)$$

where b is a scaling constant. It is easy to see that Eq. (3.123) minimizes the following energy function by gradient descent, just as in the heat diffusion process in the section above,

$$U(\mathbf{I}) = \int \int \lambda(\nabla_x \mathbf{I}) + \lambda(\nabla_y \mathbf{I}) dx dy, \qquad (3.124)$$

where $\lambda(\xi) = const. \log(1 + (\xi/b)^2)$ and $\lambda'(\xi) = const. \frac{\xi}{1+(\xi/b)^2}$ are plotted in Fig. 3.27.

Although the anisotropic diffusion equations can be adopted for removing noise and enhancing edges [186], $\mathbf{I}(z, y, t)$ converges to a flat image as $t \to \infty$ in the Perona–Malik equation.

GRADE: Gibbs Reaction and Diffusion Equations

The above sections introduced (very similar) prior PDE paradigms in synthesizing textures. However, the methods mentioned above directly follow PDE models used in chemistry and physics. Directly following Gibbs distribution as used in FRAME, we derive a family of PDEs called Gibbs Reaction and Diffusion Equation

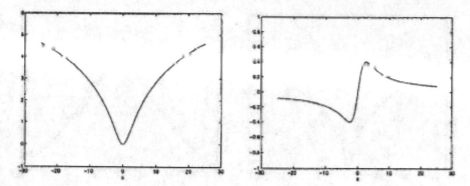

Fig. 3.27 Reprinted with permission from [278]. On the left is $\lambda(\xi) = const. \log(1 + (\xi/b)^2)$ and on the right is $\lambda'(\xi) = const. \frac{\xi}{1+(\xi/b)^2}$

(GRADE). We also believe that many PDE paradigms for image processing can be unified under this common framework by using the same approach.

Suppose image $\mathbf{I}$ is defined on an $N \times N$ lattice D. Consider the Gibbs distribution as derived in FRAME,

$$p(\mathbf{I}; \lambda, S) = \frac{1}{Z} \exp\{-U(\mathbf{I}; \Lambda, S)\}, \qquad (3.125)$$

where Z is the normalization constant independent of $\mathbf{I}$, $S = \{F_1, F_2, \ldots, F_n\}$ is a set of filters to characterize the essential features of the images, and $\Lambda = \{\lambda_1(\cdot), \ldots, \lambda_n(\cdot)\}$ is a set of potential functions on feature statistics extracted by S (using convolution). The potential is

$$U(\mathbf{I}; \Lambda, S) = \sum_{i=1}^{n} \sum_{(x,y) \in D} \lambda_i(F_i * \mathbf{I}(x, y)), \qquad (3.126)$$

where $F_i * \mathbf{I}(x, y)$ is the filter response of F_i at (x, y).

In practice, S is chosen by minimizing the entropy of $p(\mathbf{I})$ from a bank of filters such as the Gabor filters at various scales and orientations [4, 21] and wavelet Transform [13, 36]. In general, these filters can be nonlinear functions of the image intensities. In the rest of this section, we shall only study the following linear filters:

1. An intensity filter $\delta(\cdot)$ and gradient filters ∇_x, ∇_y
2. The Laplacian of Gaussian filters

$$LoG(x, y, s) = const. (x^2 + y^2 - s^2) e^{-\frac{x^2+y^2}{s^2}}, \qquad (3.127)$$

where $s = \sqrt{2}\sigma$ stands for the scale of the filter. We denote these filters by $LoG(s)$. A special filter is $LoG(\frac{\sqrt{2}}{2})$, which has a 3×3 window $[0, \frac{1}{4}, 0; \frac{1}{4}, -1, \frac{1}{4}; 0, \frac{1}{4}, 0]$.

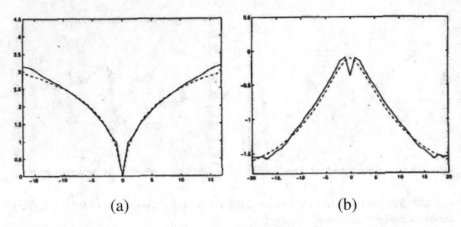

(a)   (b)

Fig. 3.28 Reprinted with permission from [278]. This figure shows two classes of functions. (a) Shows diffusion functions. (b) Shows reaction functions. Dotted lines show the fitted ϕ and ψ functions, respectively

3. Gabor filters with both sine and cosine components:

$$G(x, y, s, \theta) = const. \cdot Rot(\theta) \cdot e^{\frac{1}{2s^2}(4x^2+y^2)} e^{-i\frac{2\pi}{s}x}. \tag{3.128}$$

It is a sine wave at frequency $\frac{2\pi}{s}$ modulated by an elongated Gaussian Function and rotated at angle θ. We denote the real and imaginary parts of $G(z, y, s, \theta)$ by $G\cos(s, \theta)$ and $G\sin(s, \theta)$.

Using the filters above to learn the potential function in Eq. (3.126) in discretized bins using FRAME, we found that texture patterns generally exhibit two families of functions similar to reaction-diffusion in chemical processes as shown in Fig. 3.28. We use the following two families of functions to fit our discretized findings, and the fitted curves are shown as dashed lines in Fig. 3.28,

$$\phi(\xi) = a \left(1 - \frac{1}{1 + (|\xi - \xi_0|/b)^\gamma}\right), \quad a > 0, \tag{3.129}$$

$$\psi(\xi) = a \left(1 - \frac{1}{1 + (|\xi - \xi_0|/b)^\gamma}\right), \quad a < 0, \tag{3.130}$$

where ξ_0 and b are the translation and scaling constants, respectively, and $\|a\|$ weights the contribution of the filter. $\phi(\xi)$, the diffusion function, assigns the lowest energy (or highest probability) to filter responses around ξ_0 (and $\xi_0 = 0$ in most cases), and $\psi(\xi)$, the reaction function, has the lowest energy at the two tails which represent salient image features such as object boundaries. These inverted potential functions are in contrast to all previous image models, and they are essential for modeling realistic images. The forming of the two potential functions is closely

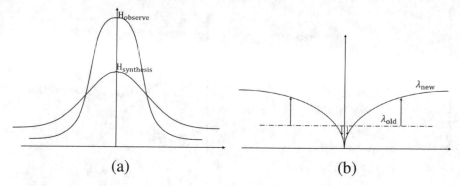

Fig. 3.29 Note the forming of the two types of potential functions. Suppose our current histogram of the synthesized image is more dispersed than the observed histogram (shown in the left image (**a**)). Then the updating of λ will push the λ on the tail to be bigger and push the λ in the center to be smaller, which forms a diffusion function (**b**). In the following sampling process, this diffusion function will push the corresponding feature to zero, making the synthesized histogram more concentrated to the center and thus closer to the observed one

related to our training process. Recalling Eq. (3.64), we have shown the gradient of λ is equal to the difference between the histogram of the current synthesized image and the histogram of the observed image. Then if the synthesized image contains larger components of a certain feature than the original one, the update will shape the new λ to be more like the diffusion type so that the larger responses are inhibited. On the other hand, if the histogram of the synthesized image is too concentrated on a certain response, the λ will be shaped to reaction type so that this feature is encouraged to appear. This process is illustrated in Fig. 3.29. Recently, especially in the deep neural network cases, instead of learning a nonlinear potential function, people use the ReLU function to get similar results. The ReLU function has a linear response on the positive half axis and zero on the negative half axis. As shown in Fig. 3.30, by setting the coefficient of ReLU function to be greater or less than zero, we can get the same results of encouraging or preventing a certain feature to appear.

Now we can design our potential function in Eq. (3.126) to be

$$U(\mathbf{I}; \Lambda, S) = \sum_{i=1}^{n_d} \sum_{(x,y)\in D} \phi_i(F_i * \mathbf{I}(x, y)) + \sum_{i=n_d}^{n} \sum_{(x,y)\in D} \phi_i(F_i * \mathbf{I}(x, y)). \quad (3.131)$$

Note that the filter set is divided into two parts $S = S_d \cup S_r$, with $S_d = \{F_i, i = 1, 2, \ldots, n_d\}$ and $S_r = \{F_i, i = n_d + 1, \ldots, n\}$. In most cases, S_d consists of filters such as gradient filters and Laplacian/Gaussian filters which capture general smoothness appearances of real-world images, and S_r contains filters such as Gabor filters at various orientations and scales which characterize salient features of images.

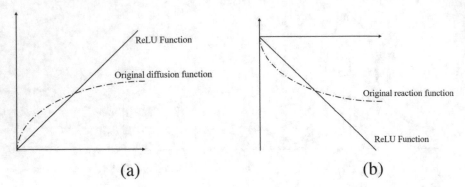

Fig. 3.30 We can use the ReLU function to replace the original nonlinear diffusion (**a**) and reaction (**b**) functions

Our PDE can therefore be designed by maximizing Gibbs distribution, which is equivalent to minimizing the above $U(\mathbf{I}; \Lambda, S)$ by gradient descent. We obtain the following nonlinear parabolic partial differential equation:

$$\frac{d\mathbf{I}}{dt} = \sum_{i=1}^{n_d} F_i^- * \phi_i'(F_i * \mathbf{I}) + \sum_{i=n_d}^{n} F_i^- * \psi_i'(F_i * \mathbf{I}), \tag{3.132}$$

where $F_i^-(x, y) = -F_i(-x, -y)$. Thus starting from an input image $\mathbf{I}(x, y, t) = \mathbf{I}$, the first term diffuses the image by reducing the gradients while the second term forms patterns as the reaction forces favor large filter responses. We call Eq. (3.132) the Gibbs Diffusion And Reaction Equation (GRADE).

Properties of GRADE

Property 1: A General Statistical Framework

GRADE as in Eq. (3.132) can be considered as an extension to the heat diffusion equation, as in Eq. (3.122), on a discrete lattice by defining a vector field

$$\vec{V} = (\phi_1'(\cdot), \ldots, \phi_{n_d}'(\cdot), \psi_{n_d+1}'(\cdot), \ldots, \psi_n'(\cdot)), \tag{3.133}$$

and the divergence operator can be generalized to

$$\text{div} = F_1^- * + \ldots + F_n^- * . \tag{3.134}$$

Equation 3.132 can be written as

$$\frac{d\mathbf{I}}{dt} = \text{div}(\vec{V}). \tag{3.135}$$

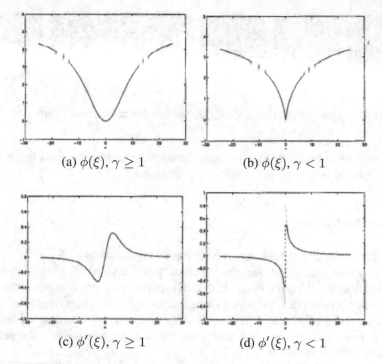

(a) $\phi(\xi),\ \gamma \geq 1$ (b) $\phi(\xi),\ \gamma < 1$

(c) $\phi'(\xi),\ \gamma \geq 1$ (d) $\phi'(\xi),\ \gamma < 1$

Fig. 3.31 Reprinted with permission from [278]. Here $\phi(\xi) = a(1 - \frac{1}{1+(|\xi-\xi_0|/b)^{\gamma}})$, $a > 0$, and its derivative $\phi'(\xi)$ for (**a**), (**c**) is $\gamma = 2.0$ and (**b**), (**d**) $\gamma = 0.8$

Compared to Eq. (3.122) which transfers "heat" among adjacent pixels, Eq. (3.135) transfers the "heat" in many directions on a graph and the conductivities are defined as functions of the local patterns instead of just the local gradients. Note that in the discrete lattice, choosing $S_d = \nabla_x, \nabla_y$, $S_r = \emptyset$, we have $F_1 = F_1^- = \nabla_x$, $F_2 = F_2^- = \nabla_y$ and div $= \nabla_x + \nabla_y$, and thus Eqs. 3.122 and 3.123 are just special cases of Eq. (3.135).

Property 2: Diffusion

Figure 3.31a,b show two best-fit potential functions. Figure 3.31a shows a round tip at $\xi = 0$, and Fig. 3.31b shows a cusp at $\xi = 0$. Interestingly, real-world objects typically show potential function with a cusp. This is because large parts of real-world objects have flat intensity, encouraging piece-wisely constant regions. Intuitively, with $\gamma < 1$ and $\xi = 0$ at location (x, y), $\phi'(0)$ forms an attractive basin in the neighborhood $\mathcal{N}_i(x, y)$ specified by filter window F_i at (x, y). For a pixel $(u, v) \in \mathcal{N}_i(x, y)$, the depth of attractive basin is $\|\omega F_i^-(u - x, v - y)\|$. If a pixel is involved in multiple zero filter responses, it will accumulate the depth of the attractive basin generated by each filter. Thus unless the absolute value of

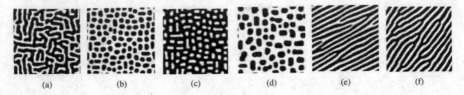

(a)   (b)   (c)   (d)   (e)   (f)

Fig. 3.32 Reprinted with permission from [278]. Generated texture patterns **(a)**–**(f)**

the driving force from other well-defined terms is larger than the total depth of the attractive basin at (u, v), $\mathbf{I}(u, v)$ will stay unchanged.

Property 3: Reaction

The other class of potential function is the reaction function. We refer back to Fig. 3.28 where gradients of reaction function "push" activations away from origin. Different from the diffusion function, this class of potential function creates features and brings pixels out of local attraction basins set up by diffusion functions.

With both reaction and diffusion, we can sample a wide variety of textures. Starting from random uniform noise, we use Langevin equations inspired by Brownian motion as below:

$$d\mathbf{I} = -\nabla U(\mathbf{I})dt + \sqrt{2T(t)}dw_t, \tag{3.136}$$

where $dw_t = \sqrt{dt}N(0, 1)$, T is the "temperature" which controls magnitude of random fluctuations. In the above figure (Fig. 3.32), we use one diffusion filter, the Laplacian of Gaussian filter, and several reaction filters: isotropic center-surround filters and Gabor filters with different orientations.

3.7 Conclusion

In this chapter, we started by asking the question,

What features and statistics are characteristic of a texture pattern, so that texture pairs that share the same features and statistics cannot be told apart by pre-attentive human visual perception?

In finding suitable features for texture modeling, we have walked through several models in history including clique-based models and filter-based models. Clique-based models specify their energy function as feature statistics but such models are shown to be inadequate in modeling texture patterns. Filter-based models more closely align with human vision as accepted in neurophysiology. Convolution of filters with an image (filter responses) and histograms of filter activations (marginal statistics projecting an image onto a filter) seems to be a reasonable choice for features.

With features selected, we then introduced Filters, Random field, and Maximum Entropy (FRAME) model as a unifying view over both clique-based and filter-based module. We presented a pursuit process involving the Minimax Entropy Principle for pursuing a set of suitable filters and learning suitable potential functions under a Gibbs distribution. Texture ensemble is then introduced and connected to statistical physics (micro-canonical ensembles and canonical ensembles). Equivalence between the FRAME model and Julesz ensemble on an infinite lattice is proved.

Finally, we studied PDE paradigms in synthesizing images and presented two classes of potential functions empirically found in texture modeling—diffusion and reaction. Diffusion is found to destroy features while the reaction is found to create features. We also derived our PDE for texture synthesis, introducing Gibbs Reaction and Diffusion Equation (GRADE) as a general statistical framework under which lie many other PDEs used in texture formation.

Chapter 4
Textons

Textons refer to fundamental micro-structures in natural images and are considered the atoms of pre-attentive human visual perception (Julesz [124]). Unfortunately, the term "texton" remains a vague concept in the literature for lacking a good mathematical model. In this chapter, we present various generative image models for textons.

4.1 Textons and Textures

Julesz's Discovery

In psychophysics, Julesz [124] and colleagues discovered that pre-attentive vision is sensitive to some basic image features while ignoring other features. He conjectured that pre-attentive human vision is sensitive to local patterns called textons. Figure 4.1 illustrates the first batch of experiments for texture discrimination.

In the second batch of experiments, Julesz measured the response time of human subjects in detecting a target element among a number of distractors in the background. For example, Fig. 4.2 shows two pairs of elements in comparison. The response time for the upper pair is instantaneous (100–200 ms) and independent of the number of distractors. In contrast, for the lower pair, the response time increases linearly with the number of distractors. This discovery was very important in psychophysics and motivated Julesz to conjecture a pre-attentive stage that detects some atomic structures, such as elongated blobs, bars, crosses, and terminators, which he called "textons" for the first time.

The early texton studies were limited by their exclusive focus on artificial texture patterns instead of natural images. It was shown that the perceptual textons could be adapted through training. Thus the dictionary of textons must be associated with or learned from the ensemble of natural images. Despite the significance of Julesz's

© Springer Nature Switzerland AG 2023
S.-C. Zhu, Y. N. Wu, *Computer Vision*, https://doi.org/10.1007/978-3-030-96530-3_4

Fig. 4.1 Pre-attentive vision is sensitive to local patterns called textons

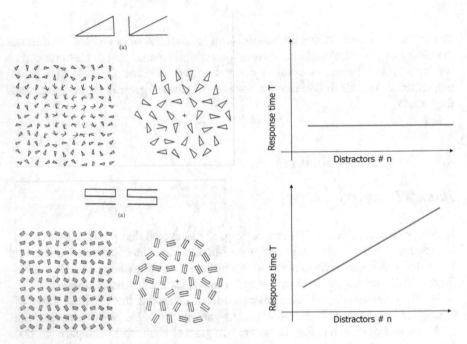

Fig. 4.2 Julesz textures show that pre-attentive vision is sensitive to local image structures such as edges, bars, and endpoints

experiments, there have been no rigorous mathematical definitions for textons. Later in this chapter, we explain that textons can be defined in the context of a generative model of images.

In natural images, textures and textons are often interwoven where the image patches are considered to be textures or textons which are, respectively, from manifolds of different dimensions in the image space. As we have discussed in the previous chapter, the texture patches are from high-dimensional manifolds defined by implicit functions, i.e., the statistical constraints. In this chapter, we will show that the texton patches are from low-dimensional manifolds defined by explicit functions, i.e., generative models.

Neural Coding Schemes

Julesz's experiments are inherently related to how neurons respond to different kinds of stimuli. Studies in neuroscience reveal that neurons propagate signals by generating characteristic electrical pulses called action potentials. With the presence of external stimuli like light and specific image patterns, sensory neurons fire sequences of action potentials in various temporal patterns, in which information about the stimulus is encoded and transmitted to the brain. This process leads to the different perceptions of textures and textons and justifies the pre-attentive stage proposed by Julesz. Yet, how exactly the neurons represent the signals is still a topic of debate. Here, we present three hypothesized coding schemes and their relations to texton and texture modeling.

Population coding is a method that represents stimuli with the joint activities of multiple neurons. In this coding scheme, each neuron has a distribution of responses over some set of inputs. The responses of many neurons are combined to determine a final value about the inputs, which triggers further reactions in the signal propagation process. In our discussion of image modeling, we can draw a parallel between the filters and the neurons. Indeed, the FRAME model combines the potentials of filter responses, i.e., the histogram statistics, and outputs a value to represent an image's likelihood of belonging to a concept. In this sense, both the histograms and the neuron responses encode the impression that we have on certain visual stimuli.

Another coding scheme is the grandmother cell coding scheme. In this formulation, high-level concepts are represented by a single neuron. It activates when a person perceives a specific entity, such as one's grandma. To be more concise, there is a single cell in your brain that responds multi-modally to your grandmother, granting you the percept of a single entity when you hear the sounds "grandmother," "grandma," and "grams," capture the sight of her smiling, wrinkly face, and recall the tasty cookies she once made as well as the stories she once told you. The hypothesis is supported by an observation in 2004, when an epilepsy patient at the UCLA Medical Center whose brain was being monitored showed vigorous neural responses to several pictures of Jennifer Aniston, but not to other celebrities. Though the theory itself is still controversial, we can nonetheless incorporate it into our study of computer vision. Borrowing the idea of the grandmother cells, we can develop object templates that are invariant under various transformations. These templates, such as cars, represent high-level concepts as a whole rather than individual parts that make up the concept, such as the wheels and front doors of the cars.

The last coding scheme that we are interested in is sparse coding, where an object is encoded by the strong activations of a relatively small set of neurons. In fact, this method is the foundation of the texton models that we will introduce later in this chapter.

4.2 Sparse Coding

Image Representation

The purpose of vision, biological and machine, is to compute a hierarchy of increasingly abstract interpretations of the observed images (or image sequences). Therefore, it is of fundamental importance to know what are the descriptions used at each level of interpretation. By analogy to physics concepts, we wonder what are the visual "electrons," visual "atoms," and visual "molecules" for visual perception. The pursuit of basic images and perceptual elements is not just for intellectual curiosity but has important implications for a series of practical problems. For example,

1. *Dimension reduction*. Decomposing an image into its constituent components reduces information redundancy and leads to lower dimensional representations. As we will show in later examples, an image of 256×256 pixels can be represented by about 500 basis functions (which are local image patches), which are, in turn, reduced to 50–80 texton elements. The dimension of representation is thus reduced by about 100 folds. Further reductions are achieved in motion sequences and lighting models.
2. *Variable decoupling*. The decomposed image elements become more and more independent of each other and thus are spatially nearly decoupled. This facilitates image modeling which is necessary for visual tasks such as segmentation and recognition.
3. *Biologic modeling*. Micro-structures in natural images provide ecological clues for understanding the functions of neurons in the early stages of biologic vision systems [10].

In the literature, there are several threads of research investigating fundamental image structures from different perspectives, with many questions left unanswered.

First, in neurophysiology, the cells in the early visual pathway (retina, LGN, and V1) of primates are found to compute some basic image structures at various scales and orientations [113]. This motivated some image pyramid representations including Laplacian of Gaussians (LoGs), Gabor functions, and their variants. However, very little is known about how V1 cells are grouped into larger structures at higher levels (say, V2 and V4). Similarly, it is unclear what are the generic image representations beyond the image pyramids in image analysis.

Second, in harmonic analysis, one treats images as 2D functions, and then it can be shown that some classes of functions (such as Sobolev, Hölder, Besov spaces) can be decomposed into basis functions, for example, Fourier basis and wavelets.

But the natural image ensemble is known to be very different from those classic mathematical functional spaces.

The third perspective, and the most direct attack on the problem, is the study of natural image statistics and image component analysis. The most important work was done by Olshausen and Field [190] who learned over-complete image basis from natural image patches (12×12 pixels) based on the principle of sparse coding. In contrast to the orthogonal and complete basis in the Fourier analysis, the learned basis functions are highly correlated, and a given image is coded by a sparse population in the over-complete dictionary. Added to the sparse coding idea is independent component analysis (ICA) which decomposes images into a linear superposition of some basis functions minimizing a measure of dependence between the coefficients of these basis elements [15].

Basis and Frame

In linear algebra, a set **B** of vectors in a vector space V is called a basis if all elements in V can be written as a unique linear combination of vectors in **B**. A vector space may have several different bases. Yet all the bases have the same number of elements equal to the dimension of the vector space. For a typical image with 1024×1024 pixels, the dimension of the image space to which it belongs is 2^{20}. Consequently, a basis must have 2^{20} vectors to reconstruct any image of the same size. However, as we have learned earlier, natural images only occupy a small subset of the entire image space, making it trickier to select a set of vectors to represent natural images more efficiently.

If the inner product operation is well defined in a vector space V, then V is an inner product space. We call a basis $\mathbf{B} = \{b_1, b_2, ..., b_n\}$ for V orthonormal if the elements are all unit vectors and orthogonal to each other. Parseval's identity for orthonormal basis states that

$$\forall x \in V, \ \|x\|^2 = \sum_{i=1}^{n} |\langle x, b_i \rangle|^2 . \tag{4.1}$$

We can slightly loosen the constraint so that a collection of vectors $\mathbf{F} = \{f_1, \ldots, f_m\}$ satisfies

$$\forall x \in V, \ A \|x\|^2 \leq \sum_{i=1}^{n} |\langle x, f_i \rangle|^2 \leq B \|x\|^2 ; \tag{4.2}$$

then, we call **F** a frame for the vector space with the corresponding frame bounds A and B. Furthermore, a frame is a tight frame if $A = B$. In finite dimensional vector spaces, the frames are exactly the spanning sets. Therefore, a vector can be expressed as a linear combination of the frame vectors in a redundant way. Using a frame, it is possible to create a simpler, more sparse representation of a signal compared to representing it strictly with linearly independent vectors.

In general, if a set of vectors is still a basis after removing some elements, then it is called over-complete. In other words, the number of vectors for an over-complete basis is greater than the dimension of the input vector. Practically speaking, over-completeness can help us to achieve a more stable, robust, and compact decomposition of input vectors. We shall see its importance in the sparse coding model of texton representation.

As a matter of terminology, if the input vector is an image, we may consider the image as a two-dimensional function, so that the basis vectors become basis functions. These basis functions form a basis to represent the input images. Thus we shall use the term "basis functions" when we talk about representing images.

Olshausen–Field Model

In image coding, one starts with a dictionary of basis functions

$$\Psi = \{\psi_\ell(u, v), \ell = 1, \ldots, L_\psi\}. \tag{4.3}$$

For example, some commonly used basis functions are Gabor, Laplacian of Gaussian (LoG), and other wavelets (Fig. 4.3).

Let $A = (x, y, \tau, \sigma)$ denote the translation, rotation, and scaling transform of a basis function and $G_A \ni A$ the orthogonal transform space (group), then we obtain a set of basis functions Δ,

$$\Delta = \{\psi_\ell(u, v, A) : A = (x, y, \tau, \sigma) \in G_A, \ell = 1, \ldots, L_\psi\}. \tag{4.4}$$

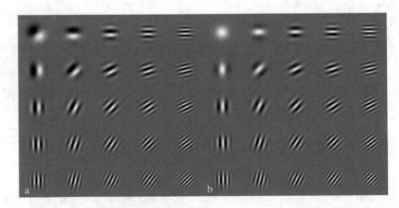

Fig. 4.3 Gabor wavelets are sine and cosine waves multiplied by Gaussian functions

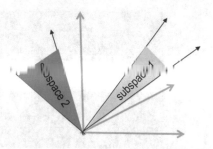

Fig. 4.4 The sparse coding model assumes that the observed signals lie on the low-dimensional subspaces

A simple generative image model, adopted in almost all image coding schemes, assumes that an image $\mathbf{I}$ is a linear superposition of some basis functions selected from Δ plus a Gaussian noise image $\mathbf{n}$,

$$\mathbf{I} = \sum_i^{n_B} \alpha_i \cdot \boldsymbol{\psi}_i + \mathbf{n}, \qquad \boldsymbol{\psi}_i \in \Delta, \ \forall i, \tag{4.5}$$

where n_B is the number of basis functions and α_i is the coefficient of the i-th base $\boldsymbol{\psi}_i$ (Fig. 4.4).

As Δ is over-complete,[1] the variables $(\ell_i, \alpha_i, x_i, y_i, \tau_i, \sigma_i)$ indexing a basis function ψ_i are treated as latent (hidden) variables and must be inferred probabilistically, in contrast to deterministic transforms such as the Fourier transform. All the hidden variables are summarized in a basis map,

$$\mathbf{B} = (n_B, \{b_i = (\ell_i, \alpha_i, x_i, y_i, \tau_i, \sigma_i) : i = 1, \ldots, n_B\}). \tag{4.6}$$

If we view each $\boldsymbol{\psi}_i$ as an attributed point with attributes $b_i = (\ell_i, \alpha_i, x_i, y_i, \tau_i, \sigma_i)$, then $\mathbf{B}$ is an attributed spatial point process.

In the image coding literature, the basis functions are assumed to be independently and identically distributed (i.i.d.), and the locations, scales, and orientations are assumed to be uniformly distributed, so

$$p(\mathbf{B}) = p(n_B) \prod_{i=1}^{n_B} p(b_i), \tag{4.7}$$

$$p(b_i) = p(\alpha_i) \cdot \text{unif}(\ell_i) \cdot \text{unif}(x_i, y_i) \cdot \text{unif}(\tau_i) \cdot \text{unif}(\sigma_i). \tag{4.8}$$

It was well known that responses of image filters on natural images have high kurtosis histograms. This means that most of the time the filters have nearly zero responses (i.e., they are silent) and they are activated with large responses

---

[1] The number of basis functions in Δ is often 100 times larger than the number of pixels in an image.

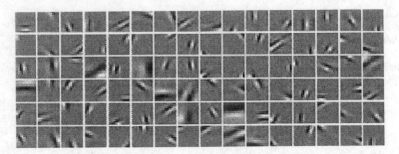

Fig. 4.5 Reprinted with permission from [190]. The above image patches are the basis functions learned from natural image patches by the sparse coding model

occasionally. This leads to the sparse coding idea by Olshausen and Field [190].[2] For example, $p(\alpha)$ is chosen to be a Laplacian distribution or a mixture of two Gaussians with σ_1 close to zero. For all $i = 1, \ldots, n_B$,

$$p(\alpha_i) \sim \exp\{-|\alpha_i|/c\} \quad \text{or} \quad p(\alpha_i) = \sum_{j=1}^{2} \omega_j N(0, \sigma_j). \tag{4.9}$$

In fact, as long as $p(\alpha)$ has high kurtosis, the exact form of $p(\alpha)$ is not so crucial. For example, one can choose a mixture of two uniform distributions on a range $[-\sigma_j, \sigma_j]$, $j = 1, 2$, respectively, with σ_1 close to zero, $p(\alpha_i) = \sum_{j=1}^{2} \omega_j \text{unif}[-\sigma_j, \sigma_j]$.

In the above image model, the basis map **B** includes the hidden variables, and the dictionary Ψ are parameters. For example, Olshausen and Field used $L_\psi = 144$ base functions, each being a 12×12 image patch. Following an EM-like learning algorithm, they learned Ψ from a large number of natural image patches. Figure 4.5 shows some of the 144 base functions. Such basis functions capture some image structures and are believed to bear resemblance to the responses of simple cells in V1 of primates. In their experiments, the training images are chopped into 12×12 pixel patches, and therefore they did not really infer the hidden variables for the transformation A_i. Thus the learned basis functions are not aligned at the centers.

A Three-Level Generative Model

We may extend the previously introduced simple form of a generative image model to a three-level generative model as shown in Fig. 4.6. In this model, an image **I** is

---

[2] Note that the filter responses are convolutions of a filter with image in a deterministic way and are different from the coefficients of the basis functions.

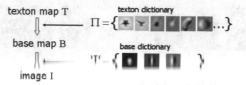

Fig. 4.6 Reprinted with permission from [277]. A three-level generative model: an image **I** is a linear superposition of some basis functions selected from a base dictionary Ψ, such as Gabor or Laplacian of Gaussians. The basis map is further generated by a smaller number of textons selected from a texton dictionary Π. Each texton consists of a number of basis functions in certain deformable configurations

generated by a basis map **B** as in image coding, and the basis functions are selected from a dictionary Ψ with some transforms.

The basis map **B** is, in turn, generated by a texton map **T**. The texton elements are selected from a texton dictionary Π with some transforms. Each texton element in **T** consists of a few basis functions with a deformable geometric configuration. So we have

$$\mathbf{T} \xrightarrow{\Pi} \mathbf{B} \xrightarrow{\Psi} \mathbf{I}, \tag{4.10}$$

with

$$\Psi = \{\psi_\ell, \ell = 1, 2, \ldots, L_\psi\}, \quad \text{and} \quad \Pi = \{\pi_\ell; \ell = 1, 2, \ldots, \mathbf{L}_\pi\}. \tag{4.11}$$

By analogy to the waveform–phoneme–word hierarchy in speech, the pixel–basis–texton hierarchy presents an increasingly abstract visual description. This representation leads to dimension reduction and the texton elements account for spatial co-occurrence of the basis functions.

To clarify terminology, a basis function $\psi \in \Psi$ is like a mother wavelet, and an image base b_i in the basis map **B** is an instance under certain transforms of a basis function. Similarly, a "texton" in a texton dictionary $\pi \in \Pi$ is a deformable template, while a "texton element" is an instance in the texton map **T** which is a transformed and deformed version of a texton in Π.

For natural images, it is reasonable to guess that the number of basis functions is about $|\Psi| = O(10)$, and the number of textons is in the order of $|\Pi| = O(10^3)$ for various combinations. Intuitively, textons are meaningful objects viewed at distance (i.e., small scale), such as stars, birds, cheetah blobs, snowflakes, beans, etc.

In this chapter, we fix the basis dictionary to three common basis functions: Laplacian of Gaussian, Gabor cosine, and Gabor sine, i.e.,

$$\Psi = \{\psi_1, \psi_2, \psi_3\} = \{\text{LoG, Gcos, Gsin}\}. \tag{4.12}$$

These basis functions are not enough for patterns like hair or water, etc. But we fix them for simplicity and focus on the learning of texton dictionary Π. This

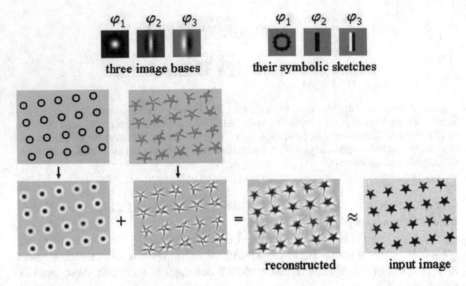

Fig. 4.7 Reprinted with permission from [277]. Reconstructing a star pattern by two layers of basis functions. An individual star is decomposed into a LoG basis function in the upper layer for the body of the star plus a few other basis functions (mostly Gcos, Gsin) in the lower layer for the angles

chapter is also limited to learning textons for each individual texture pattern instead of generic natural images, and therefore $|\Pi|$ is a small number for each texture.

Before we formulate the problem, we show an example of a simple star pattern to illustrate the generative texton model. In Fig. 4.7, we first show the three basis functions in Ψ (the first row) and their symbolic sketches. Then for an input image, a matching pursuit algorithm is adopted to compute the basis map $\mathbf{B}$ in a bottom-up fashion. This map will be modified later by stochastic inference. It is generally observed that the map $\mathbf{B}$ can be divided into two sub-layers. One sub-layer has relatively large ("heavy") coefficients α_i and captures some larger image structures. For the star pattern, these are the LoG basis functions shown in the first column. We show both the symbolic sketch of these LoG basis functions (above) and the image generated by these basis functions (below). The heavy basis functions are usually surrounded by a number of "light" basis functions with relatively small coefficients α_i. We put these secondary basis functions in another sub-layer (see the second column of Fig. 4.7.) When these image basis functions are superposed, they generate a reconstructed image (see the third column in Fig. 4.7.) The residues of reconstruction are assumed to be Gaussian noise.

By analogy to the physics model, we call the heavy basis functions the "nucleus basis functions" as they have heavy weights like protons and neutrons and the light basis functions the "electron basis functions." Figure 4.7 displays an "atomic" model for the star texton. It is a LoG surrounded by 5 electron basis functions.

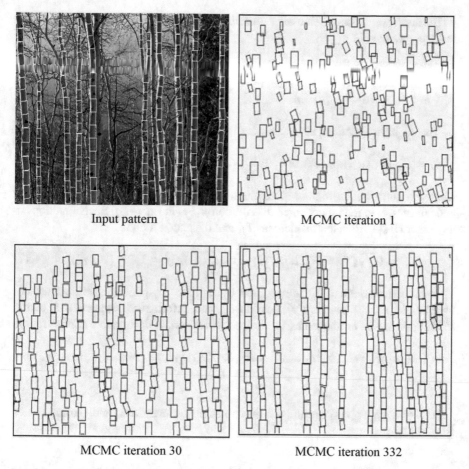

Fig. 4.8 Reprinted with permission from [89]. The spatial arrangement of textons can be modeled by a point process model

The three-level generative model is governed by a joint probability distribution specified with parameters $\Theta = (\Psi, \Pi, \kappa)$.

$$p(\mathbf{I}, \mathbf{B}, \mathbf{T}; \Theta) = p(\mathbf{I}|\mathbf{B}; \Psi) p(\mathbf{B}|\mathbf{T}; \Pi) p(\mathbf{T}; \kappa), \qquad (4.13)$$

where Ψ and Π are dictionaries for two generating processes, and $p(\mathbf{T}; \kappa)$ is a descriptive (Gibbs) model for the spatial distribution of the textons as a stochastic attributed point process (Fig. 4.8).

We rewrite the basis map as

$$\mathbf{B} = (n_B, \{b_i = (\ell_i, \alpha_i, x_i, y_i, \tau_i, \sigma_i) : i = 1, 2, \ldots, n_B\}). \qquad (4.14)$$

Because we assume Gaussian distribution $N(0, \sigma_o^2)$ for the reconstruction residues, we have

$$p(\mathbf{I}|\mathbf{B}; \boldsymbol{\Psi}) \propto \exp\left\{ - \sum_{(u,v)\in D} (\mathbf{I}(u, v) - \sum_{i=1}^{n_B} \alpha_i \psi_{\ell_i}(u, v; x_i, y_i, \tau_i, \sigma_i))^2 / 2\sigma_o^2 \right\}.$$

(4.15)

The n_B basis functions in map $\mathbf{B}$ are divided into $n_T + 1$ groups ($n_T < n_B$).

$$\{b_i = (\ell_i, \alpha_i, x_i, y_i, \tau_i, \sigma_i) : i = 1, 2, \ldots, n_B\} = \varpi_0 \cup \varpi_1 \cup \cdots \cup \varpi_{n_T}. \quad (4.16)$$

The basis functions in ϖ_0 are "free electrons" which do not belong to any texton and are subject to the independent distribution $p(b_j)$ in Eq. (4.8). Basis functions in any other class form a texton element T_j, and the texton map is

$$\mathbf{T} = (n_T, \{T_j = (\ell_j, \alpha_j, x_j, y_j, \tau_j, \sigma_j, \delta_j) : j = 1, 2, \ldots, n_T\}). \quad (4.17)$$

Each texton element T_j is specified by its type ℓ_j, photometric contrast α_j, translation (x_j, y_j), rotation τ_j, scaling σ_j, and deformation vector δ_j. A texton $\pi \in \Pi$ consists of m image basis functions with a certain deformable configuration

$$\pi = (\, (\ell_1, \alpha_1, \tau_1, \sigma_1), \ (\ell_2, \alpha_2, \delta x_2, \delta y_2, \delta \tau_2, \delta \sigma_2), \ldots, \ (\ell_m, \alpha_m, \delta x_m, \delta y_m, \delta \tau_m, \delta \sigma_m) \,).$$

(4.18)

The $(\delta x, \delta y, \delta \tau, \delta \sigma)$ are the relative positions, orientations, and scales. Therefore, we have

$$p(\mathbf{B}|\mathbf{T}; \boldsymbol{\Pi}) = p(|\varpi_0|) \prod_{b_j \in \varpi_0} p(b_j) \prod_{c=1}^{n_T} p(\varpi_c | T_c; \boldsymbol{\pi}_{\ell_c}). \quad (4.19)$$

$p(\mathbf{T}; \kappa)$ is another distribution that accounts for the number of textons n_T and the spatial relationship among them. It can be a Gibbs model for the attributed point process. For simplicity, we assume the textons are independent at this moment as a special Gibbs model.

By integrating out the hidden variables, we obtain the likelihood for any observable image $\mathbf{I}^{\text{obs}}$,

$$p(\mathbf{I}^{\text{obs}}; \Theta) = \int p(\mathbf{I}^{\text{obs}}|\mathbf{B}; \boldsymbol{\Psi}) p(\mathbf{B}|\mathbf{T}; \boldsymbol{\Pi}) p(\mathbf{T}; \kappa) \ d\mathbf{B} \, d\mathbf{T}. \quad (4.20)$$

In $p(\mathbf{I}; \Theta)$ above, the parameters Θ (dictionaries, etc.) characterize the entire image ensemble, like the vocabulary for English or Chinese languages. In contrast, the hidden variables $\mathbf{B}$ and $\mathbf{T}$ are associated with an individual image $\mathbf{I}$ and correspond to the parsing tree in language.

Our goal is to learn the parameters $\Theta = (\Psi, \Pi, \kappa)$ by maximum likelihood estimation, or equivalently minimizing a Kullback–Leibler divergence between an underlying data distribution of images $f(\mathbf{I})$ and the model distribution $p(\mathbf{I}; \Theta)$

$$\Theta^* = (\Psi, \Pi, \kappa)^* = \arg\min \mathrm{KL}(f(\mathbf{I}) \| p(\mathbf{I}; \Theta)) = \arg\max \sum_m \log p(\mathbf{I}_m^{\mathrm{obs}}; \Theta) + \epsilon,$$

(4.21)

where ϵ is an approximation error that diminishes as sufficient data are available for training. In practice, ϵ may decide the complexity of the model and thus the number of basis functions L_ψ and textons L_π. For simplicity, we use only one large $\mathbf{I}^{\mathrm{obs}}$ for training, because multiple images can be considered just patches of a larger image. For motion and lighting models, $\mathbf{I}^{\mathrm{obs}}$ is extended to image sequence and image set with illumination variations.

By fitting the generative model to observed images, we can learn the texton dictionary as parameters of the generative model.

With such a model, we may study the geometric, dynamic, and photometric structures of the texton representation by further extending the generative model to account for motion and illumination variations. (1) For the geometric structures, a texton consists of a number of basis functions with deformable spatial configurations. The geometric structures are learned from static texture images (Figs. 4.9 and 4.10). (2) For the dynamic structures, the motion of a texton is characterized by a Markov chain model in time which sometimes can switch geometric configurations during the movement. We call the moving textons as "motons." The dynamic models are learned using the trajectories of the textons inferred from a video sequence

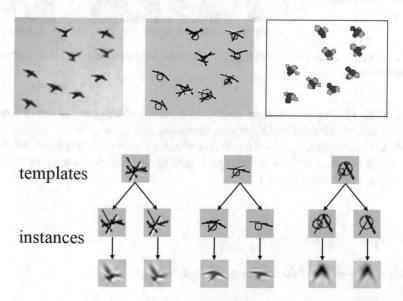

Fig. 4.9 Reprinted with permission from [247]. A texton template is a deformable composition of basis functions

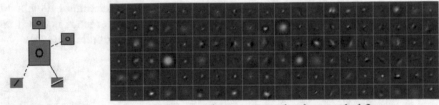

a texton template π   many texton instances randomly sampled from π

Fig. 4.10 Reprinted with permission from [247]. A texton template is a deformable composition of basis functions

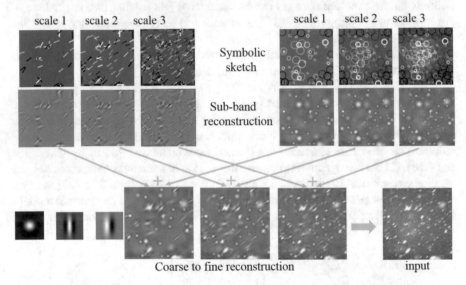

Fig. 4.11 Reprinted with permission from [247]. The image can be decomposed into sub-bands at multiple scales

(Fig. 4.11). (3) For photometric structures, a texton represents the set of images of a 3D surface element under varying illuminations and is called a "lighton." We adopt an illumination cone representation where a lighton is a texton triplet. For a given light source, a lighton image is generated as a linear combination of the three texton basis functions (Fig. 4.12).

4.3 Active Basis Model

Olshausen–Field Model for Sparse Coding

The active basis model is based on the sparse coding model of Olshausen and Field [188]. Olshausen and Field proposed that the role of simple V1 cells is to compute

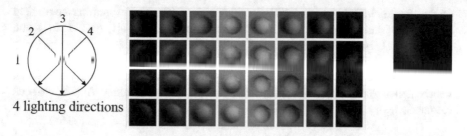

4 lighting directions

Fig. 4.12 Reprinted with permission from [277]. The concept of textons can also be generalized to incorporate lighting variations

sparse representations of natural images. Let $\{\mathbf{I}_m, m = 1, \ldots, M\}$ be a set of small image patches. For example, they may be 12×12 patches, in which case $\mathbf{I}_m \in \mathbb{R}^{12 \times 12}$. We may think of each $\mathbf{I}_m$ as a two-dimensional function defined on the 12×12 lattice. The Olshausen–Field model seeks to represent these images by

$$\mathbf{I}_m = \sum_{i=1}^{N} c_{m,i} B_i + U_m, \tag{4.22}$$

where $(B_i, i = 1, \ldots, N)$ is a dictionary of basis functions defined on the same image lattice (e.g., 12×12) as $\mathbf{I}_m$ and $c_{m,i}$ are the coefficients, and U_m is the unexplained residual image. N is often assumed to be greater than the number of pixels in $\mathbf{I}_m$, so the dictionary is said to be over-complete and is therefore redundant. However, the number of coefficients $(c_{m,i}, i = 1, \ldots, N)$ that are non-zero (or significantly different from zero) is assumed to be small (e.g., less than 10) for each image $\mathbf{I}_m$.

One may also assume that the basis functions in the dictionary are translated, rotated, and dilated versions of one another, so that each B_i can be written as $B_{x,s,\alpha}$, where x is the location (a two-dimensional vector), s is the scale, and α is the orientation. We call such a dictionary self-similar, and we call (x, s, α) the geometric attribute of $B_{x,s,\alpha}$.

Model (4.22) then becomes

$$\mathbf{I}_m = \sum_{x,s,\alpha} c_{m,x,s,\alpha} B_{x,s,\alpha} + U_m, \tag{4.23}$$

where $B_{x,s,\alpha}$ are translated, rotated, and dilated copies of a single basis function, e.g., $B = B_{x=0,s=1,\alpha=0}$, and (x, s, α) are properly discretized (default setting: α is discretized into 16 equally spaced orientations). B can be learned from training images $\{\mathbf{I}_m\}$.

From now on, we assume that the dictionary of basis functions is self-similar, and $\{B_{x,s,\alpha}, \forall(x, s, \alpha)\}$ is already given. In the following, we assume that $B_{x,s,\alpha}$ is a Gabor wavelet, and we also assume that $B_{x,s,\alpha}$ is normalized to have unit ℓ_2 norm so that $\|B_{x,y,\alpha}\|^2 = 1$. $B_{x,s,\alpha}$ may also be a pair of Gabor sine and cosine wavelets, so that for each Gabor wavelet B, $B = (B_0, B_1)$. The corresponding coefficient $c = (c_0, c_1)$, and $cB = c_0 B_0 + c_1 B_1$. The projection $\langle \mathbf{I}, B \rangle = (\langle \mathbf{I}, B_0 \rangle, \langle \mathbf{I}, B_1 \rangle)$, and $|\langle \mathbf{I}, B \rangle|^2 = \langle \mathbf{I}, B_0 \rangle^2 + \langle \mathbf{I}, B_1 \rangle^2$.

Given the dictionary $(B_{x,s,\alpha}, \forall(x, s, \alpha))$, the encoding of an image $\mathbf{I}_m$ amounts to inferring the coefficients $(c_{m,x,s,\alpha}, \forall(x, s, \alpha))$ in (4.23) under the sparsity constraint, which means that only a small number of $(c_{m,x,s,\alpha})$ are non-zero. That is, we seek to encode $\mathbf{I}_m$ by

$$\mathbf{I}_m = \sum_{i=1}^{n} c_{m,i} B_{x_{m,i}, s_{m,i}, \alpha_{m,i}} + U_m, \tag{4.24}$$

where $n \ll N$ is a small number, and $(x_{m,i}, s_{m,i}, \alpha_{m,i}, i = 1, \ldots, n)$ are the geometric attributes of the selected basis functions whose coefficients $(c_{m,i})$ are non-zero. The attributes $(x_{m,i}, s_{m,i}, \alpha_{m,i}, i = 1, \ldots, n)$ form a spatial point process (we continue to use i to index the basis functions, but here i only runs through the n selected basis functions instead of all the N basis functions as in (4.22)).

Active Basis Model for Shared Sparse Coding of Aligned Image Patches

The active basis model was proposed for modeling deformable templates formed by basis functions. Suppose we have a set of training image patches $\{\mathbf{I}_m, m = 1, \ldots, M\}$. This time we assume that they are defined on the same bounding box, and the objects in these images come from the same category. In addition, these objects appear at the same location, scale, and orientation, and in the same pose. See Fig. 4.13 for 9 image patches of deer. We call such image patches aligned.

The active basis model is of the following form:

$$\mathbf{I}_m = \sum_{i=1}^{n} c_{m,i} B_{x_i + \Delta x_{m,i}, s, \alpha_i + \Delta \alpha_{m,i}} + U_m, \tag{4.25}$$

where $\mathbf{B} = (B_{x_i, s, \alpha_i}, i = 1, \ldots, n)$ forms the nominal template of an active basis model (sometimes we simply call $\mathbf{B}$ an active basis template). Here we assume that the scale s is fixed and given. $\mathbf{B}_m = (B_{x_i + \Delta x_{m,i}, s, \alpha_i + \Delta \alpha_{m,i}}, i = 1, \ldots, n)$ is the deformed version of the nominal template $\mathbf{B}$ for encoding $\mathbf{I}_m$, where $(\Delta x_{m,i}, \Delta \alpha_{m,i})$ are the perturbations of the location and orientation of the i-th basis function from its nominal location x_i and nominal orientation α_i, respectively. The perturbations are introduced to account for shape deformation. Both $\Delta x_{m,i}$ and $\Delta \alpha_{m,i}$ are assumed to vary within limited ranges (default setting: $\Delta x_{m,i} \in [-3, 3]$ pixels, and $\Delta \alpha_{m,i} \in \{-1, 0, 1\} \times \pi/16$).

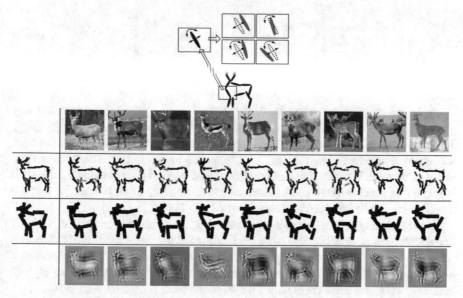

Fig. 4.13 First published in the Quarterly of Applied Mathematics in Volume 72:373–406, 2014, published by Brown University. Reprinted with permission from [107]. (**a**) An active basis model is a composition of a small number of basis functions, such as Gabor wavelets at selected locations and orientations. Each basis function can perturb its location and orientation within limited ranges. (**b**) Supervised learning of active basis model from aligned images. In this example, two active basis models are learned using Gabor wavelets at two different scales. The first row displays the 9 training images. The second row: the first plot is the nominal template formed by 50 basis functions. The rest of the plots are the deformed templates matched to the images by perturbing the basis functions. The third row: the same as the second row, except that the scale of the Gabor wavelets is about twice as large, and the number of wavelets is 14. The last row displays the linear reconstruction of each training image from 100 selected and perturbed basis functions

Prototype Algorithm

Given the dictionary of basis functions $\{B_{x,s,\alpha}, \forall x, s, \alpha\}$, the learning of the active basis model from the aligned image patches $\{\mathbf{I}_m\}$ involves the sequential selection of B_{x_i,s,α_i} and the inference of its perturbed version $B_{x_i + \Delta x_{m,i}, s, \alpha_i + \Delta \alpha_{m,i}}$ in each image $\mathbf{I}_m$. We call the learning supervised because the bounding boxes of the objects are given and the images are aligned. See Fig. 4.13 for an illustration of the learning results.

In this subsection, we consider a prototype version of the shared matching pursuit algorithm, which is to be revised in the following subsections. The reason we start from this prototype algorithm is that it is simple and yet captures the key features of the learning algorithm.

We seek the maximal reduction of the least squares reconstruction error in each iteration (recall that the basis functions are normalized to have unit ℓ_2 norm):

$$\sum_{m=1}^{M} \left\| \mathbf{I}_m - \sum_{i=1}^{n} c_{m,i} B_{x_i+\Delta x_{m,i},s,\alpha_i+\Delta \alpha_{m,I}} \right\|^2. \tag{4.26}$$

The prototype algorithm is a greedy algorithm that minimizes the reconstruction error:

In Eq. (4.29), the perturbed basis function $B_{x_i+\Delta x_{m,i},s,\alpha_i+\Delta \alpha_{m,i}}$ explains away part of U_m. As a result, nearby basis functions that overlap with $B_{x_i+\Delta x_{m,i},s,\alpha_i+\Delta \alpha_{m,i}}$ tend not to be selected in future iterations. So the basis functions selected for each deformed template $\mathbf{B}_m = (B_{x_i+\Delta x_{m,i},s,\alpha_i+\Delta \alpha_{m,i}}, i = 1, \cdots, n)$ usually have little overlap with each other. For computational and modeling convenience, we shall assume that these selected basis functions are orthogonal to each other, so that the coefficients can be obtained by projection: $c_{m,i} = \langle \mathbf{I}_m, B_{x_i+\Delta x_{m,i},s,\alpha_i+\Delta \alpha_{m,i}} \rangle$.

Correspondingly, the explaining-away step can then be carried out by local inhibition. Specifically, after we identify the perturbed basis function $B_{x_i+\Delta x_{m,i},s,\alpha_i+\Delta \alpha_{m,i}}$, we simply prohibit nearby basis functions that are correlated with $B_{x_i+\Delta x_{m,i},s,\alpha_i+\Delta \alpha_{m,i}}$ from being included in the deformed template $\mathbf{B}_m$. In practice, we allow small correlations between the basis functions in each $\mathbf{B}_m$.

0. Initialize $i \leftarrow 0$. For $m = 1, \ldots, M$, initialize the residual image $U_m \leftarrow \mathbf{I}_m$.
1. $i \leftarrow i + 1$. Select the next basis function by

$$(x_i, \alpha_i) = \arg\max_{x,\alpha} \sum_{m=1}^{M} \max_{\Delta x, \Delta \alpha} |\langle U_m, B_{x+\Delta x,s,\alpha+\Delta \alpha} \rangle|^2, \tag{4.27}$$

where $\max_{\Delta x, \Delta \alpha}$ is the local maximum pooling within the small ranges of $\Delta x_{m,i}$ and $\Delta \alpha_{m,i}$.
2. For $m = 1, \ldots, M$, given (x_i, α_i), infer the perturbations in location and orientation by retrieving the arg-max in the local maximum pooling of step 1:

$$(\Delta x_{m,i}, \Delta \alpha_{m,i}) = \arg\max_{\Delta x, \Delta \alpha} |\langle U_m, B_{x_i+\Delta x,s,\alpha_i+\Delta \alpha} \rangle|^2. \tag{4.28}$$

Let $c_{m,i} \leftarrow \langle U_m, B_{x_i+\Delta x_{m,i},s,\alpha_i+\Delta \alpha_{m,i}} \rangle$, and update the residual image by explaining away:

$$U_m \leftarrow U_m - c_{m,i} B_{x_i+\Delta x_{m,i},s,\alpha_i+\Delta \alpha_{m,i}}. \tag{4.29}$$

3. Stop if $i = n$, else go back to step 1.

Algorithm 4: Prototype Algorithm

Statistical Modeling

The above algorithm guided by (4.26) implicitly assumes that the unexplained background image U_m is Gaussian white noise. This assumption can be problematic because the unexplained background may contain salient structures such as edges, and the Gaussian white noise distribution clearly cannot account for such structures. This is why we need to revise the above algorithm which is based on the Gaussian white noise assumption. A better assumption is to assume that U_m follows the same distribution as that of natural images.

More precisely, the distribution of $\mathbf{I}_m$ given the deformed template $\mathbf{B}_m = (B_{x_i + \Delta x_{m,i}, s, \alpha_i + \Delta \alpha_{m,i}}, i = 1, \ldots, n)$, i.e., $p(\mathbf{I}_m \mid \mathbf{B}_m)$, is obtained by modifying the distribution of natural images $q(\mathbf{I}_m)$ in such a way that we only change the distribution of $C_m = (c_{m,i} = \langle \mathbf{I}_m, B_{x_i + \Delta x_{m,i}, s, \alpha_i + \Delta \alpha_{m,i}} \rangle, i = 1, \ldots, n)$ from $q(C_m)$ to $p(C_m)$, while leaving the conditional distribution of U_m given C_m unchanged. Here $p(C_m)$ and $q(C_m)$ are the distributions of C_m under $p(\mathbf{I}_m \mid \mathbf{B}_m)$ and $q(\mathbf{I}_m)$, respectively. Thus the model is in the form of foreground $p(C_m)$ popping out from background $q(\mathbf{I}_m)$. Specifically, $p(\mathbf{I}_m \mid \mathbf{B}_m) = q(\mathbf{I}_m)p(C_m)/q(C_m)$.

The reason for such a form is as follows. C_m is the projection of $\mathbf{I}_m$ into $\mathbf{B}_m$. Let U_m be the projection of $\mathbf{I}_m$ into the remaining subspace that is orthogonal to $\mathbf{B}_m$. Then $p(\mathbf{I}_m \mid \mathbf{B}_m)/q(\mathbf{I}_m) = p(C_m, U_m)/q(C_m, U_m) = p(C_m)/q(C_m)$. The second equality follows from the assumption that $p(U_m \mid C_m) = q(U_m \mid C_m)$, i.e., we keep the conditional distribution of U_m given C_m fixed.

For computational simplicity, we further assume $(c_{m,i} = \langle \mathbf{I}_m, B_{x_i + \Delta x_{m,i}, s, \alpha_i + \Delta \alpha_{m,i}} \rangle, i = 1, \ldots, n)$ are independent given $\mathbf{B}_m$, under both p and q, so

$$p(\mathbf{I}_m \mid \mathbf{B}_m) = q(\mathbf{I}_m) \prod_{i=1}^{n} \frac{p_i(c_{m,i})}{q(c_{m,i})}, \tag{4.30}$$

where $q(c)$ is assumed to be the same for $i = 1, \ldots, n$ because $q(\mathbf{I}_m)$ is translation and rotation invariant. $q(c)$ can be pooled from natural images in the form of a histogram of Gabor filter responses. This histogram is heavy-tailed because of the edges in natural images.

For parametric modeling, we model $p_i(c_{m,i})/q(c_{m,i})$ in the form of exponential family model. Specifically, we assume the following exponential family model $p_i(c) = p(c; \lambda_i)$, which is in the form of exponential tilting of the reference distribution $q(c)$:

$$p(c; \lambda) = \frac{1}{Z(\lambda)} \exp\{\lambda h(|c|^2)\} q(c), \tag{4.31}$$

so that $p(c; \lambda)/q(c)$ is in the exponential form. We assume $\lambda_i > 0$, and $h(r)$ is a sigmoid-like function of the response $r = |c|^2$ that saturates for large r (recall that the Gabor filter response $c = (c_0, c_1)$ consists of responses from the pair of

Gabor sine and cosine wavelets, and $|c|^2 = c_0^2 + c_1^2$). Specifically, we assume that $h(r) = \xi[2/(1 + e^{-2r/\xi}) - 1]$, so $h(r) \approx r$ for small r, and $h(r) \to \xi$ as $r \to \infty$ (default setting: $\xi = 6$). The reason we want $h(r)$ to approach a fixed constant for large r is that there can be strong edges in both the foreground and the background, albeit with different frequencies. $p(c; \lambda)/q(c)$ should approach the ratio between these two frequencies for large $r = |c|^2$. In (4.31),

$$Z(\lambda) = \int \exp\{\lambda h(r)\} q(c) dc = E_q[\exp\{\lambda h(r)\}] \tag{4.32}$$

is the normalizing constant.

$$\mu(\lambda) = E_\lambda[h(r)] = \int h(r) p(c; \lambda) dc \tag{4.33}$$

is the mean parameter. Both $Z(\lambda)$ and $\mu(\lambda)$ can be computed beforehand from a set of natural images.

The exponential family model can be justified by the maximum entropy principle. Given the deformed template $\mathbf{B}_m = (B_{x_i + \Delta x_{m,i}, s, \alpha_i + \Delta\alpha_{m,i}}, i = 1, \ldots, n)$, consider the coefficients obtained by projection: $(c_{m,i}(\mathbf{I}_m) = \langle \mathbf{I}_m, B_{x_i + \Delta x_{m,i}, s, \alpha_i + \Delta\alpha_{m,i}} \rangle, i = 1, \ldots, n)$. Suppose we want to find a probability distribution $p(\mathbf{I}_m \mid \mathbf{B}_m)$ so that $E[h(|c_{m,i}(\mathbf{I}_m)|^2)] = \mu_i$ for some fixed μ_i, $i = 1, \ldots, n$, where μ_i can be estimated from the training images. Then among all the distributions that satisfy the constraints on $E[h(|c_{m,i}(\mathbf{I}_m)|^2)]$, the distribution that is closest to $q(\mathbf{I}_m)$ in terms of the Kullback–Leibler divergence is given by

$$p(\mathbf{I}_m \mid \mathbf{B}_m) = \frac{1}{Z(\Lambda)} \exp\left\{ \sum_{i=1}^{n} \lambda_i h(|c_{m,i}(\mathbf{I}_m)|^2) \right\} q(\mathbf{I}_m), \tag{4.34}$$

where $\Lambda = (\lambda_i, i = 1, \ldots, n)$, $Z(\Lambda) = E_q[\exp\{\sum_{i=1}^{n} \lambda_i h(|c_{m,i}(\mathbf{I}_m)|^2)\}]$ is the normalizing constant, and Λ is chosen to satisfy the constraints on $E[h(|c_{m,i}(\mathbf{I}_m)|^2)]$. If we further assume that $c_{m,i}(\mathbf{I}_m)$ are independent of each other for $i = 1, \ldots, n$ under $q(\mathbf{I}_m)$, then $c_{m,i}(\mathbf{I}_m)$ are also independent under $p(\mathbf{I}_m \mid \mathbf{B}_m)$, and their distributions are of the form (4.31).

In order to choose the nominal template $\mathbf{B}$ and the deformed templates $\{\mathbf{B}_m, m = 1, \ldots, M\}$, we want $p(\mathbf{I}_m \mid \mathbf{B}_m)$ to be farthest from $q(\mathbf{I}_m)$ in terms of the Kullback–Leibler divergence. From a classification point of view, we want to choose $\mathbf{B}$ and $\{\mathbf{B}_m\}$ so that the features $\{h(|c_{m,i}|^2), i = 1, \ldots, n\}$ lead to the maximal separation between training images (e.g., images of deer) and generic natural images.

The log-likelihood ratio between the current model $p(\mathbf{I}_m | \mathbf{B}_m)$ and the reference model $q(\mathbf{I}_m)$ is

$$l(\{\mathbf{I}_m\} \mid \mathbf{B}, \{\mathbf{B}_m\}, \Lambda) = \sum_{m=1}^{M} \log \frac{p(\mathbf{I}_m \mid \mathbf{B}_m)}{q(\mathbf{I}_m)} \tag{4.35}$$

$$= \sum_{m=1}^{M} \sum_{i=1}^{n} \left[\lambda_i h(|\langle \mathbf{I}_m, B_{x_i + \Delta x_{m,i}, s, \alpha_i + \Delta \alpha_{m,i}} \rangle|^2) - \log Z(\lambda_i) \right]. \tag{4.36}$$

The expectation of the above log-likelihood ratio is the Kullback–Leibler divergence between $p(\mathbf{I}_m \mid \mathbf{B}_m)$ and $q(\mathbf{I}_m)$.

Given the training images $\{\mathbf{I}_m, m = 1, \ldots, M\}$, $\sum_{m=1}^{M} \log q(\mathbf{I}_m)$ is a constant. Thus maximizing the log-likelihood ratio $\sum_{m=1}^{M} \log p(\mathbf{I}_m \mid \mathbf{B}_m, \Lambda)/q(\mathbf{I}_m)$ is equivalent to maximizing the log-likelihood $\sum_{i=1}^{M} \log p(\mathbf{I}_m \mid \mathbf{B}_m, \Lambda)$.

Shared Matching Pursuit

We revise the prototype algorithm in Sect. 4.3 so that each iteration seeks the maximal increase of the log-likelihood ratio (4.36) instead of the maximum reduction of the least squares reconstruction error (4.26) as in Sect. 4.3. The revised version of the shared matching pursuit algorithm is as follows.

0. Initialize $i \leftarrow 0$. For $m = 1, \ldots, M$, initialize the response maps
   $R_m(x, \alpha) \leftarrow \langle \mathbf{I}_m, B_{x,s,\alpha} \rangle$ for all (x, α).
1. $i \leftarrow i + 1$. Select the next basis function by finding

$$(x_i, \alpha_i) = \arg \max_{x, \alpha} \sum_{m=1}^{M} \max_{\Delta x, \Delta \alpha} h(|R_m(x + \Delta x, \alpha + \Delta \alpha)|^2), \tag{4.37}$$

   where $\max_{\Delta x, \Delta \alpha}$ is again the local maximum pooling.
2. For $m = 1, \ldots, M$, given (x_i, α_i), infer the perturbations by retrieving the arg-max in the local maximum pooling of step 1:

$$(\Delta x_{m,i}, \Delta \alpha_{m,i}) = \arg \max_{\Delta x, \Delta \alpha} |R_m(x_i + \Delta x, \alpha_i + \Delta \alpha)|^2. \tag{4.38}$$

   Let $c_{m,i} \leftarrow R_m(x_i + \Delta x_{m,i}, \alpha_i + \Delta \alpha_{m,i})$, and update $R_m(x, \alpha) \leftarrow 0$ if the correlation

$$\mathrm{corr}[B_{x,s,\alpha}, B_{x_i + \Delta x_{m,i}, s, \alpha_i + \Delta \alpha_{m,i}}] > \epsilon \tag{4.39}$$

   (default setting: $\epsilon = .1$). Then compute λ_i by solving the maximum likelihood equation $\mu(\lambda_i) = \sum_{m=1}^{M} h(|c_{m,i}|^2)/M$.
3. Stop if $i = n$, else go back to step 1.

Algorithm 5: Revised Prototype Algorithm

For each candidate (x_i, α_i), the maximum likelihood equation $\mu(\lambda_i) = \sum_{m=1}^{M} h(|c_{m,i}|^2)/M$ is obtained by taking the derivative of the log-likelihood ratio, where $\mu(\lambda_i) = E_{\lambda_i}[h(|c|^2)] = \int h(|c|^2)p(c; \lambda_i)dc$ is the mean parameter and is a monotonically increasing function of $\lambda_i > 0$. So its inverse $\mu^{-1}()$ is also a monotonically increasing function. λ_i is solved so that $\mu(\lambda_i)$ matches the empirical average of $h(|c_{m,i}|^2), m = 1, \ldots, M$. The function $\mu()$ can be computed and stored over a discrete set of equal-spaced values so that λ_i can be solved by looking up these values with linear interpolations between them.

Because $h()$ is monotonically increasing, the maximized log-likelihood ratio is monotone in the estimated λ_i. The estimated λ_i is in turn monotone in the average $\sum_{m=1}^{M} h(|c_{m,i}|^2)/M$. So the maximized log-likelihood ratio is monotone in $\sum_{m=1}^{M} h(|c_{m,i}|^2)/M$. Therefore, in step [1], (x_i, α_i) is chosen by maximizing the sum $\sum_{m=1}^{M} \max_{\Delta x, \Delta \alpha} h(|R_m(x + \Delta x, \alpha + \Delta \alpha)|^2)$ over all possible (x, α).

In step 2, the arg-max basis function inhibits nearby basis functions to enforce the approximate orthogonality constraint. The correlation is defined as the square of the inner product between the basis functions and can be computed and stored beforehand.

After learning the template from training images $\{\mathbf{I}_m\}$, we can use the learned template to detect the object in a testing image $\mathbf{I}$.

1. For every pixel X, compute the log-likelihood ratio $l(X)$, which serves as the template matching score at putative location X:

$$l(X) = \sum_{i=1}^{n} \left[\lambda_i \max_{\Delta x, \Delta \alpha} h(|\langle \mathbf{I}, B_{X+x_i+\Delta x, s, \alpha_i+\Delta \alpha} \rangle|^2) - \log Z(\lambda_i) \right]. \quad (4.40)$$

2. Find maximum likelihood location $\hat{X} = \arg \max_X l(X)$. For $i = 1, \ldots, n$, inferring perturbations by retrieving the arg-max in the local maximum pooling in step 1:

$$(\Delta x_i, \Delta \alpha_i) = \arg \max_{\Delta x, \Delta \alpha} |\langle \mathbf{I}, B_{\hat{X}+x_i+\Delta x, s, \alpha_i+\Delta \alpha} \rangle|^2. \quad (4.41)$$

3. Return the location $\hat{X}$, and $(B_{\hat{X}+x_i+\Delta x_i, s, \alpha_i+\Delta \alpha_i}, i = 1, \ldots, n)$, which is the translated and deformed template.

Algorithm 6: Object detection

We can rotate the template and scan the template over multiple resolutions of the original image, to account for uncertainties about the orientation and scale of the object in the testing image.

4.4 Sparse FRAME Model

One generalization of the FRAME model is a sparse FRAME model [261, 265] where the potential functions are location specific, and they are non zero only in selected locations. This model is intended to model image patterns that are non-stationary in the spatial domain, such as object patterns. The model can be written as a shared sparse coding model, where the observed images are represented by a commonly shared set of wavelets selected from a dictionary. In this shared sparse coding model, the original linear filters for bottom-up computation (from image to filter responses) become linear basis functions for top-down representation (from coefficients to image).

Dense FRAME

We start from the non-stationary or spatially inhomogeneous FRAME model [254, 261, 265] based on a dictionary of basis functions or wavelets $\{B_{k,x}, \forall k, x\}$ (we assume that the dictionary of wavelets, such as the Gabor and DoG wavelets, has been given or has been learned by sparse component analysis [3, 22, 190]). The model is a random field of the following form:

$$p(\mathbf{I}; w) = \frac{1}{Z(w)} \exp\left\{ \sum_{k=1}^{K} \sum_{x \in \mathcal{D}} w_{k,x} h(\langle \mathbf{I}, B_{k,x} \rangle) \right\} q(\mathbf{I}). \tag{4.42}$$

The above model is a simple generalization of FRAME model, where $\langle \mathbf{I}, B_{k,x} \rangle$ is the filter response, which can also be written as $[F_k * \mathbf{I}](x)$. The parameter $w_{k,x}$ depends on position x, so the model is non-stationary. $w = (w_{k,x}, \forall k, x)$. Again $Z(w)$ is the normalizing constant. $h()$ is a pre-specified rectification function. In [261], $h(r) = |r|$, i.e., the model is insensitive to the signs of filter responses. $q(\mathbf{I})$ is a reference distribution, such as the Gaussian white noise model

$$q(\mathbf{I}) = \frac{1}{(2\pi\sigma^2)^{D/2}} \exp\left\{ -\frac{1}{2\sigma^2} ||\mathbf{I}||^2 \right\}, \tag{4.43}$$

where again D counts the number of pixels in the image domain $\mathcal{D}$.

Sparse Representation

Assume we are given a dictionary of wavelets or basis functions $\{B_{k,x}\}$, where k may index a finite collection of prototype functions $\{B_k, k = 1, \ldots, K\}$, and where

$B_{k,x}$ is a spatially translated copy of B_k to position x. We can represent an image $\mathbf{I}$ by

$$\mathbf{I} = \sum_{k,x} c_{k,x} B_{k,x} + \epsilon, \qquad (4.44)$$

where $c_{k,x}$ are the coefficients, and ϵ is the residual image. It is often assumed that the representation is sparse, i.e., most of the $c_{k,x}$ are equal to zero. The resulting representation is also called sparse coding [49, 190].

The sparsification of $c_{k,x}$, i.e., the selection of the basis functions, can be accomplished by matching pursuit [167] or basis pursuit/Lasso [26, 230]. Using a Lasso-like objective function, the dictionary of basis functions $\{B_k\}$ can be learned from a collection of training images [3, 190]. It is sometimes called sparse component analysis [46]. It can be considered a generalization of factor analysis. For natural images, the learned basis functions resemble the Gabor and DoG wavelets.

Maximum Likelihood Learning

The basic learning algorithm estimates the parameters $w = (w_{k,x}, \forall k, x)$ from a set of aligned training images $\{\mathbf{I}_i, i = 1, \dots, n\}$ that come from the same category, where n is the total number of training images. The algorithm can be extended to learn from non-aligned images from mixed categories. The basic learning algorithm seeks to maximize the log-likelihood

$$L(w) = \frac{1}{n} \sum_{i=1}^{n} \log p(\mathbf{I}_i; w), \qquad (4.45)$$

whose partial derivatives are

$$\frac{\partial L(w)}{\partial w_{k,x}} = \frac{1}{n} \sum_{i=1}^{n} h(\langle \mathbf{I}_i, B_{k,x} \rangle) - \mathrm{E}_w \left[h(\langle \mathbf{I}, B_{k,x} \rangle) \right], \qquad (4.46)$$

where E_w denotes expectation with respect to $p(\mathbf{I}; w)$ in (4.42). This expectation can be approximated by the Monte Carlo integration. Thus, w can be computed by the stochastic gradient ascent algorithm [205, 274]

$$w_{k,x}^{(t+1)} = w_{k,x}^{(t)} + \gamma_t \left[\frac{1}{n} \sum_{i=1}^{n} h(\langle \mathbf{I}_i, B_{k,x} \rangle) - \frac{1}{\tilde{n}} \sum_{i=1}^{\tilde{n}} h(\langle \tilde{\mathbf{I}}_i, B_{k,x} \rangle) \right], \qquad (4.47)$$

where γ_t is the step size or the learning rate, and $\{\tilde{\mathbf{I}}_i, i = 1, \dots, \tilde{n}\}$ are the synthetic images sampled from $p(\mathbf{I}; w^{(t)})$ using MCMC, such as Hamiltonian Monte Carlo [179] or the Gibbs sampler [74]. $\tilde{n}$ is the total number of independent parallel Markov chains that sample from $p(\mathbf{I}; w^{(t)})$.

Generative Boosting

Model (4.42) is a dense model in that all the wavelets (or filters) in the dictionary are included in the model. We can sparsify the model by forcing most of the $w_{k,x}$ to be zero so that only a small number of wavelets are included in the model. This can be achieved by a generative version [265] of the epsilon-boosting algorithm [64, 67] (see also [37, 59, 250, 257]). The algorithm starts from $w = 0$, the zero vector. At the t-th iteration, let

$$\Delta_{k,x} = \frac{1}{n} \sum_{i=1}^{n} h(\langle \mathbf{I}_i, B_{k,x} \rangle) - \frac{1}{\tilde{n}} \sum_{i=1}^{\tilde{n}} h(\langle \tilde{\mathbf{I}}_i, B_{k,x} \rangle) \qquad (4.48)$$

be the Monte Carlo estimate of $\partial L(w)/\partial w_{k,x}$, where again $\{\tilde{\mathbf{I}}_i, i = 1, \ldots, \tilde{n}\}$ are the synthetic images sampled from the current model. We select

$$(\hat{k}, \hat{x}) = \arg \max_{k,x} |\Delta_{k,x}| \qquad (4.49)$$

and update $w_{\hat{k},\hat{x}}$ by

$$w_{\hat{k},\hat{x}} \leftarrow w_{\hat{k},\hat{x}} + \gamma_t \Delta_{\hat{k},\hat{x}}, \qquad (4.50)$$

where γ_t is the step size, assumed to be sufficiently small (thus the term "epsilon" in the epsilon-boosting algorithm). We call this algorithm generative epsilon boosting because the derivatives are estimated by images generated from the current model. See Fig. 4.14 for an illustration.

The selected wavelet $B_{\hat{k},\hat{x}}$ reveals the dimension along which the current model is most conspicuously lacking in reproducing the statistical properties of the training images. By including $B_{\hat{k},\hat{x}}$ into the model and updating the corresponding parameter $w_{\hat{k},\hat{x}}$, the model receives the most needed boost. The process is like an artist making a painting, where $B_{\hat{k},\hat{x}}$ is the stroke that is most needed to make the painting look more similar to the observed objects.

The epsilon boosting algorithm [67, 96] has an interesting relationship with the ℓ_1 regularization in the Lasso [230] and basis pursuit [26]. As pointed out by [206], under a monotonicity condition (e.g., the components of w keep increasing), such an algorithm approximately traces the solution path of the ℓ_1 regularized minimization of

$$- L(w) + \rho \|w\|_{\ell_1}, \qquad (4.51)$$

where the regularization parameter ρ starts from a big value so that all the components of w are zero and gradually lowers itself to allow more components to be non-zero so that more wavelets are induced into the model.

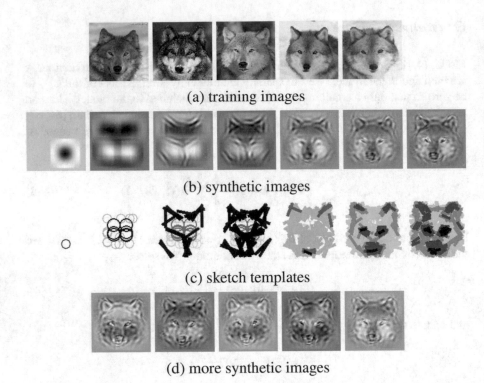

(a) training images

(b) synthetic images

(c) sketch templates

(d) more synthetic images

Fig. 4.14 Reprinted with permission from [256]. Learning process of generative boosting. (**a**) Observed training images (100×100 pixels) from which the random field model is learned. (**b**) A sequence of synthetic images generated by the learned model as more and more wavelets are induced into the model. The numbers of the selected wavelets are $1, 20, 65, 100, 200, 500$, and 800, respectively. (**c**) A sequence of sketch templates that illustrate the wavelets selected from the given dictionary. The dictionary includes 4 scales of Gabor wavelets, illustrated by bars of different sizes, and 2 scales of Difference of Gaussian (DoG) wavelets, illustrated by circles. In each template, smaller scale wavelets appear darker than larger ones. (**d**) More synthetic images independently generated from the final learned model

Sparse Model

After selecting m wavelets, we have the following sparse FRAME model:

$$p(\mathbf{I}; \mathbf{B}, w) = \frac{1}{Z(w)} \exp\left\{\sum_{j=1}^{m} w_j h(\langle \mathbf{I}, B_{k_j, x_j}\rangle)\right\} q(\mathbf{I}), \qquad (4.52)$$

where $\mathbf{B} = (B_j = B_{k_j, x_j}, j = 1, \ldots, m)$ is the set of wavelets selected from the dictionary, and $w_j = w_{k_j, x_j}$.

In model (4.52), m is much smaller than D, the number of pixels. Thus, we can represent $\mathbf{I}$ by

$$\mathbf{I} = \sum_{j=1}^{m} c_j B_{k_j, x_j} + \epsilon, \tag{4.53}$$

where $C = (c_j, j = 1, \ldots, m)^\top$ are the least square regression coefficients of $\mathbf{I}$ on $\mathbf{B} = (B_j, j = 1, \ldots, m)$, i.e., $C = (\mathbf{B}^\top \mathbf{B})^{-1} \mathbf{B}^\top \mathbf{I}$, and ϵ is the residual image. The distribution of C under $p(\mathbf{I}; \mathbf{B}, w)$ is

$$p_C(C; w) = \frac{1}{Z(w)} \exp\left\{ \langle w, h(\mathbf{B}^\top \mathbf{B} C) \rangle \right\} q_C(C), \tag{4.54}$$

where $q_C(C)$ is the distribution of C under $q(\mathbf{I})$, and the transformation $h()$ is applied element-wise. Thus, $p(\mathbf{I}; B, w)$ in (4.52) can be written as a wavelet sparse coding model (4.53) and (4.54). The forms of (4.52) and (4.53) show that the selected wavelets $\{B_j\}$ serve both as filters and as basis functions. The sparse coding form of the model (4.53) and (4.54) is used for sampling $\{\tilde{\mathbf{I}}_i\}$ from $p(\mathbf{I}; \mathbf{B}, w)$ by first sampling $C \sim p_C(C; w)$ using the Gibbs sampler [74] and then generating $\tilde{\mathbf{I}}_i$ according to (4.53).

Model (4.53) suggests that we can also select the wavelets by minimizing

$$\sum_{i=1}^{n} \left\| \mathbf{I}_i - \sum_{j=1}^{m} c_{i,j} B_{k_j, x_j} \right\|^2, \tag{4.55}$$

using a shared matching pursuit method [261]. See Fig. 4.15 for an illustration. We can also allow the selected wavelets to perturb their locations and orientations to account for deformations [254].

The sparse FRAME model can be used for unsupervised learning tasks such as model-based clustering [57]. Extending the learning algorithm, one can learn a codebook of multiple sparse FRAME models from non-aligned images. The learned models can be used for tasks such as transfer learning [107, 261].

The sparse FRAME model merges two important research themes in image representation and modeling, namely, Markov random fields [19, 75] and wavelet sparse coding [3, 190].

The wavelets can be mapped to the first layer filters of a ConvNet [144] to be described later. The sparse FRAME models can be mapped to the second layer nodes of a ConvNet, except that the sparse FRAME versions of the second layer nodes are selectively and sparsely connected to the first layer nodes.

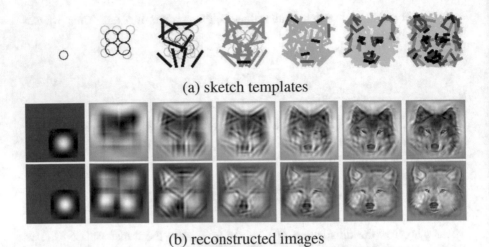

(a) sketch templates

(b) reconstructed images

Fig. 4.15 Reprinted with permission from [256]. The shared matching pursuit for the purpose of wavelet selection. (**a**) Sequence of sketch templates that illustrate the wavelets selected sequentially in order to reconstruct all the training images simultaneously. The selected wavelets are shared by all the training images (100 × 100) in their reconstructions. The numbers of selected wavelets in the sequence are 2, 20, 60, 100, 200, 500, and 800, respectively. (**b**) Sequences of reconstructed images by the selected wavelets for the 1st and 3rd training images in Fig. 4.14a

4.5   Compositional Sparse Coding

Sparsity and Composition

The goal of this section is to develop a compositional sparse code for natural images. Figure 4.16 illustrates the basic idea. We start with a dictionary of Gabor wavelets centered at a dense collection of locations and tuned to a collection of scales and orientations. In Fig. 4.16, each Gabor wavelet is illustrated by a bar at the same location and with the same length and orientation as the corresponding wavelet. Figure 4.16a displays the training image. (b) Displays a mini-dictionary of 2 compositional patterns of wavelets learned from the training image. Each compositional pattern is a template formed by a group of a small number of wavelets at selected locations and orientations. The learning is unsupervised in the sense that the images are not labeled or annotated. The number of templates in the dictionary is automatically determined by an adjusted Bayesian information criterion. The 2 templates are displayed in different colors, so that it can be seen clearly how the translated, rotated, scaled, and deformed copies of the 2 templates are used to represent the training image, as shown in (b). In (c), the templates are overlaid on the original image. In our current implementation, we allow some overlap between the bounding boxes of the templates. The templates learned from the training image can be generalized to testing images, as shown in (d) and (e).

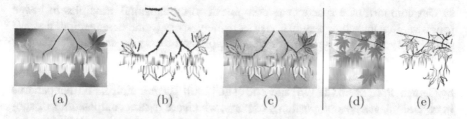

(a)   (b)   (c)   (d)   (e)

Fig. 4.16 First published in the Quarterly of Applied Mathematics in Volume 72:373–406, 2014, published by Brown University. Reprinted with permission from [107]. Unsupervised learning of compositional sparse code (**a,b,c**) and using it for recognition and segmentation (**d,e**). (**a**) Training image of 480 × 768 pixels. (**b**) Above: 2 compositional patterns (twig and leaf) in the form of shape templates learned from the training image. Each constituent Gabor wavelet (basis function) of a template is illustrated by a bar at the same location and with the same orientation and length as the corresponding wavelet. The size of each template is 100 × 100 pixels. The number of basis functions in each template is no more than 40 and is automatically determined. Below: representing the training image by translated, rotated, scaled, and deformed copies of the 2 templates. (**c**) Superposing the deformed templates on the original image. (**d**) Testing image. (**e**) Representation (recognition) of the testing image by the 2 templates

Fig. 4.17 First published in the Quarterly of Applied Mathematics in Volume 72:373–406, 2014, published by Brown University. Reprinted with permission from [107]. The compositional patterns (templates) are learned from 20 training images (only 6 of them are shown in this figure). The training images are not registered or otherwise annotated. The size of each template is 100 × 100 pixels. The number of basis functions in each template is no more than 40 and is automatically determined

Figure 4.17 shows another example, where part templates of egrets and templates of water waves and grasses are learned from 20 training images without supervision. That is, the training images are not registered, in that we do not assume that the objects in the training images appear at the same location and scale. It is interesting to observe that in this example, unsupervised learning also accomplishes image segmentation, object detection, and perceptual grouping (e.g., grass pattern), which are important tasks in vision.

Our compositional sparse code combines two fundamental principles in image representation and computational vision, namely, sparsity and compositionality. We shall briefly review these two principles below and then give an overview of our methodology.

The compositionality principle was proposed in the context of computer vision by Geman, Potter, and Chi [76] and Zhu et al. [280]. The principle holds that patterns in natural images are compositions of parts, which are in turn compositions of sub-parts, and so on. An interesting example cited by Geman et al. is Laplace's remark that one strongly prefers to view the string CONSTANTINOPLE as a single word, rather than 14 individual letters. This is also the case with the basis functions in the sparse coding of natural images. Like letters forming the words, the basis functions in the sparse representations of natural images also form various compositional patterns in terms of their spatial arrangements. We call such sparsity compositional sparsity, which is a special form of structured sparsity.

Any hierarchical compositional model will necessarily end with constituent elements that cannot be further decomposed, and such elements may be called "atoms." Interestingly, the basis functions are commonly referred to as atoms in sparse coding literature, and the sparse representation based on atoms is usually called "atomic decomposition." Compositionality enables us to compose atoms into composite representational units, which leads to much sparser and thus more meaningful representations of the signals.

The current form of our model consists of two layers of representational units: basis functions and shape templates. It is possible to extend it to multiple layers of hierarchy.

Compositional Sparse Coding Model

We will write down our model in an analogous form as the Olshausen–Field model $\mathbf{I}_m = \sum_{i=1}^{n} c_{m,i} B_{x_{m,i}, s_{m,i}, \alpha_{m,i}} + U_m$, by making the notation compact.

As the first step, let us slightly generalize the active basis model by assuming that the template may appear at location X_m in image $\mathbf{I}_m$, and then we can write the representation in the following form:

$$\mathbf{I}_m = \sum_{i=1}^{n} c_{m,i} B_{X_m + x_i + \Delta x_{m,i}, s, \alpha_i + \Delta \alpha_{m,i}} + U_m$$

$$= C_m \mathbf{B}_{X_m} + U_m, \tag{4.56}$$

where $\mathbf{B}_{X_m} = (B_{X_m + x_i + \Delta x_{m,i}, s, \alpha_i + \Delta \alpha_{m,i}}, i = 1, \ldots, n)$ is the deformed template spatially translated to X_m, $C_m = (c_{m,i}, i = 1, \ldots, n)$, and $C_m \mathbf{B}_{X_m} = \sum_{i=1}^{n} c_{m,i} B_{X_m + x_i + \Delta x_{m,i}, s, \alpha_i + \Delta \alpha_{m,i}}$ by definition.

$\mathbf{B}_{X_m}$ explains the part of $\mathbf{I}_m$ that is covered by $\mathbf{B}_{X_m}$. For each image $\mathbf{I}_m$ and each X_m, we can define the log-likelihood ratio similar to (4.40)

$$l(\mathbf{I}_m \mid \mathbf{B}_{X_m}) = \log \frac{p(\mathbf{I}_m \mid \mathbf{B}_{X_m})}{q(\mathbf{I}_m)}$$

$$= \sum_{i=1}^{n} \left[\lambda_i \max_{\Delta x, \Delta \alpha} h(|\langle \mathbf{I}_m, B_{X_m + x_i + \Delta x, s, \alpha_i + \Delta \alpha} \rangle|^2) - \log Z(\lambda_i) \right] \quad (4.57)$$

As the next step of this modeling procedure, in addition to spatial translation and deformation, we can also rotate and scale the template. So a more general version of (4.56) is

$$\mathbf{I}_m = C_m \mathbf{B}_{X_m, S_m, A_m} + U_m, \quad (4.58)$$

where X_m is the location, S_m is the scale, and A_m is the orientation of the translated, rotated, scaled, and deformed template. The scaling of the template is implemented by changing the resolution of the original image. We adopt the convention that whenever the notation $\mathbf{B}$ appears in image representation, it always means the deformed template, where the perturbations of the basis functions can be inferred by local max pooling. The log-likelihood ratio $l(\mathbf{I}_m \mid \mathbf{B}_{X_m, S_m, A_m})$ can be similarly defined as in (4.57). Figure 4.18 illustrates the basic idea of representation (4.58). In addition to spatial translation, dilation, and rotation of the template, we may also allow mirror reflection as well as the change of aspect ratio.

Fig. 4.18 First published in the Quarterly of Applied Mathematics in Volume 72:373–406, 2014, published by Brown University. Reprinted with permission from [107]. Objects appear at different locations, scales, and orientations in the training images. In each row, the first plot displays the nominal active basis template. The rest of the row displays some examples of training images and the suppositions of the spatially translated, scaled, rotated, and deformed versions of the nominal template

Now suppose we have a dictionary of T active basis templates, $\{\mathbf{B}^{(t)}, t = 1, \ldots, T\}$, where each $\mathbf{B}^{(t)}$ is a compositional pattern of basis functions. Then we can represent the image $\mathbf{I}_m$ by K_m templates that are spatially translated, rotated, scaled, and deformed versions of these T templates in the dictionary:

$$\mathbf{I}_m = \sum_{k=1}^{K_m} C_{m,k} \mathbf{B}^{(t_{m,k})}_{X_{m,k}, S_{m,k}, A_{m,k}} + U_m, \tag{4.59}$$

where each $\mathbf{B}^{(t_{m,k})}_{X_{m,k}, S_{m,k}, A_{m,k}}$ is obtained by translating the template of type $t_{m,k}$, i.e., $\mathbf{B}^{(t_{m,k})}$, to location $X_{m,k}$, scaling it to scale $S_{m,k}$, rotating it to orientation $A_{m,k}$, and deforming it to match $\mathbf{I}_m$. Note that according to (4.59), the images represented by the dictionary are no longer assumed to be aligned.

If the K_m templates do not overlap with each other, then the log-likelihood ratio is

$$\sum_{m=1}^{M} \sum_{k=1}^{K_m} \left[l(\mathbf{I}_m \mid \mathbf{B}^{(t_{m,k})}_{X_{m,k}, S_{m,k}, A_{m,k}}) \right]. \tag{4.60}$$

The above representation is in analogy to model (4.24) in Sect. 4.3, which we copy here: $\mathbf{I}_m = \sum_{i=1}^{n} c_{m,i} B_{x_{m,i}, s_{m,i}, \alpha_{m,i}} + U_m$. The difference is that each $\mathbf{B}^{(t_{m,k})}_{X_{m,k}, S_{m,k}, A_{m,k}}$ is a composite representational unit, which is itself a group of basis functions that follow a certain compositional pattern of type $t_{m,k}$. Because of such grouping or packing, the number of templates K_m needed to encode $\mathbf{I}_m$ is expected to be much smaller than the total number of basis functions needed to represent $\mathbf{I}_m$, thus resulting in sparser representation. Specifically, if each template is a group of g basis functions, then the number of basis functions in the representation (4.59) is $K_m g$. In fact, we can unpack model (4.59) into the representation (4.24). The reason that it is advantageous to pack the basis functions into groups is that these groups exhibit T types of frequently occurring spatial grouping patterns, so that when we encode the image $\mathbf{I}_m$, for each selected group $\mathbf{B}^{(t_{m,k})}_{X_{m,k}, S_{m,k}, A_{m,k}}$, we only need to code the overall location, scale, orientation, and type of the group, instead of the locations, scales, and orientations of the individual constituent basis functions.

It is desirable to allow some limited overlap between the bounding boxes of the K_m templates that encode $\mathbf{I}_m$. Even if the bounding boxes of two templates have some overlap with each other, their constituent basis functions may not overlap much. If we do not allow any overlap between the bounding boxes of the templates, some salient structures of $\mathbf{I}_m$ may fall through the cracks between the templates. Also, it is possible that the frequently occurring patterns may actually overlap with each other. For instance, in a string "ABABABA," the pattern "AB" is frequently occurring, but at the same time, the pattern "BA" is as frequent as "AB," and these two patterns overlap with each other. So it can be desirable to allow some overlap between the patterns in order to recover all the important recurring patterns. On the other hand, we do not want to allow excessive overlap between the templates.

Otherwise, the learned templates will be too redundant, and we will need a lot of them in order to describe the training images. In practice, we assume the following limited overlap constraint: for each template $\mathbf{B}_{X_{m,k},S_{m,k},A_{m,k}}^{(t_{m,k})}$ centered at $X_{m,k}$, let D be the side length of its squared bounding box, then no other templates are allowed to be centered within a distance of ρD from $X_{m,k}$ (default setting: $\rho = .4$).

Such an assumption naturally leads to an inhibition step when we use a dictionary of templates to encode a training or testing image. Specifically, when a template is chosen to encode an image, this template will prevent overlapping templates from being selected. The template matching pursuit algorithm to be described below adopts such an inhibition scheme.

The following are the details of the two steps.

Step (I): Image encoding by template matching pursuit. Suppose we are given the current dictionary $\{\mathbf{B}^{(t)}, t = 1, \ldots, T\}$. Then for each $\mathbf{I}_m$, the template matching pursuit process seeks to represent $\mathbf{I}_m$ by sequentially selecting a small number of templates from the dictionary. Each selection seeks to maximally increase the log-likelihood ratio (4.60).

[I.0] Initialize the maps of template matching scores for all (X, S, A, t):

$$\mathbf{R}_m^{(t)}(X, S, A) \leftarrow l(\mathbf{I}_m \mid \mathbf{B}_{X,S,A}^{(t)}) - n^{(t)}\gamma, \qquad (4.61)$$

where $n^{(t)}$ is the number of basis functions in the t-th template in the dictionary and γ is a constant controlling model complexity as explained above. This can be accomplished by first rotating the template $\mathbf{B}^{(t)}$ to orientation A and then scanning the rotated template over the image zoomed to the resolution that corresponds to scale S. The larger the S is, the smaller the resolution is. Initialize $k \leftarrow 1$.

[I.1] Select the translated, rotated, scaled, and deformed template by finding the global maximum of the response maps:

$$(X_{m,k}, S_{m,k}, A_{m,k}, t_{m,k}) = \arg \max_{X,S,A,t} \mathbf{R}_m^{(t)}(X, S, A). \qquad (4.62)$$

[I.2] Let the selected arg-max template inhibit overlapping candidate templates to enforce limited overlap constraint. Let D be the side length of the bounding box of the selected template $\mathbf{B}_{X_{m,k},S_{m,k},A_{m,k}}^{(t_{m,k})}$, then for all (X, S, A, t), if X is within a distance ρD from $X_{m,k}$, then set the response $\mathbf{R}_m^{(t)}(X, S, A) \leftarrow -\infty$ (default setting: $\rho = .4$).

[I.3] Stop if all $\mathbf{R}_m^{(t)}(X, S, A, t) < 0$. Otherwise let $k \leftarrow k + 1$, and go to [I.1].

The template matching pursuit algorithm implements a hard inhibition to enforce the limited overlap constraint. In a more rigorous implementation, we may update the residual image by $U_m \leftarrow U_m - C_m \mathbf{B}_{X_{m,k},S_{m,k},A_{m,k}}^{(t_{m,k})}$ as in the original version of matching pursuit. But the current simplified version is more efficient.

Step (II): Dictionary re-learning by shared matching pursuit. For each $t = 1, \ldots, T$, we re-learn $\mathbf{B}^{(t)}$ from all the image patches that are currently covered by $\mathbf{B}^{(t)}$.

[II.0] Image patch cropping. For each $\mathbf{I}_m$, go through all the selected templates $\{\mathbf{B}^{(t_{m,k})}_{X_{m,k},S_{m,k},A_{m,k}}, \forall k\}$ that encode $\mathbf{I}_m$. If $t_{m,k} = t$, then crop the image patch of $\mathbf{I}_m$ (at the resolution that corresponds to $S_{m,k}$) covered by the bounding box of the template $\mathbf{B}^{(t_{m,k})}_{X_{m,k},S_{m,k},A_{m,k}}$.

[II.1] Template re-learning. Re-learn template $\mathbf{B}^{(t)}$ from all the image patches covered by $\mathbf{B}^{(t)}$ that are cropped in [II.0], with their bounding boxes aligned. The learning is accomplished by the shared matching pursuit algorithm of Sect. 4.3.

This dictionary re-learning step re-learns each compositional pattern from the re-aligned raw image patches, where the sparse representations and the correspondences between the selected basis functions are obtained simultaneously by the shared matching pursuit algorithm.

As mentioned above, the learning algorithm is initialized by learning from image patches randomly cropped from the training images. As a result, the initially learned templates are rather meaningless, but meaningful templates emerge very quickly after a few iterations.

In the beginning, the differences among the initial templates are small. However, as the algorithm proceeds, the small differences among the initial templates trigger a polarizing or specializing process, so that the templates become more and more different, and they specialize in encoding different types of image patches.

Chapter 5
Gestalt Laws and Perceptual Organization

5.1 Gestalt Laws for Perceptual Organization

The word Gestalt, as used in Gestalt Psychology, is often understood as "pattern." The school of Gestaltism, which emerged in Austria and Germany in the early twentieth century, focuses on the study of how organisms perceive entire patterns or configurations from an image, rather than just individual components. As is commonly cited, "the whole is more than the sum of its parts" best describes the idea of Gestaltism.

Many people share the experience that when we look at the world, we tend to decompose complex scenes into groups of objects against a background and perceive the objects as a composition of parts, and sometimes even those parts have sub-parts. But what actually enables us to do this? Does that surprise you that we manage to do such a remarkable thing when what we see is, to some extent, just a distribution of colored points? The Gestalt Psychology believes that we are able to do it because our vision system favors a set of principles when we understand the world. The set of principles is later formulated based on regularities of wholes, sub-wholes, groups, or Gestalten and called Gestalt laws. Gestalt laws, proximity, similarity, figure-ground, continuity, closure, and connection, determine how humans perceive visuals in connection with different objects and environments.

Specifically, Gestalt laws include the following:

- Law of Similarity: The law of similarity suggests that similar things tend to show up together. The grouping can occur in various modalities, including visual and auditory stimuli.
- Law of Pragnanz: "Pragnanz" in German means "good figure." Therefore, this law is sometimes referred to as the law of good figure or the law of simplicity. It states that we perceive figures in the simplest way possible, say a composition of simple shapes.
- Law of Proximity: The law of proximity suggests that when we perceive an image, closer objects tend to be grouped together. This law could be particularly

© Springer Nature Switzerland AG 2023
S.-C. Zhu, Y. N. Wu, *Computer Vision*, https://doi.org/10.1007/978-3-030-96530-3_5

helpful when we describe a set of objects and explain how we separate them into several smaller groups.

- Law of Continuity: We often have the impression that points connected by lines or curves form a smooth path together instead of segmented lines and angles. The Gestaltism explains this phenomenon using the law of continuity.
- Law of Closure: If things, when grouped as a whole, would become a simple entity, we usually ignore contradictory evidence and choose to fill in the missing pieces to treat it as a group. This law of closure helps us understand the segmented arcs of a circle to be a whole.
- Law of Common Region: According to this law, elements in the same region of space tend to be grouped together.

Gestalt laws provide a way for us to understand some important perception heuristics and psychological research continues to offer insights into it. But from a modeling point of view, despite long-standing observations in the psychology literature, there were no explicit mathematical models that can account for these Gestalt laws and weight them properly when multiple laws are working together or in competition.

5.2   Texton Process Embedding Gestalt Laws

In this section, we discuss a way to model visual patterns based on Gestalt laws by integrating descriptive and generative methods.

In particular, we present a mathematical framework for visual learning that integrates two popular statistical learning paradigms in the literature:

1. Descriptive methods, such as Markov random fields and minimax entropy learning [281].
2. Generative methods, such as principal component analysis, independent component analysis [15], transformed component analysis [60], wavelet coding [26, 165], and sparse coding [152, 188].

The integrated framework creates richer classes of probabilistic models for visual patterns. In this section, we demonstrate the integrated framework by learning a class of hierarchical models for texton patterns. At the bottom level of the model, we assume that an observed texture image is generated by multiple hidden "texton maps," and textons on each map are translated, scaled, and oriented versions of a window function, like mini-templates or wavelet basis function. The texton maps generate the observed image by occlusion or linear superposition. This bottom level of the model is generative in nature. At the top level of the model, the spatial arrangements or global organizations of the textons in the texton maps are characterized by the minimax entropy principle, which leads to the Gibbs point process models [27]. The top level of the model is descriptive in nature.

The learning framework achieves four goals:

1. Computing the window functions (or appearances) of different types of textons.
2. Inferring the hidden texton maps that generate the image.
3. Learning Gibbs point process models for the texton maps.
4. Verifying the learned window functions and Gibbs models through texture synthesis by stochastic sampling.

We use a stochastic gradient algorithm for inferential computation. We demonstrate the learning framework through a set of experiments.

Introduction

What a vision algorithm can accomplish depends crucially upon how much it "understands" the contents of the observed images. Thus in computer vision and especially Bayesian image analysis, an important research theme is visual learning whose objective is to construct parsimonious and general models that can realistically characterize visual patterns in natural scenes. Due to the stochastic nature of visual patterns, visual learning is posed as statistical modeling and inference problem, and existing methods in the literature can be generally divided into two categories. In this book, we call one category the *descriptive methods* and the other category the *generative methods*.[1]

Descriptive methods model a visual pattern by imposing statistical constraints on features extracted from signals. Descriptive methods include Markov random fields, minimax entropy learning [281], deformable models, etc. For example, many methods of texture modeling fall into this category [36, 98, 197, 281]. These models are built on pixel intensities or some deterministic transforms of the original signals, such as linear filtering. The shortcomings of descriptive methods are twofold. First, they do not capture high-level semantics in visual patterns, which are often very important in human perception. For example, a descriptive model of texture can realize a cheetah skin pattern with impressive synthesis results, but it does not have an explicit notion of individual blobs. Second, as descriptive models are built directly on the original signals, the resulting probability densities are often of very high dimensions and the sampling and inference are computationally expensive. It is desirable to have dimension reduction so that the models can be built in a low-dimensional space that often better reflects the intrinsic complexity of the pattern.

In contrast to descriptive methods, generative methods postulate hidden variables as the causes for the complicated dependencies in raw signals, and thus the models are hierarchical in nature. Generative methods are widely used in vision and image analysis. For example, principal component analysis (PCA), independent component analysis (ICA) [15], transformed component analysis (TCA) [60],

---

[1] There is a third category of methods which are discriminative. The goal of discriminative methods is not for modeling visual patterns explicitly but for classification. For example, pattern recognition, feed-forward neural networks, and classification trees, etc.

wavelet image representation [26, 165], sparse coding [152, 188], and the random collage model for generic natural images [146]. Despite their simplicity, these generative models suffer from an over-simplified assumption that hidden variables are independent and identically distributed (i.i.d.).[2] As a result, they are not sophisticated enough to generate realistic visual patterns. For example, a wavelet image coding model can easily reconstruct an observed image, but it cannot synthesize a texture pattern through i.i.d. random sampling because the spatial relationships between the wavelet coefficients are not characterized.

The two learning paradigms were developed almost independently by somewhat disjoint communities working on different problems, and their relationship has yet to be studied. Therefore, we present a visual learning framework that integrates both descriptive and generative methods and extends them to a richer class of probabilistic models for computer vision.

The integrated learning framework makes contributions to visual learning in the following four aspects.

First, it combines the advantages of both descriptive and generative methods and provides a general visual learning framework for modeling complex visual patterns. In computer vision, a fundamental observation, stated in Marr's primal sketch paradigm [169], is that natural visual patterns consist of multiple layers of stochastic processes. For example, Fig. 5.1 displays two natural images. When we look at the ivy-wall image, we not only perceive the texture "impression" in terms of pixel intensities but also see the repeated elements in the ivy and bricks. To capture the hierarchical notion, we propose a multi-layer generative model as shown in Fig. 5.2. We assume that a texture image is generated by a few layers of stochastic processes and each layer consists of a finite number of distinct but similar elements, called "textons" (following the terminology of Julesz). In experiments, each texton covers more than 100 pixels on average, so the layered representation

Fig. 5.1 Two examples of natural patterns

---

[2] Interested readers are referred to the paper [208] for discussion of the problem with existing generative models.

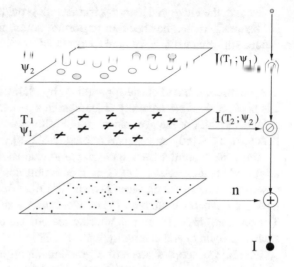

Fig. 5.2 Reprinted with permission from [89]. A generative model for an image **I** consists of multiple layers of texton maps $\mathbf{I}(\mathbf{T}_l; \Psi_l), l = 1, \ldots, L$ superimposed with occlusion plus an additive noise image **n**

achieves a nearly 100-fold dimension reduction.[3] The spatial arrangements of the textons at each layer are characterized by Markov random field (MRF) models through the minimax entropy learning [281], and previous MRF texture models can be considered special cases where the models have only one layer and each "texton" is just a pixel.

It is our belief that descriptive models are precursors of generative models and both are ingredients of the integrated learning process. In visual learning, the model can be initially built on image intensities via some features computed deterministically from the image intensities. Then we can replace the features with hidden causes, and such a process would incrementally discover more abstract elements or concepts such as textons, curves, flows, and so on, where elements at the more abstract levels become causes for the elements of lower abstractions. For instance, the flows generate curves, and the curves generate textons, which in turn generate pixel intensities. At each stage, the elements at the most abstract level have no further hidden causes, and thus they have to be characterized by a descriptive model based on some deterministic features, and such models can be derived by the minimax entropy principle as demonstrated in [255]. When a new hidden level of elements is introduced, it replaces the current descriptive model with a simplified one. The learning process evolves until the descriptive model for the most abstract elements becomes simple enough for a certain vision purpose. By analogy, the learning process is very similar to the situation in physics, where experimental observations are explained by a hierarchy of elements (say from quarks, electrons, atoms, to molecules) and their interactions.

---

[3] A texton has to be described by a few variables for location, scale, orientation, etc.

Second, the integrated learning framework provides a representational definition of "textons." Texton has been an important notion in texture perception and early vision. Unfortunately, it was only expressed vaguely in psychology [124], and a precise definition of texton has yet to be found. In this chapter, we argue that the definition of "texton" is possible in the context of a generative model. In contrast to the constraint-based clustering method by Malik, Leung, etc. [150, 151, 163], in this book, textons are naturally embedded in a generative model and are inferred as hidden variables of the generative model. This is consistent with the philosophy of ICA [15], TCA [60], and sparse coding [152, 188].

Third, we present a *Gestalt ensemble* to characterize the hidden texton maps as attributed point processes. The Gestalt ensemble corresponds to the grand canonical ensemble in statistical physics [24], and it differs from traditional Gibbs models by having an unknown number of textons whose neighborhood changes dynamically. The relationships between neighboring textons are captured by some Gestalt laws, such as proximity and continuity, etc.

Fourth, we adopt a stochastic gradient algorithm [88] for effective learning and inference, in contrast to the conventional EM algorithm [38]. In the adapted algorithm, we simplify the original likelihood function and solve the simplified maximum likelihood problem first. Starting from the initial solution, we then use the stochastic gradient algorithm to find refined solutions.

We demonstrate the proposed learning method on texture images. For an input texture image, the learning algorithm achieves the following four objectives:

1. Learning the appearance of textons for each stochastic process. Textons of the same stochastic process are translated, scaled, and oriented versions of a window function, like mini-templates or wavelet basis functions.
2. Inferring the hidden texton maps, each of which consists of an unknown number of similar textons which are related to each other by affine transformations.
3. Learning the minimax entropy models for the stochastic processes that generate the textons maps.
4. Verifying the learned window functions and generative models through stochastic sampling.

Background on Descriptive and Generative Learning

Given a set of images $\mathcal{I} = \{\mathbf{I}_1^{\text{obs}}, \ldots, \mathbf{I}_M^{\text{obs}}\}$, where $\mathbf{I}_m^{\text{obs}}, m = 1, \ldots, M$ are considered realizations of some underlying stochastic process governed by a probability distribution $f(\mathbf{I})$. The objective of visual learning is to estimate a probabilistic model $p(\mathbf{I})$ based on $\mathcal{I}$ so that $p(\mathbf{I})$ approaches $f(\mathbf{I})$ by minimizing the Kullback–Leibler divergence $\text{KL}(f \| p)$ from f to p [29],

$$\text{KL}(f \| p) = \int f(\mathbf{I}) \log \frac{f(\mathbf{I})}{p(\mathbf{I})} d\mathbf{I} = \text{E}_f[\log f(\mathbf{I})] - \text{E}_f[\log p(\mathbf{I})]. \tag{5.1}$$

In practice, the expectation $E_f[\log p(\mathbf{I})]$ is replaced by the sample average. Thus we have the standard maximum likelihood estimator (MLE),

$$p^* = \arg\min_{p\in\Omega_p} \text{KL}(f\|p) \approx \arg\max_{p\in\Omega_p} \sum_{m=1}^{M} \log p(\mathbf{I}_m^{obs}), \qquad (5.2)$$

where Ω_p is the family of distributions. One general procedure is to search for p in a sequence of nested probability families,

$$\Omega_0 \subset \Omega_1 \subset \cdots \subset \Omega_K \to \Omega_f \ni f, \qquad (5.3)$$

where K indexes the dimensionality of the space. For example, K could be the number of free parameters in a model. As K increases, the probability family should be general enough to approach f to arbitrary precision.

There are two choices of families for Ω_p in the literature and both are general enough for approximating any distribution f.

The first choice is the exponential family. The exponential family can be derived by the descriptive method through maximum entropy and has its root in statistical mechanics [24]. A descriptive method extracts a set of K feature statistics as deterministic transforms of an image $\mathbf{I}$, denoted by $\phi_k(\mathbf{I}), k = 1, \ldots, K$. Then it constructs a model p by imposing descriptive constraints so that p reproduces the observed statistics $\mathbf{h}_k^{obs}$ extracted from $\mathcal{I}$,

$$E_p[\phi_k(\mathbf{I})] = \frac{1}{M} \sum_{m=1}^{M} \phi_k(\mathbf{I}_m^{obs}) \approx E_f[\phi_k(\mathbf{I})] = \mathbf{h}_k, \quad k = 1, \ldots, K. \qquad (5.4)$$

One may consider $\mathbf{h}_k$ as a projected statistics of $f(\mathbf{I})$, and thus when M is large enough, p and f will have the same projected (marginal) statistics on the K chosen dimensions. By the maximum entropy principle [120], this leads to the Gibbs model,

$$p(\mathbf{I}; \boldsymbol{\beta}) = \frac{1}{Z(\boldsymbol{\beta})} \exp\left\{-\sum_{k=1}^{K} \beta_k \phi_k(\mathbf{I})\right\}. \qquad (5.5)$$

The parameters $\boldsymbol{\beta} = (\beta_1, \ldots, \beta_K)$ are Lagrange multipliers and they are computed by solving the constraint equations (5.4). The K features are chosen by a minimum entropy principle [281].

The descriptive learning method augments the dimension of the space Ω_p by increasing the number of feature statistics and generates a sequence of exponential family models,

$$\Omega_1^d \subset \Omega_2^d \subset \cdots \Omega_K^d \to \Omega_f. \qquad (5.6)$$

This family includes all the MRF and minimax entropy models for texture [281]. For example, a type of descriptive model for texture chooses $\phi_j(\mathbf{I})$ as the histograms of responses from some Gabor filters.

The second choice is the mixture family, which can be derived by integration or summation over some hidden variables $W = (w_1, \ldots, w_K)$,

$$p(\mathbf{I}; \Theta) = \int p(\mathbf{I}, W; \Theta)dW = \int p(\mathbf{I}|W; \Psi)p(W; \boldsymbol{\beta})dW. \tag{5.7}$$

The parameters of a generative model include two parts $\Theta = (\Psi, \boldsymbol{\beta})$. It assumes a joint probability distribution $p(\mathbf{I}, W; \Theta)$, and that W generates $\mathbf{I}$ through a conditional model $p(\mathbf{I}|W; \Psi)$ with parameters Ψ. The hidden variables are characterized by a model $p(W; \boldsymbol{\beta})$. W should be inferred from $\mathbf{I}$ in a probabilistic manner, and this is in contrast to the deterministic features $\phi_k(\mathbf{I}), k = 1, \ldots, K$ in descriptive models. The generative method incrementally adds hidden variables to augment the space Ω_p and thus generates a sequence of mixture families,

$$\Omega_1^g \subset \Omega_2^g \subset \cdots \subset \Omega_K^g \to \Omega_f \ni f. \tag{5.8}$$

For example, principal component analysis, wavelet image coding [26, 165], and sparse coding [152, 188] all assume a linear additive model where an image $\mathbf{I}$ is the result of linear superposition of some window functions $\Psi_k, k = 1, \ldots, K$, plus a Gaussian noise process $\mathbf{n}$.

$$\mathbf{I} = \sum_{k=1}^{K} a_k \Psi_k + \mathbf{n}, \tag{5.9}$$

where $a_k, k = 1, \ldots, K$, are the coefficients, Ψ_i are the eigenvectors in PCA, wavelets in image coding, or over-complete basis for sparse coding. The hidden variables are the K coefficients of basis functions plus the noise, so $W = (a_1, \ldots, a_K, \mathbf{n})$.[4] The coefficients are assumed to be independently and identically distributed,

$$a_k \sim p(a_k) \propto \exp^{-\lambda_o|a_k|\rho}, \quad k = 1, \ldots, K. \tag{5.10}$$

The norm $\rho = 1$ for sparse coding [152, 188] and basis pursuit [26], and $\rho = 2$ for principal component analysis. Thus we have a simple distribution for W,

$$p(W; \boldsymbol{\beta}) \propto \prod_{k=1}^{k} \exp^{-\lambda_o|a_k|\rho} \prod_{(x,y)} \exp^{-\frac{\mathbf{n}^2(x,y)}{2\sigma_o^2}}. \tag{5.11}$$

---

[4] In PCA, since the basis functions are orthogonal, a_k can be computed as transform, but for an over-complete basis, the a_k has to be inferred.

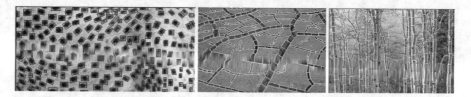

Fig. 5.3 Reprinted with permission from [89]. Texture images with texton processes

In this model $p(W; \boldsymbol{\beta})$ is from the exponential family. However, in the literature, hidden variables $a_k, k = 1, \ldots, K$, are assumed to be i.i.d. Gaussian or Laplacian distributed. Thus the concept of descriptive models is trivialized.

A Multi-layered Generative Model for Images

We focus on a multi-layer generative model for images with mainly texture content and we believe that the same learning framework can be applied to other patterns such as object shapes. An image $\mathbf{I}$ is assumed to be generated by L layers of stochastic processes, and each layer consists of a finite number of distinct but similar elements, called "textons." Figure 5.3 shows three typical examples of texture images, and each texton is represented by a rectangular window. A layered model is shown in Fig. 5.2.

Textons at layer l are image patches transformed from a square template Ψ_l. The j-th texton in layer l is identified by six transformation variables,

$$t_{lj} = (x_{lj}, y_{lj}, \sigma_{lj}, \tau_{lj}, \theta_{lj}, A_{lj}), \tag{5.12}$$

where (x_{lj}, y_{lj}) represents the texton center location, σ_{lj} the scale (or size), τ_{lj} the "shear" (aspect ratio of height versus width), θ_{lj} the orientation, and A_{lj} for photometric transforms such as lighting variability.

t_{lj} defines an affine transform denoted by $G[t_{lj}]$, and the pixels covered by a texton t_{lj} is denoted by D_{lj}. Thus the image patch $\mathbf{I}_{D_{lj}}$ of a texton t_{lj} is

$$\mathbf{I}_{D_{lj}} = G[t_{lj}] \odot \Psi_l, \quad \forall j, \ \forall l, \tag{5.13}$$

where $\odot$ denotes the transformation operator. Texton examples of a circular template at different scales, shears, and orientations are shown in Fig. 5.4.

We define the collection of all textons in layer l as a texton map,

$$\mathbf{T}_l = (n_l, \{t_{lj}, j = 1 \ldots n_l\}), l = 1 \ldots L, \tag{5.14}$$

where n_l is the number of textons in layer l.

ψ scale   shear   scale/shear/rotation

Fig. 5.4 Reprinted with permission from [89]. A template Ψ (leftmost) and its three transformed copies

In each layer, the texton map $\mathbf{T}_l$ and the template Ψ_l generate an image $\mathbf{I}_l = \mathbf{I}(\mathbf{T}_l; \Psi_l)$ deterministically. If several texton patches overlap at site (x, y) in $\mathbf{I}_l$, the pixel value is taken as average,

$$\mathbf{I}_l(x, y) = \frac{\sum_{j=1}^{n_l} \delta((x, y) \in D_{lj}) \mathbf{I}_{D_{lj}}(x, y)}{\sum_{j=1}^{n_l} \delta((x, y) \in D_{lj})}, \tag{5.15}$$

where $\delta(\cdot) = 1$ if $\bullet$ is true, otherwise $\delta(\cdot) = 0$. In image $\mathbf{I}_l$, pixels not covered by any texton patches are transparent. The image $\mathbf{I}$ is generated in the following way:

$$\mathbf{I}(\mathbf{T}; \Psi) = \mathbf{I}(\mathbf{T}_1; \Psi_1) \oslash \mathbf{I}(\mathbf{T}_2; \Psi_2) \oslash \cdots \oslash \mathbf{I}(\mathbf{T}_L; \Psi_L), \quad \text{and} \quad \mathbf{I}^{\text{obs}} = \mathbf{I}(\mathbf{T}; \Psi) + \mathbf{n}. \tag{5.16}$$

The symbol $\oslash$ denotes occlusion (or linear addition), i.e., $\mathbf{I}_1 \oslash \mathbf{I}_2$ means $\mathbf{I}_1$ occludes $\mathbf{I}_2$. $\mathbf{I}(\mathbf{T}; \Psi)$ is called a reconstructed image and $\mathbf{n}$ is a Gaussian noise process $\mathbf{n}(x, y) \sim N(0, \sigma_0^2), \forall (x, y)$. Thus pixel value at site (x, y) in the image $\mathbf{I}$ is the same as the top layer image at that point, while uncovered pixels are only modeled by noises.

In this generative model, the hidden variables are

$$\mathbf{T} = (L, \{(\mathbf{T}_l, d_l) : l = 1, \ldots, L\}, \mathbf{n}), \tag{5.17}$$

where d_l indexes the order (or relative depth) of the l-th layer.

To simplify computation, we assume that $L = 2$ and the two stochastic layers are called "background" and "foreground," respectively. The two texton processes $\mathbf{T}_l, l = 1, 2$, are assumed to be independent of each other. We find that this assumption holds true for most of the texture patterns. Otherwise one has to treat L as an unknown complexity parameter in the model.

Thus the likelihood for an observable image $\mathbf{I}$ can be computed

$$p(\mathbf{I}; \Theta) = \int p(\mathbf{I}|\mathbf{T}; \Psi) p(\mathbf{T}; \boldsymbol{\beta}) d\mathbf{T}, \tag{5.18}$$

$$= \int p(\mathbf{I}|\mathbf{T}_1, \mathbf{T}_2; \Psi) \prod_{l=1}^{2} p(\mathbf{T}_l; \boldsymbol{\beta}_l) d\mathbf{T}_1 d\mathbf{T}_2. \tag{5.19}$$

Let $\Psi = (\Psi_1, \Psi_2)$ be texton templates, $\beta = (\beta_1, \beta_2)$ the parameters for the two texton processes which we shall discuss in the next section, and σ^2 the variance of the noise. The generative part of the model is a conditional probability $p(\mathbf{I}|\mathbf{T}_1, \mathbf{T}_2, \Psi)$,

$$p(\mathbf{I}^{obs}|\mathbf{T}_1, \mathbf{T}_2; \Psi) \propto \exp\left\{\frac{-\|\mathbf{I}^{obs} - \mathbf{I}(\mathbf{T}_1, \mathbf{T}_2; \Psi)\|^2}{2\sigma^2}\right\}, \qquad (5.20)$$

where $\mathbf{I}(\mathbf{T}_1, \mathbf{T}_2; \Psi)$ is the reconstructed image from the two hidden layers without noise (see (5.16)). As the generative model is very simple, the texture pattern should be captured by the spatial arrangements of textons in models $p(\mathbf{T}_l; \beta_l), l = 1, 2$, which are in much lower dimensional spaces and are more semantically meaningful than previous Gibbs models on pixels [281].

In the next section, we discuss the model $p(\mathbf{T}_l; \beta_l), l = 1, 2$, for the texton processes.

A Descriptive Model of Texton Processes

As the texton processes $\mathbf{T}_l$ are not generated by further hidden layers in the model,[5] they must be characterized by descriptive models in exponential families. In this section, we first review some background on three physical ensembles and then introduce a Gestalt ensemble for the texton process. Finally, we show some experiments for realizing the texton processes.

Background: Physics Foundation for Visual Modeling

There are two main differences between a texton process $\mathbf{T}_l$ and a conventional texture defined on a lattice $D \subset Z^2$.

- A texton process has an unknown number of elements and each element has many attributes t_{lj}, while a texture image has a fixed number of pixels and each pixel has only one variable for intensity.
- The neighborhood of a texton can change depending on their relative positions, scales, and orientations, while pixels always have fixed neighborhoods.

Although a texton process is more complicated than a texture image, they share a common property that they all have a large number of elements and global patterns arise from simple local interactions between elements. Thus a well-suited theory

---

[5] We may introduce additional layers of hidden variables for curve processes that render the textons. But our model stops at the texton level.

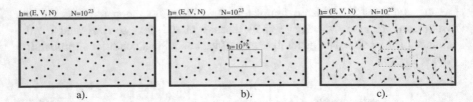

Fig. 5.5 Three typical ensembles in statistical mechanics. (**a**) Micro-canonical ensemble. (**b**) Canonical ensemble. (**c**) Grand-canonical ensemble

for studying these patterns is statistical physics—a subject studying macroscopic properties of a system involving a huge number of elements [24].

To understand the intuitive ideas behind various texture and texton models, we find it revealing to discuss three physical ensembles which are shown in Fig. 5.5.

1. *Micro-canonical ensemble.* Figure 5.5a is an insulated system of N elements. The elements could be atoms or molecules in systems such as solid ferromagnetic material, fluid, or gas. N is nearly infinity, say $N = 10^{23}$. The system is decided by a configuration $S = (\mathbf{x}^N, \mathbf{m}^N)$, where $\mathbf{x}^N$ describes the coordinates of the N elements and $\mathbf{m}^N$ their momenta. The system is subject to some global constraints $\mathbf{h}_o = (N, E, V)$. That is, the number of elements N, the total system energy E, and the total volume V are fixed. When it reaches equilibrium, this insulated system is characterized by a so-called *micro-canonical ensemble*,

$$\Omega_{mcn} = \{S : \mathbf{h}(S) = \mathbf{h}_o, \ f(S; \mathbf{h}_o) = 1/|\Omega_{mcn}|\}. \tag{5.21}$$

S is a microscopic state or instance, and $\mathbf{h}(S)$ is the macroscopic summary of the system. The state S is assumed to be uniformly distributed within Ω_{mcn}, and thus it is associated with a probability $f(S; \mathbf{h}_o)$. The system is identified by $\mathbf{h}_o$.

2. *Canonical ensemble.* Figure 5.5b illustrates a small subsystem embedded in a micro-canonical ensemble. The subsystem has $n \ll N$ elements, fixed volume $v \ll V$, and energy e. It can exchange energy through the wall with the remaining elements which is called the "heat bath" or "reservoir." At thermodynamic equilibrium, the microscopic state $s = (\mathbf{x}^n, \mathbf{m}^n)$ for the small system is characterized by a canonical ensemble with a Gibbs model $p(s; \boldsymbol{\beta})$,

$$\Omega_{cn} = \{s; \ p(s; \boldsymbol{\beta}) = c \cdot \exp\{-\boldsymbol{\beta} e(s)\}\}. \tag{5.22}$$

In texture modeling [255], the micro-canonical ensemble is mapped to a *Julesz ensemble* where $S = \mathbf{I}$ is an infinite image on 2D plane $\mathbf{Z}^2$, and $\mathbf{h}_o$ is a collection of Gabor filtered histograms. The canonical ensemble is mapped to a FRAME model [281] with $s = \mathbf{I}_D$ being an image on a finite lattice D. Intuitively, s is a small patch of S viewed from a window D. The intrinsic relationship between the two ensembles is that the Gibbs model $p(s; \boldsymbol{\beta})$ in Ω_{cn} is derived as a conditional distribution of

$f(S; \mathbf{h}_o)$ in Ω_{mcn}. There is a duality between $\mathbf{h}_o$ and β (see [255] and the references therein).

3, *Grand-Canonical ensemble*. Figure 5.5c illustrates a third system where the subsystem is open and can exchange not only energy but also elements with the bath. So v is fixed, but n and e may change. This models liquid or gas materials. At equilibrium, the microscopic state s for this small system is governed by a distribution $p(s; \beta_o, \beta)$ with β_o controlling the density of elements in s. Thus a grand-canonical ensemble is

$$\Omega_{gd} = \{s = (n, \mathbf{x}^n, \mathbf{m}^n); \ p(s; \beta_o, \boldsymbol{\beta})\}. \tag{5.23}$$

The grand-canonical ensemble is a mathematical model for visual patterns with varying numbers of elements, thus laying the foundation for modeling texton processes. In the next subsection, we map the grand-canonical ensemble to a Gestalt ensemble in visual modeling.

Gestalt Ensemble

Without loss of generality, we represent a spatial pattern by a set of attributed elements called textons as was discussed in Sect. 5.2. To simplify notation, we consider only one texton layer on a lattice D,

$$\mathbf{T} = (n, \{t_j = (x_j, y_j, \sigma_j, \tau_j, \theta_j, A_j), j = 1, \ldots, n\}). \tag{5.24}$$

For a texton map $\mathbf{T}$, we define a neighborhood system $\partial(\mathbf{T})$.

$$\partial(\mathbf{T}) = \{\partial t : \ t \in \mathbf{T}, \partial t \subset \mathbf{T}\}, \tag{5.25}$$

where ∂t is a set of neighboring textons for each texton t. We find the nearest neighbors are often enough. Because each texton covers a 15×15 patch on average, a pair of adjacent textons captures image features at the scale of often more than 30×30 pixels.

There are a few different ways of defining $\partial(\mathbf{T})$. One may treat each texton as a point and compute a Voronoi diagram or Delaunay triangularization which provides graph structures for the neighborhood. For example, a Voronoi neighborhood was used in [5] for grouping dot patterns. However, for textons, we need to consider other attributes such as orientation in defining the neighborhood. Figure 5.6a shows a texton t. The plane is separated into four quadrants relative to the two axes of the rectangle. In each quadrant, the nearest texton is considered the neighbor texton. Unlike the Markov random field on image lattice, the texton neighborhood is no longer translation invariant.

The above neighborhood is defined deterministically. In more general settings, $\partial(\mathbf{T})$ shall be represented by a set of hidden variables that can be inferred from $\mathbf{T}$. Thus a texton may have a varying number of neighbors referenced by some indexing

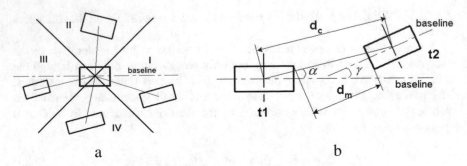

a   b

Fig. 5.6 Reprinted with permission from [89]. Texton neighborhood. (**a**) A texton has four neighbors. (**b**) Four measurements between texton t_1 and its neighbor t_2, d_c, d_m, α, and γ

(or address) variables. These address variables could be decided probabilistically depending on the relative positions, orientations, and scales or intensities. This leads to the so-called *mixed Markov random field* and is beyond the scope of this section. Mumford and Fridman discussed such cases in another context (see [62]).

For a texton t_1 and its neighbor $t_2 \in \partial t$, we measure five features shown in Fig. 5.6b, which capture various Gestalt properties:

1. d_c: Distance between two centers, which measures proximity.
2. d_m: Gap between two textons, which measures connectedness and continuation.
3. α: Angle of a neighbor relative to the main axis of the reference texton. This is mostly useful in quadrants I and III. α/d_c measures the curvature of possible curves formed by the textons or co-linearity and co-circularity in the Gestalt language.
4. γ: Relative orientations between the two textons. This is mostly useful for neighbors in quadrants II and IV and measures parallelism.
5. r: Size ratio that denotes the similarity of texton sizes. r is the width of t_2 divided by the width of t_1 for neighbors into quadrants I and III and r is the length of t_2 divided by the length of t_1 for neighbors in quadrants II and IV.

Thus a total of $4 \times 5 = 20$ pairwise features are computed for each texton plus two features of each texton itself: the orientation θ_j and a two-dimensional feature consisting of the scaling and shearing (σ_j, τ_j). Following the notation of descriptive models in Sect. 5.2, we denote these features by

$$\phi^{(k)}(t|\partial t), \quad \text{for } k = 1, \dots, 22. \tag{5.26}$$

We compute 21 one-dimensional marginal histograms and one two-dimensional histogram for (σ_j, τ_j), averaged over all textons.

$$H^{(k)}(z) = \sum_{j=1}^{n} \delta(z - \phi^{(k)}(t_j|\partial t_j)), \ \forall k. \tag{5.27}$$

We denote these histograms by

$$H(\mathbf{T}) = (H^{(1)}, \ldots, H^{(22)}), \quad \text{and} \quad \mathbf{h}(\mathbf{T}) = \frac{1}{n} H(\mathbf{T}). \tag{5.28}$$

The vector length of $\mathbf{h}(\mathbf{T})$ is the total number of bins in all histograms. One may choose other features and high-order statistics as well. In the vision literature, the work (Steven, 1978) was perhaps the earliest attempt for characterizing spatial patterns using histogram of attributes (see [169] for some examples).

The distribution of $\mathbf{T}$ is characterized by a statistical ensemble in correspondence to the grand-canonical ensemble in Fig. 5.5c. We call it a Gestalt ensemble on a finite lattice D as it is the general representation for various Gestalt patterns,

$$\text{a Gestalt ensemble} = \Omega_{gst} = \{\mathbf{T} : p(\mathbf{T}; \beta_o, \boldsymbol{\beta})\}. \tag{5.29}$$

The Gestalt ensemble is governed by a Gibbs distribution,

$$p(\mathbf{T}; \beta_o, \boldsymbol{\beta}) = \frac{1}{Z} \exp\left\{-\beta_o n - \langle \boldsymbol{\beta}, H(\mathbf{T}) \rangle\right\}. \tag{5.30}$$

Z is the partition function. β_o is the parameter controlling texton density. We can rewrite the vector-valued potential functions $\boldsymbol{\beta}$ as energy functions $\beta^{(k)}()$, and then we have

$$p(\mathbf{T}; \beta_o, \boldsymbol{\beta}) = \frac{1}{Z} \exp\left\{-\beta_o n - \sum_{j=1}^{n} \sum_{k=1}^{K=22} \beta^{(k)}(\phi^{(k)}(t_j|t_{\partial j}))\right\}. \tag{5.31}$$

This model provides a rigorous way for integrating multiple feature statistics into one probability model and generalizes existing point processes [27].

The probability $p(\mathbf{T}; \beta_o, \boldsymbol{\beta})$ is derived from the Julesz ensemble (or micro-canonical ensemble). We first define a close system with $N \gg n$ elements on a lattice D, and we assume the density of textons is fixed

$$\lim_{N \to \infty} \frac{N}{|D|} = \rho, \quad \text{as } N \to \infty, \text{ and } D \to \mathbf{Z}^2. \tag{5.32}$$

Thus we obtain a Julesz ensemble on $\mathbf{Z}^2$ [255],

$$\text{a Julesz ensemble} = \Omega_{jlz} = \{\mathbf{T}_\infty : \mathbf{h}(\mathbf{T}_\infty) = \mathbf{h}_o, N \to \infty, f(\mathbf{T}_\infty; \mathbf{h}_o)\}, \tag{5.33}$$

where $\mathbf{h}_o = (\rho, \mathbf{h})$ is the macroscopic summary of the system state $\mathbf{T}_\infty$. On any finite image, a texton process should be a conditional density of $f(\mathbf{T}_\infty; \mathbf{h}_o)$. There is a one-to-one correspondence between $\mathbf{h}_o = (\rho, \mathbf{h})$ and the parameters $(\beta_o, \boldsymbol{\beta})$.

We can learn the parameters $(\beta_o, \boldsymbol{\beta})$ and select effective features $\phi^{(k)}$ when learning the descriptive method. In the following subsection, we discuss some computational issues and experiments for learning $p(\mathbf{T}; \beta_o, \beta)$ and simulating the Gestalt ensembles.

An Integrated Learning Framework

After discussing the descriptive models for the hidden texton layers, we now return to the integrated framework presented in Sect. 5.2.

Integrated Learning

The generative model for an observed image $\mathbf{I}^{\text{obs}}$ is rewritten from Eq. (5.19),

$$p(\mathbf{I}^{\text{obs}}; \Theta) = \int p(\mathbf{I}^{\text{obs}}|\mathbf{T}_1, \mathbf{T}_2; \Psi) \prod_{l=1}^{2} p(\mathbf{T}_l; \beta_l) d\mathbf{T}_1 d\mathbf{T}_2. \tag{5.34}$$

We follow the maximum likelihood estimate in Eq. (5.2),

$$\Theta^* = \arg \max_{\Theta \in \Omega_K^g} \log p(\mathbf{I}^{\text{obs}}; \Theta). \tag{5.35}$$

The parameters Θ include the texton templates Ψ_l, the Lagrange multipliers β_l, $l = 1, 2$, for two Gestalt ensembles, and the variance of the Gaussian noise, σ^2,

$$\Theta = (\Psi, \beta, \sigma), \quad \Psi = (\Psi_1, \Psi_2), \quad \text{and} \quad \beta = (\beta_{10}, \beta_1, \beta_{20}, \beta_2). \tag{5.36}$$

To maximize the log-likelihood, we take the derivative with respect to Θ. Let $\mathbf{T} = (\mathbf{T}_1, \mathbf{T}_2)$,

$$\frac{\partial \log p(\mathbf{I}^{\text{obs}}; \Theta)}{\partial \Theta} \tag{5.37}$$

$$= \int \frac{\partial \log p(\mathbf{I}^{\text{obs}}, \mathbf{T}; \Theta)}{\partial \Theta} p(\mathbf{T}|\mathbf{I}^{\text{obs}}; \Theta) d\mathbf{T}$$

$$= \int \left[\frac{\partial \log p(\mathbf{I}^{\text{obs}}|\mathbf{T}; \Psi)}{\partial \Psi} + \sum_{l=1}^{2} \frac{\partial \log p(\mathbf{T}_l; \beta_l)}{\partial \beta_l} \right] p(\mathbf{T}|\mathbf{I}^{\text{obs}}; \Theta) d\mathbf{T}$$

$$= E_{p(\mathbf{T}|\mathbf{I}^{\text{obs}}; \Theta)} \left[\frac{\partial \log p(\mathbf{I}^{\text{obs}}|\mathbf{T}; \Psi)}{\partial \Psi} + \sum_{l=1}^{2} \frac{\partial \log p(\mathbf{T}_l; \beta_l)}{\partial \beta_l} \right]. \tag{5.38}$$

In the literature, there are two well-known methods for computation. One is the EM algorithm [38], and the other is data augmentation [229] in the Bayesian context. We propose to use a stochastic gradient algorithm [88] which is more effective for ׀׀׀׀ ׀׀׀׀׀׀׀׀ ׀׀׀

A Stochastic Gradient Algorithm

– Step 0. Initialize the hidden texton maps $\mathbf{T}$ and the templates Ψ using a simplified likelihood as discussed in the next section. Set $\beta = 0$.

Repeat Steps I and II below iteratively (like EM algorithm).

– Step I. With the current $\Theta = (\Psi, \beta, \sigma)$, obtain a sample of texton maps from the posterior probability

$$\mathbf{T}_m^{\mathrm{syn}} \sim p(\mathbf{T}|\mathbf{I}^{\mathrm{obs}}; \Theta) \propto p(\mathbf{I}^{\mathrm{obs}}|\mathbf{T}_1, \mathbf{T}_2; \Psi)p(\mathbf{T}_1; \beta_{1o}, \beta_1)p(\mathbf{T}_2; \beta_{2o}, \beta_2), \quad m = 1, \ldots, M. \tag{5.39}$$

This is Bayesian inference. The sampling process is realized by a Monte Carlo Markov chain which simulates a random walk with two types of dynamics.

- (I.a) A diffusion dynamics realized by a Gibbs sampler—sampling (relaxing) the transform group for each texton, for example, moving textons, updating their scales and rotating them, etc.
- (I.b) A jump dynamics—adding or removing a texton (death/birth) by reversible jumps [83].

— Step II. We treat $\mathbf{T}_m^{\mathrm{syn}}, m = 1, \ldots, M$, as "observations" and estimate the integration in (5.38) by importance sampling. Thus we have

$$\frac{\partial \log p(\mathbf{I}^{\mathrm{obs}}|\mathbf{T}; \Psi)}{\partial \Psi} + \sum_{l=1}^{2} \frac{\partial \log p(\mathbf{T}_l; \beta_l)}{\partial \beta_l}. \tag{5.40}$$

We learn $\Theta = (\Psi, \beta, \sigma)$ of the texton templates and Gibbs models, respectively, by gradient ascent:

- (II.a) Update the texton templates Ψ by maximizing $\sum_{m=1}^{M} \log p(\mathbf{I}^{\mathrm{obs}}|\mathbf{T}_m^{\mathrm{syn}}; \Psi)$; this is a fitting process. In our experiment, the texton templates Ψ_1 and Ψ_2 are represented by 15×15 windows, and thus there are 2×225 unknowns.[6]
- (II.b) Update $\beta_{lo}, \beta_l, l = 1, 2$, by maximizing $\sum_{m=1}^{M} \log p(\mathbf{T}_m^{\mathrm{syn}}; \beta_{lo}, \beta_l)$. This is exactly the maximum entropy learning process in the descriptive method except that the texton processes are given by Step I.
- (II.c) Update σ for the noise process.

In Step I, we choose to sample $M = 1$ example each time. If the learning rate in Steps (II.a) and (II.b) is slow enough, the expectation is estimated by importance sampling through samples $\mathbf{T}^{\mathrm{syn}}$ over time. It has been proved [88] that such an

---

[6] Each point in the window can be transparent, and thus the shape of the texton can change during the learning process.

algorithm converges to the optimal Θ if the step size in Step II satisfies some mild conditions.

The following are some useful observations:

1. Descriptive models and learning is part of the integrated learning framework, in terms of both representation and computing (Step II.b)).

2. Bayesian vision inference is a sub-task (Step I) of the integrated learning process. A vision system, machine or biological, evolves by learning generative models $p(\mathbf{I}; \Theta)$ and makes an inference about the world $\mathbf{T}$ (or W in more general generative models) using the current imperfect knowledge Θ—the Bayesian view of vision. What is missing in this learning paradigm is the "discovery process" that introduces new hidden variables.

Mathematical Definitions of Visual Patterns

Any visual learning paradigm must answer the question of conceptualization: How do we define, mathematically, a visual concept or a pattern? for example, a human face, a wood grain texture, and so on. In this section, we show how the descriptive and generative methods conceptualize a visual pattern.

The concept of a visual pattern is an abstraction (or summary) for an assembly of configurations (or instances) s that are not distinguished by human perception for a certain vision purpose. Because of the stochastic nature of the visual signal, instances in this assembly are governed by a frequency $f(s)$, and an instance can then be considered a random sample from the probability distribution $f(s)$. Thus a concept is said to be equal to an ensemble,

$$\text{a visual concept } c = \Omega_c = \{s : f(s)\}. \tag{5.41}$$

By a descriptive method, the ensemble is defined through statistical constraints (see (5.4)),

$$\text{a descriptive concept } c = \Omega(\mathbf{h}_c) = \{s : f(s; \mathbf{h}_c), \ \mathrm{E}_f[\mathbf{h}(s)] = \mathbf{h}_c\}. \tag{5.42}$$

For example, s could be a human face represented by a list of key points, or $s = \mathbf{I}$ could be a texture image. $\mathbf{h}(s)$ is the statistics extracted from s which are sufficient for a certain vision purpose. $\mathbf{h}_c$ is the vector value that identifies this concept. $\mathbf{h}_c$ corresponds to a Gibbs model $p(s; \boldsymbol{\beta})$ with parameter $\boldsymbol{\beta}$. It is accepted that two concepts may have overlapping ensembles.

When the signal is homogeneous, such as texture or texton maps, the expectation $\mathrm{E}_f[\mathbf{h}(s)]$ can be computed from a single instance through spatial average over a large enough lattice D, and thus we can define a concept as an equivalence class—called the Julesz ensemble in (Zhu et al. 2000) [224].

$$\text{a descriptive concept } c = \Omega(\mathbf{h}_c) = \{s : \mathbf{h}(s) = \mathbf{h}_c \ \ D \to Z^2, f(s; \mathbf{h}_c)\}. \tag{5.43}$$

For example, for defining a texture, $\mathbf{h}_c$ is the sufficient and necessary statistics extracted in texture perception. The ensembles on large lattice $D \to Z^2$ are disjoint and deterministic as $f(s; \mathbf{h}_c)$ is a uniform distribution.

Though the pure definition constraint and Julesz ensemble are technically sound, they are only a coarse or first-stage approximation to human perception in texture discrimination. Texture studies in psychology [119, 124] suggest that human vision is sensitive to the perception of some basic elements called textons, perhaps for some vision purpose. It was also argued and demonstrated by Malik et al. [150, 163, 164] that the detection of individual textons plays an important role in texture discrimination, segmentation, and grouping. However, it was unclear what textons really are.

We argue that the mathematical definition of textons can be guided by a generative model. So the integrated learning paradigm extends the Julesz ensemble definition to a generative concept with W being the hidden variables,

$$\text{a generative concept } c = \{s : f(s; \Theta) = \int f(s|W; \Psi) f(W; \mathbf{h}_w) dW\}, \quad (5.44)$$

where W are instances of a descriptive ensemble,

$$\{W : \mathbf{h}(W) = \mathbf{h}_w, D \to Z^2, f(W; \mathbf{h}_w)\}. \quad (5.45)$$

Thus a concept is identified by the parameters $\Theta = (\Psi, \mathbf{h}_w)$ (or $\Theta = (\Psi, \beta)$ because of the duality between $\mathbf{h}_w$ and β). In this definition, Ψ is a mathematically sound definition of texton.

Effective Inference by Simplified Likelihood

In this section, we address some computational issues in the integrated learning paradigm and propose a method for initializing the stochastic gradient algorithm (in Step 0).

Initialization by Likelihood Simplification and Clustering

The stochastic algorithm presented in Sect. 5.2 needs a long "burn-in" period if it starts from an arbitrary condition. To accelerate the computation, we use a simplified likelihood in Step 0 of the stochastic gradient algorithm. Thus given an input image $\mathbf{I}^{obs}$, our objective is to compute some good initial texton templates Ψ_1, Ψ_2 and hidden texton maps $\mathbf{T}_1, \mathbf{T}_2$, before the iterative process in Steps I and II.

A close analysis reveals that the computational complexity is largely due to the complex coupling between the textons in both the generative model $p(\mathbf{I}|\mathbf{T}_1, \mathbf{T}_2; \Psi)$

and the descriptive models $p(\mathbf{T}_1; \beta_{1o}, \boldsymbol{\beta}_1)$ and $p(\mathbf{T}_2; \beta_{2o}, \boldsymbol{\beta}_2)$. Thus we simplify both models by decoupling the textons.

First, we decouple the textons in $p(\mathbf{T}_1; \beta_{1o}, \boldsymbol{\beta}_1)$ and $p(\mathbf{T}_2; \beta_{2o}, \boldsymbol{\beta}_2)$. We fix the total number of textons $n_1 + n_2$ to an excessive number, and thus we do not need to simulate the death–birth process. We set $\boldsymbol{\beta}_1$ and $\boldsymbol{\beta}_2$ to 0, and therefore $p(\mathbf{T}_l; \beta_{lo}, \boldsymbol{\beta}_l)$ becomes a uniform distribution and the texton elements are decoupled from spatial interactions.

Second, we decouple the textons in $p(\mathbf{I}_{obs}|\mathbf{T}_1, \mathbf{T}_2; \Psi)$. Instead of using the image generating model in Eq. (5.16), which implicitly imposes couplings between texton elements through Eq. (5.20), we adopt a constraint-based model

$$p(\mathbf{I}^{\text{obs}}|\mathbf{T}, \Psi) \propto \exp\left\{ -\sum_{l=1}^{2}\sum_{j=1}^{n_l} \|\mathbf{I}^{\text{obs}}_{D_{lj}} - G[T_{lj}] \odot \Psi_l\|^2/2\sigma^2 \right\}, \tag{5.46}$$

where $\mathbf{I}^{\text{obs}}_{D_{lj}}$ is the image patch of the domain D_{lj} in the observed image. For pixels in $\mathbf{I}^{\text{obs}}$ not covered by any textons, a uniform distribution is assumed to introduce a penalty.

We run the stochastic gradient algorithm on the decoupled log-likelihood, which reduces to a conventional clustering problem. We start with two random texton maps and the algorithm iterates the following two steps.

(I) Given Ψ_1 and Ψ_2, the algorithm runs a Gibbs sampler to change each texton t_{lj}, respectively, by moving, rotating, scaling the rectangle, and changing the cluster into which each texton falls according to the simplified model of Eq. (5.46). Thus the texton windows intend to cover the entire observed image and at the same time try to form tight clusters around Ψ.

(II) Given $\mathbf{T}_1$ and $\mathbf{T}_2$, the algorithm updates the texton Ψ_1 and Ψ_2 by averaging

$$\Psi_l = \frac{1}{n_l} \sum_{j=1}^{n_l} G^{-1}[T_{lj}] \odot \mathbf{I}^{\text{obs}}_{D_{lj}}, \quad l = 1, 2, \tag{5.47}$$

where $G^{-1}[T_{lj}]$ is the inverse transformation. The layer orders d_1 and d_2 are not needed for the simplified model.

This initialization algorithm for computing $(\mathbf{T}_1, \mathbf{T}_2, \Psi_1, \Psi_2)$ resembles the transformed component analysis [60]. It is also inspired by a clustering algorithm by Leung and Malik [151], which did not engage hidden variables and thus compute a variety of textons Ψ at different scale and orientations. We also experimented with representing the texton template Ψ by a set of Gabor basis functions instead of a 15×15 window. However, the results were not as encouraging as in this generative model.

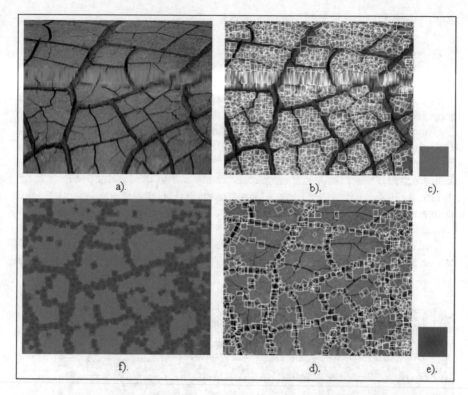

Fig. 5.7 Reprinted with permission from [89]. Result of the initial clustering algorithm. (**a**) Input image. (**b**) Cluster 1 textons T_1. (**c**) ψ_1. (**d**) Cluster 2 textons T_2. (**e**) ψ_2. (**f**) Reconstructed image

Experiment I: Texton Clustering

In the following, we demonstrate experiments for initialization and clustering as Sect. 5.2 stated.

Figure 5.7 shows an experiment on the initialization algorithm for a crack pattern. 1055 textons are used with the template size of 15×15. The number of textons is as twice as necessary to cover the whole image. In optimizing the likelihood in Eq. (5.46), an annealing scheme is utilized with the temperature decreasing from 4 to 0.5. The sampling process converges to a result shown in Fig. 5.7.

Figure 5.7a is the input image; Fig. 5.7b,d are the texton maps T_1 and T_2, respectively. Figure 5.7c,e are the cluster centers Ψ_1 and Ψ_2, shown by rectangles, respectively. Figure 5.7f is the reconstructed image. The results demonstrate that the clustering method provides a rough but reasonable starting solution for generative modeling.

Experiment II: Integrated Learning and Synthesis

Next, we show experimental results obtained by the integrated learning paradigm.
For an input image, we first do a clustering step as Sect. 5.2 showed. Then we run
the stochastic gradient algorithm on the full models to refine the clustering results.

Figure 5.8 shows the result for the crack image obtained by the stochastic
gradient algorithm, which took about 80 iterations of the two steps (Step I and Step
II), following the initial solution (Step 0) shown in Fig. 5.7. Figure 5.8b,d are the
background and foreground texton maps T_1 and T_2, respectively. Figure 5.8c,e are
the learned textons Ψ_1 and Ψ_2, respectively. Figure 5.8f is the reconstructed image
from learned Texton maps and templates. Compared to the results in Fig. 5.7, the
results in Fig. 5.8 have more precise texton maps and accurate texton templates due
to an accurate generative model. The foreground texton Ψ_2 is a bar, and one pixel at
corner of the left-top is transparent.

The integrated learning results for a cheetah skin image are shown in Fig. 5.9. It
can be seen that in the foreground template, the surrounding pixels are learned as

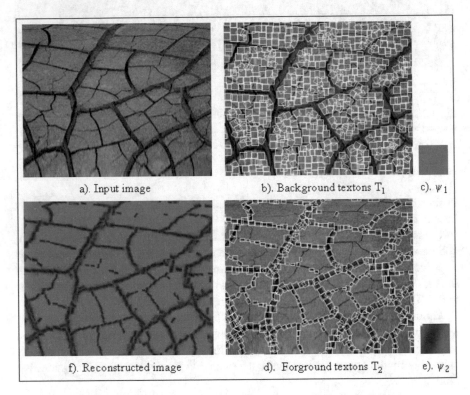

Fig. 5.8 Reprinted with permission from [89]. Generative model learning result for the crack
image. (**a**) Input image, (**b**) and (**d**) are background and foreground textons discovered by the
generative model, (**c**) and (**e**) are the templates for the generative model, and (**f**) is the reconstructed
image from the generative model

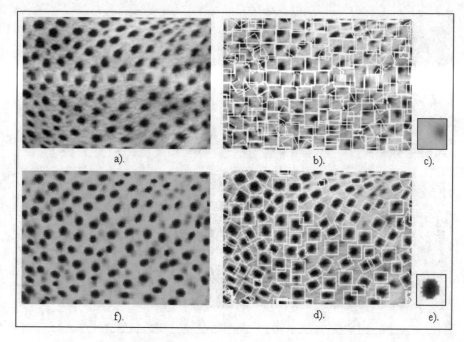

Fig. 5.9 Reprinted with permission from [89]. Generative model learning result for a cheetah skin image. (**a**) Input image. (**b**) Background textons T_1. (**c**) ψ_1. (**d**) Forground textons T_2. (**e**) ψ_2. (**f**) Reconstructed image

being transparent and the blob is exactly computed as the texton. Figure 5.10 shows the results for a brick image. No point in the template is transparent for the gap lines between bricks.

Figure 5.11 shows the learning of another short crack pattern. Figure 5.12 displays a pine corn pattern. The seeds and the black intervals are separated cleanly, and the reconstructed image keeps most of the pine structures. However, the pine corn seeds are learned as the background textons and the gaps between pine corns are treated as foreground textons.

After the parameters Ψ and β of the generative model are discovered for a type of texture images, new random samples can be drawn from the generative model. This proceeds in three steps: first, texton maps are sampled from the Gibbs models $p(\mathbf{T}_1; \boldsymbol{\beta}_1)$ and $p(\mathbf{T}_2; \boldsymbol{\beta}_2)$, respectively. Second, background and foreground images are synthesized from the texton maps and texton templates. Third, the final image is generated by combining these two images according the occlusion model.

We show synthesis experiments on three patterns.

1. Figures 5.13 and 5.14 are two synthesis examples of the two-layer model synthesis for the cheetah skin pattern. The templates used here are the learned results in Fig. 5.9.
2. Figure 5.15 shows texture synthesis for the crack pattern computed in Fig. 5.11.

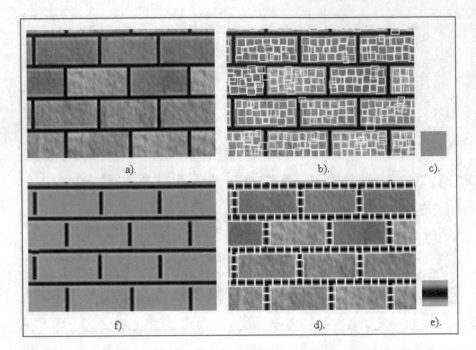

Fig. 5.10 Reprinted with permission from [89]. Generative model learning result for a brick image. (**a**) Input image. (**b**) Background textons T_1. (**c**) ψ_1. (**d**) Forground textons T_2. (**e**) ψ_2. (**f**) Reconstructed image

3. Figure 5.16 displays texture synthesis for the brick pattern in Fig. 5.10.

Note that, in these texture synthesis experiments, the Markov chain operates with meaningful textons instead of pixels.

Discussion

We present a visual learning paradigm that integrates and extends descriptive and generative models and which also provides a framework for visual conceptualization and for defining textons. The hierarchical model for textures has advantages over the previous pure descriptive method with Markov random fields on pixel intensities.

First, from the representational perspective, the neighborhood in the texton map is much smaller than the pixel neighborhood in a FRAME model [281]. The generative method captures more semantically meaningful elements on the texton maps.

Second, from the computational perspective, the Markov chain operating on the texton maps can move textons according to affine transforms and can add or delete a texton by birth–death dynamics, and thus it is much more effective than the Markov

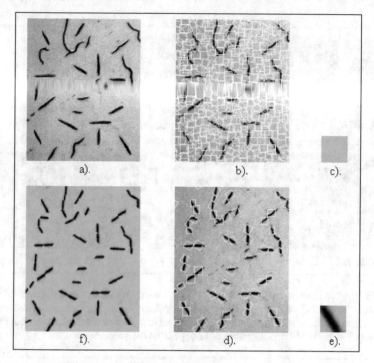

Fig. 5.11 Reprinted with permission from [89]. Generative model learning result for a crack image. (**a**) Input image. (**b**) Background textons T_1. (**c**) ψ_1. (**d**) Forground textons T_2. (**e**) ψ_2. (**f**) Reconstructed image

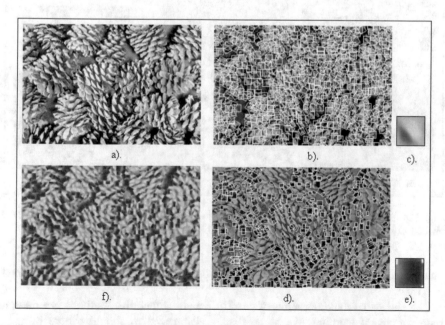

Fig. 5.12 Reprinted with permission from [89]. Generative model learning result for a pine corn image. (**a**) Input image. (**b**) Background textons T_1. (**c**) ψ_1. (**d**) Forground textons T_2. (**e**) ψ_2. (**f**) Reconstructed image

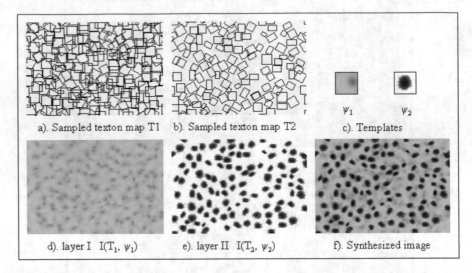

Fig. 5.13 Reprinted with permission from [89]. An example of a randomly synthesized cheetah skin image. (**a**) and (**b**) are the background and foreground texton maps sampled from $p(\mathbf{T}_l; \beta_l)$, (**d**) and (**e**) are synthesized background and foreground images from the texton map and templates in (**c**), and (**f**) is the final random synthesized image from the generative model

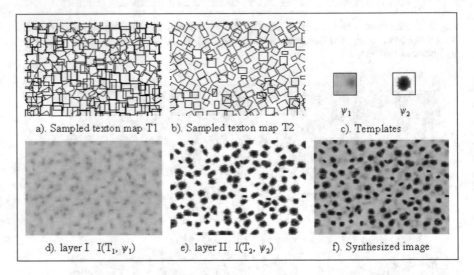

Fig. 5.14 Reprinted with permission from [89]. The second example of a randomly synthesized cheetah skin image. Notation is the same as in Fig. 5.13

chain used in traditional Markov random fields which flips the intensity of one pixel at a time.

We show that the integration of descriptive and generative methods is a natural path for visual learning. We argue that a vision system should evolve by pro-

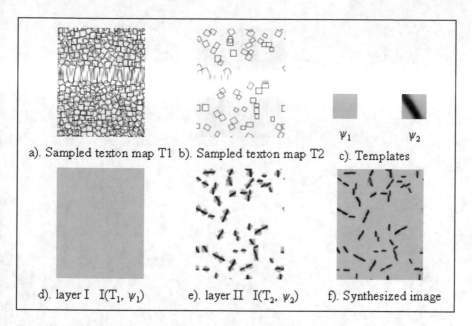

Fig. 5.15 Reprinted with permission from [89]. An example of a randomly synthesized crack image. Notations are the same as in Fig. 5.13

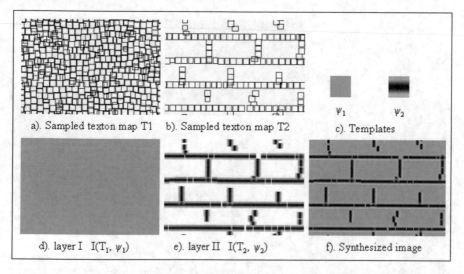

Fig. 5.16 Reprinted with permission from [89]. An example of a randomly synthesized brick image. Notation is the same as in Fig. 5.13

gressively replacing descriptive models with generative models, which realizes a transition from empirical and statistical models to physical and semantical models. The work presented here provides a step toward this goal.

Chapter 6
Primal Sketch: Integrating Textures and Textons

In his monumental book [169], Marr inherited Julesz's texton [124] notion and proposed the concept of image primitives as basic perceptual tokens, such as edges, bars, junctions, and terminators. Inspired by the Nyquist sampling theorem in signal processing, Marr went a step further and asked for a token representation which he named "primal sketch" as a perceptually lossless conversion from the raw image. He tried to reconstruct the image with zero-crossings unsuccessfully and his effort was mostly limited by the lack of proper models of texture.

6.1 Marr's Conjecture on Primal Sketch

In the early stage of visual perception, an image may be divided into two components—the structural part with noticeable elements called "textons" by Julesz or "image primitives" by Marr and the textural part without distinguishable elements in pre-attentive vision. The structural part is often composed of objects, such as tree twigs and trunks at a near distance whose positions and shapes can be clearly perceived. In contrast, the textural part is composed of objects at a far distance whose structures become indistinguishable and thus yield various texture impressions.

The modeling of texture and structure has been a long-standing puzzle in the study of early vision. In the 1960s, Julesz first proposed a texture theory and conjectured that a texture is a set of images sharing some common statistics on some features related to human perception. Later he switched to a texton theory and identified bars, edges, and terminators as textons—the atomic elements in early vision. Marr summarized Julesz's theories along with experimental results and proposed a primal sketch model in his book as a "symbolic" or "token" representation in terms of image primitives. Marr argued that this symbolic representation should be parsimonious and sufficient to reconstruct the original image without much perceivable distortion.

© Springer Nature Switzerland AG 2023
S.-C. Zhu, Y. N. Wu, *Computer Vision*, https://doi.org/10.1007/978-3-030-96530-3_6

In fact, Andrew Glennerster [42] shows that the primal sketch representation provides a stable coordinate frame under rapid eye rotation, which is biologically plausible. Once a certain task is defined, the visual system extracts "raw" visual information from the primal sketch representation to higher level information.

Despite many inspiring observations, Marr's description provided neither an explicit mathematical formulation nor a rigorous definition of the primal sketch model.

Since the 1980s, the studies of early image modeling followed two distinct paths which represent two prevailing mathematical theories for generative image modeling, respectively. In fact, the two theories are two distinct ways of learning image manifolds residing in different entropy regimes, respectively. In the following, we should briefly review the two theories. Interested readers are recommended to review the last chapter for a more complete account of generative and descriptive models.

The first theory is a two-layer generative model originating from computational harmonic analysis which represents images by a linear superposition of basis functions selected from a dictionary—often over-complete like various wavelets, image pyramids, and sparse coding. Each image base is supposed to represent some image features with hidden variables describing their locations, orientations, scales, and intensity contrast. The image is reconstructed with minimum error on the pixel intensity.

The second theory is the Markov random fields (MRFs) originated from statistical mechanics. It represents a visual pattern by pooling the responses of a bank of filters over all locations into some statistical summary like the histograms which are supposed to represent our texture impressions. On large image lattices, a Julesz ensemble is defined as a perceptual equivalence class where all images in the set share identical statistics. The statistics or texture impression is the macroscopic properties and the differences between microscopic states (i.e., image instances in the Julesz ensemble) are ignored. In other words, all images in this equivalence class are perceptually the same, replacing one with the other does not cause perceivable distortion, although the two images have large differences in pixel-by-pixel comparison.

6.2   The Two-Layer Model

According to the model, the image is generated as a mosaic as follows: the image lattice is divided into two disjoint parts: a structured domain, or a sketchable part, and a textured domain, or a non-sketchable part (Figs. 6.1 and 6.2),

$$D = D_{sk} \cup D_{nsk}, \ D_{sk} \cap D_{nsk} = \phi. \tag{6.1}$$

The image intensities on the structure domain are represented by a set of coding functions for edges and ridges. The image intensities on the texture domain are

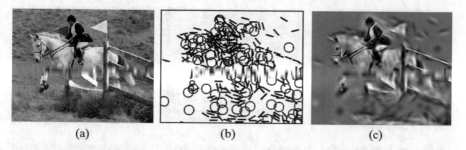

Fig. 6.1 Reprinted with permission from [90]. A sparse coding example is computed by matching pursuit. (b) is a symbolic representation where each base B_k is represented by a bar at the same location, with the same elongation and orientation. The isotropic LOG basis functions are represented by a circle. (**a**) Original image. (**b**) LoG and Gabor tokens. (**c**) Reconstructed image

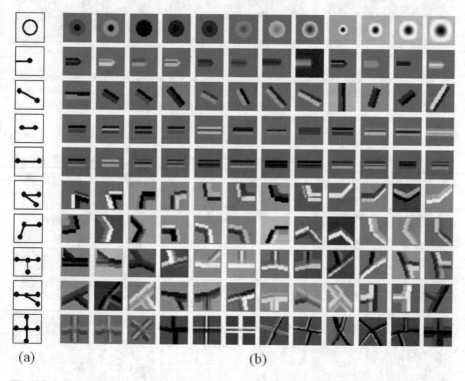

Fig. 6.2 Reprinted with permission from [91]. A collection of local structure elements employed by the model. There are eight types of elements: blobs, endpoints, edges, ridges, multi-ridges, corners, junctions, and crosses. (**a**) The symbolic representation. (**b**) The photometric representation

characterized by Markov random fields that interpolate the structure domain of the image. See Figs. 6.3 and 6.4 for two examples of primal sketch model.

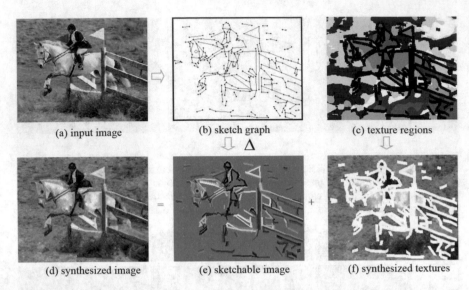

(a) input image   (b) sketch graph   (c) texture regions

(d) synthesized image   (e) sketchable image   (f) synthesized textures

Fig. 6.3 Reprinted with permission from [91]. Primal sketch model. (**a**) Observed image. (**b**) "Sketchable" part is described by a geometric sketch graph. (**c**) The texture regions of the image. (**d**) Fill in the "non-sketchable" part by matching feature statistics. (**e**) The sketchable part of the image. (**f**) The non-sketchable part of the image

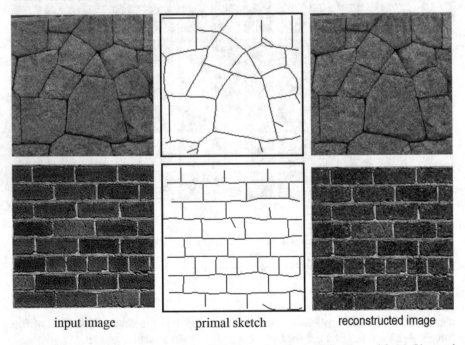

input image   primal sketch   reconstructed image

Fig. 6.4 Reprinted with permission from [91]. Examples of primal sketch model. (**a**) Observed image. (**b**) Sketch graph. (**c**) Synthesized image from the fitted model

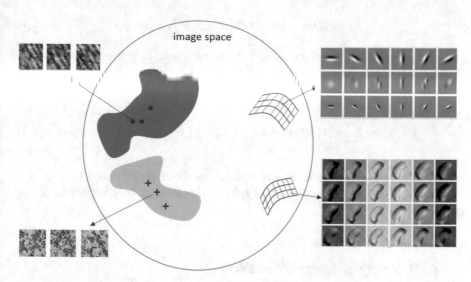

Fig. 6.5 In the space of image patches, there are simple shape primitives, and there are also stochastic texture patterns

The motivation is that in the space of image patches, there are simple geometric primitives such as edges, but there are also stochastic texture patterns, as illustrated in Fig. 6.5. We need both to describe an image.

Structure Domain

The structured domain D_{sk} is further divided into a number of disjoint patches with each patch being fitted by an image primitive

$$D_{sk} = \cup_{i=1}^{K} D_{sk,i}; \; D_{sk,i} \cap D_{sk,j} = \phi, i \neq j. \tag{6.2}$$

Some examples of the image primitive are shown in Fig. 6.2. These primitives are aligned through their landmarks to form a sketch graph S_{sk}. Specifically, we index the selected image primitives by $i = 1, \ldots, n$ and denote image patch for primitive i as $B(x, y|\theta_i)$. Here

$$\theta_i = (\theta_{\text{topological},i}, \theta_{\text{geometric},i}, \theta_{\text{photometric},i}), \tag{6.3}$$

where $\theta_{\text{topological},i}$ is the type (degree of arms) of the primitive (blob, terminator, corner, junctions, etc.), $\theta_{\text{geometric},i}$ collects the locations of the landmarks of the primitive, and $\theta_{\text{photometric},i}$ collects the intensity profiles of the arms of the primitive. The sketch graph is a layer of hidden representation, which has to be inferred from the image

$$S_{sk} = ((D_{sk,i}, B(x, y|\theta_i), a_i), i = 1, \ldots, n), \tag{6.4}$$

where S_{sk} decides the structure domain of the image, and a_i is the address variable pointing to the neighbors of the vertex $S_{sk,i} = (D_{sk,i}, B(x, y|\theta_i))$.

The model for the structure domain of the image is

$$\mathbf{I}(x, y) = \sum_{i=1}^{n} B(x, y|\theta_i) + \epsilon(x, y), \quad (x, y) \in D_{sk}, \quad i = 1, \ldots, n, \tag{6.5}$$

where $\epsilon(x, y)$ denotes the noise term. Since the set of pixels coded by $B(x, y \mid \theta_i)$ does not overlap each other, $B(x, y|\theta_i)$ is similar to coding vectors in vector quantization.

The Dictionary of Image Primitives

An edge segment is modeled by a 2D function that is constant along the edge and has a profile across the edge. Specifically,

$$B(x, y \mid \theta) = f(-(x - u)\sin\alpha + (y - v)\cos\alpha), \tag{6.6}$$

where

$$-l < (x - u)\cos\alpha + (y - v)\sin\alpha \le l, \tag{6.7}$$

$$-w \le -(x - u)\sin\alpha + (y - v)\cos\alpha \le w. \tag{6.8}$$

That is, the function $B(x, y \mid \theta)$ is supported on a rectangle centered at (u, v), with length $2l + 1$, width $2w + 1$, and orientation α.

For the profile function $f(x)$, let $f_0(x) = -1/2$ for $x < 0$ and $f_0(x) = 1/2$ for $x \ge 0$, and let $g_s()$ be a Gaussian function of standard deviation s. Then $f(x) = a + bf_0(x) * g_s(x)$. This is the model proposed by Elder and Zucker [52]. The convolution with Gaussian kernel is used to model the blurred transition of intensity values across the edge, caused by the three-dimensional shape of the underlying physical structure, as well as the resolution and focus of the camera. As proposed by Elder and Zucker, the parameter s can be determined by the distance between the two extrema of the second derivative $f''(x)$.

Thus in the coding function $B(x, y \mid \theta)$ for an edge segment, $\theta = (t, u, v, \alpha, l, w, s, a, b)$, namely, type (which is edge in this case), center, orientation, length, width, sharpness, average intensity, intensity jump. θ captures geometric and photometric aspects of an edge explicitly, and the coding function is nonlinear in θ.

A ridge segment has the same functional form, where the profile $f(x)$ is a composition of two edge profiles. The profile of a multi-ridge is a composition of two or more ridge profiles. A blob function is modeled by rotating an edge profile, more specifically, $B(x, y \mid \theta) = f(\sqrt{(x-u)^2 + (y-v)^2} - r)$, where $(x-u)^2 + (y-v)^2 \leq R^2$, and again $f(x) = a + bf_0(x) * g_s(x)$ being a step edge convolved with a Gaussian kernel. This function is supported on a disk area centered at (u, v) with radius R. The transition of intensity occurs at the circle of radius $r < R$.

The corners and junctions are important structures in images. They are modeled as compositions of edge or ridge functions. When a number of such coding functions join to form a corner or a junction, the image intensities of the small number of overlapping pixels are modeled as averages of these coding functions. The endpoint of a ridge is modeled by a half blob.

See Fig. 6.2 for a sample of local structure elements, which are the coding functions and their combinations. There are eight types of elements: blobs, endpoints, edges, ridges, multi-ridges, corners, junctions, and crosses. Figure 6.2a shows the symbolic representations of these elements. Figure 6.2b displays the image patches of these elements.

Let S_{sk} be the sketch graph formed by these coding functions. The graph has a set of nodes or vertices $V = \cup_{d=0}^4 V_d$, where V_d is the set of nodes with degree d, i.e., the nodes with d arms. For instance, a blob node has a degree 0, an endpoint has a degree 1, a corner has a degree 2, a T-junction has a degree 3, and across has a degree 4. We do not allow nodes with more than 4 arms. S_{sk} is regularized by a simple spatial prior model:

$$p(S_{sk}) \propto \exp\left\{ -\sum_{d=0}^4 \lambda_d |V_d| \right\}, \tag{6.9}$$

where $|V_d|$ is the number of nodes with d arms. The prior probability or the energy term $\gamma_{sk}(S_{sk}) = \sum_{d=0}^4 \lambda_d |V_d|$ penalizes free endpoints by setting λ_{sk} at a large value.

Texture Domain

The texture domain D_{nsk} is segmented into m regions of homogenous texture patterns,

$$D_{nsk} = \cup_{j=1}^m D_{nsk,j}; \; D_{nsk,i} \cap D_{nsk,j} = \phi, i \neq j. \tag{6.10}$$

Within each region j, we pool the marginal histograms of the responses from the K filters, $h_j = (h_{j,k}, k = 1, \ldots, K)$, where

$$h_{j,k,z} = \frac{1}{|D_{\mathrm{nsk},j}|} \sum_{(x,y)\in D_{\mathrm{nsk},j}} \delta(z; F_k * \mathbf{I}(x, y)), \tag{6.11}$$

where z indexes the histogram bins, and $\delta(z; x) = 1$ if x belongs to bin z, and $\delta(z; x) = 0$ otherwise. This yields a Markov random field model, also known as the descriptive model, for each texture region:

$$p(\mathbf{I}_{D_{\mathrm{nsk},j}}) \propto \exp\left\{ -\sum_{(x,y)\in D_{\mathrm{nsk},j}} \sum_{k=1}^{K} \phi_{j,k}(F_k * \mathbf{I}(x, y)) \right\}. \tag{6.12}$$

These Markov random fields have the structure domain as boundary conditions, because when we apply filters F_k on the pixels in D_{nsk}, these filters may also cover some pixels in D_{sk}. These Markov random fields in-paint the texture domain D_{nsk} while interpolating the structure domain D_{sk}, and the in-painting is guided by the marginal histograms of linear filters within each region.

Let $S_{\mathrm{nsk}} = ((D_{\mathrm{nsk},j}, h_{j,k}, \phi_{j,k}), j = 1, \ldots, m, k = 1, \ldots, K)$ denote the segmentation of the texture domain. S_{nsk} follows a prior model $p(S_{\mathrm{nsk}}) \propto \exp\{-\gamma_{\mathrm{nsk}}(S_{\mathrm{nsk}})\}$, for instance, $\gamma_{\mathrm{nsk}}(S_{\mathrm{nsk}}) = \rho m$ to penalize the number of regions.

Integrated Model

Formally, we can integrate the structure model (6.5) and the texture model (6.12) into a probability distribution. Our inspiration for such integration comes from the model of Mumford and Shah [177]. In their method, the prior model for the noiseless image can be written as

$$p(\mathbf{I}, S) = \frac{1}{Z} \exp\left\{ -\sum_{(x,y)\in D/S} \lambda |\nabla \mathbf{I}(x, y)|^2 - \gamma|S| \right\}, \tag{6.13}$$

where S is a set of pixels of discontinuity that correspond to the boundaries of objects and $|S|$ is the number of pixels in S. In model (6.13), S is the structure domain of the image, and the remaining part is the texture domain.

Our model can be viewed as an extension of the Mumford–Shah model. Let $S = (S_{\mathrm{sk}}, S_{\mathrm{nsk}})$, and we have

$$p(\mathbf{I}, S) = \frac{1}{Z} \exp\left\{ -\sum_{i=1}^{n} \sum_{(x,y)\in D_{\mathrm{sk},i}} \frac{1}{2\sigma^2} (\mathbf{I}(x, y) - B_i(x, y \mid \theta_i))^2 - \gamma_{\mathrm{sk}}(S_{\mathrm{sk}}) \right.$$

$$-\sum_{j=1}^{m} \sum_{(x,y) \in D_{\mathrm{nsk},j}} \sum_{k=1}^{K} \phi_{j,k}(F_k * \mathbf{I}(x, y)) - \gamma_{\mathrm{nsk}}(S_{\mathrm{nsk}}) \Bigg\}. \quad (6.14)$$

Compared to Mumford–Shah model, model (6.14) is more sophisticated in both the structure part and the texture part.

The Sketch Pursuit Algorithm

To learn the integrated model, the traditional maximum likelihood algorithm requires the MCMC method for global inference. Instead, a greedy algorithm, sketch pursuit algorithm, is proposed. It consists of the following phases:

1. *Phase 0*: an edge and ridge detector based on linear filters are run to give an initialization for the sketch graph.
2. *Phase 1*: a greedy algorithm is used to determine the sketch graph but without using the spatial prior model.
3. *Phase 2*: a greedy algorithm based on a set of graph operators is used to edit the sketch graph to achieve good spatial organization as required by the spatial prior model.
4. *Phase 3*: the remaining portion of the image is segmented into homogeneous texture regions by clustering the local marginal histograms of filter responses. The inference algorithm yields two outputs:

   (a) A sketch graph for the image, with edge and ridge segments, as well as corners and junctions.
   (b) A parameterized representation of the image which allows the image to be re-synthesized and to be encoded efficiently.

6.3 Hybrid Image Templates

The primal sketch model is developed for generic scenes. In this section, we study a more specific framework for learning representations for objects from images, the hybrid image templates (HITs).

First, just as the primal sketch, the appearances of objects have two types of descriptors: local sketch and texture gradient. For instance, the image in Fig. 6.6 consists of both the patches of geometric primitives on the object boundary and the patches of textures on the object surface. In addition to sketch features and texture features, we also add flatness features and color features, so that the templates give complete descriptions of the images. Figure 6.7 shows some examples of such

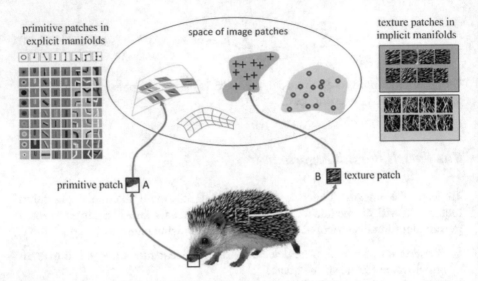

Fig. 6.6 Reprinted with permission from [217]. An image of an object consists of patches of simple primitives and patches of textures

templates, with local sketch (edge or bar), texture gradients (with orientation field), flatness regions (smooth surface and lighting), and colors.

The modeling and learning strategy of the hybrid template is similar to the active basis. For each type of feature, we pool a background histogram q from natural images and then estimate the distribution p by exponential tilting. We select the features sequentially as in active basis. After each feature is selected, it will inhibit nearby features of the same type.

Naturally, there are large variations in the representations of different classes, for example, teapots may have a common shape outline but do not have a common texture or color. The hedgehog in Fig. 6.6 has a distinct texture and shape, but its color is often less distinguishable from its background. The essence of our learning framework is to automatically select, in a principled way, informative patches from a large pool and compose them into a template with a normalized probability model.

Representation

Let D be the image lattice for the object template which is typically 150×150 pixels. This template will undergo a similarity transform to align with object instances in images. The lattice is decomposed into a set of K patches $\{D_k, k = 1, 2 \ldots, K\}$ selected from a large pool in the learning process through feature pursuit. These patches belong to four bands: sketch, texture/gradient field, flatness, and color, respectively, and they do not form a partition of the lattice D for two reasons:

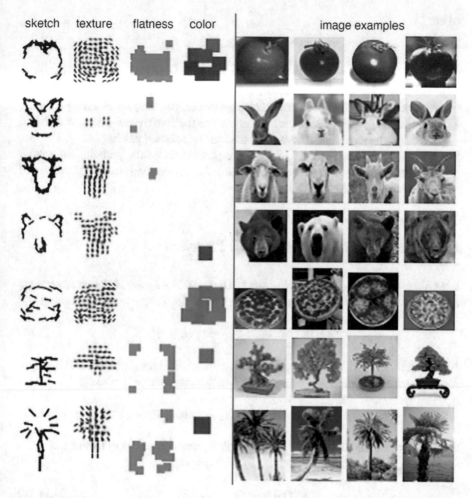

sketch	texture	flatness	color	image examples

Fig. 6.7 Reprinted with permission from [217]. The hybrid templates consist of sketch, texture, flatness, and color features

- Certain pixels on D are left unexplained due to inconsistent image appearances at these positions.
- Two selected patches from different bands may overlap each other in position. For example, a sketch patch and a color patch can occupy the same region, but we make sure the sketch feature descriptor and color descriptor extracted from them would represent largely uncorrelated information.

The hybrid image template consists of the following components:

$$\text{HIT} = (\{D_k, \ell_k, \{B_k \text{or} \mathbf{h}_k\}, \delta_k :, k = 1, 2, \ldots, K\}, \Theta), \tag{6.15}$$

where

1. $D_k \subset D$ is the k-th patch lattice described above.
2. $\ell_k \in \{$ skt, txt, flt, clr$\}$ is the type of the patch.
3. B_k or $\mathbf{h}_k$ is the feature prototype for the k-th patch. If $\ell_k =$ skt, then the patch is described by a basis function B_k for the image primitive; otherwise, it is described by a histogram $\mathbf{h}_k$ for texture gradients, flatness, or color, respectively.
4. $\delta_k = (\delta_k x, \delta_k y, \delta_k \theta)$: the latent variables for the local variabilities of the k-th patch, i.e., the local translations and rotations of selected patches.
5. $\Theta = \{\lambda_k, z_k : k = 1, 2, \ldots, K\}$ are the parameters of the probabilistic model p (to be discussed in the later subsection). λ_k, z_k are the linear coefficient and normalizing constant for the k-th patch.

Prototypes, ϵ-Balls, and Saturation Function

Let $\mathbf{I}_{D_k}$ be the image defined on the patch $D_k \subset D$. For $\ell_k =$ skt, the prototype B_k defines a subspace through an explicit function for $\mathbf{I}_{D_k}$ (a sparse coding model),

$$\Omega(B_k) = \{\mathbf{I}_{D_k} : \mathbf{I}_{D_k} = c_k B_k + \epsilon\}. \tag{6.16}$$

For $\ell_k \in \{$txt, flt, clr$\}$, the prototype defines a subspace through an implicit function for $\mathbf{I}_{D_k}$ which constrains the histogram (a Markov random field model),

$$\Omega(\mathbf{h}_k) = \{\mathbf{I}_{D_k} : H(\mathbf{I}_{D_k}) = \mathbf{h}_k + \epsilon\}. \tag{6.17}$$

$H(\mathbf{I}_{D_k})$ extracts the histogram (texture gradient, flatness, or color) from $\mathbf{I}_{D_k}$.

In $\Omega(B_k)$, the distance is measured in the image space,

$$\rho^{\text{ex}}(\mathbf{I}_{D_k}) = \|\mathbf{I}_{D_k} - c B_k\|^2, \tag{6.18}$$

while in $\Omega(h_k)$, the distance is measured in the projected histogram space with L1 or L2 norms,

$$\rho^{\text{im}}(\mathbf{I}_{D_k}) = \|H(\mathbf{I}_{D_k}) - \mathbf{h}_k\|^2. \tag{6.19}$$

Intuitively, we may view $\Omega(B_k)$ and $\Omega(\mathbf{h}_k)$ as ϵ-balls centered at the prototypes B_k and $\mathbf{h}_k$, respectively, with different metrics. Each ϵ-ball is a set of image patches that are perceptually equivalent. Thus the image space of HIT is the product space of these heterogeneous subspaces: $\Omega(HIT) = \prod_{k=1}^{K} \Omega_k$, on which a probability model is concentrated. Due to statistical fluctuations in small patches, these ϵ-balls have soft boundaries. Thus we use a sigmoid function to indicate whether a patch $\mathbf{I}_{D_k}$ belongs to a ball $\Omega(B_k)$ or $\Omega(\mathbf{h}_k)$.

$$r(\mathbf{I}_{D_k}) = S(\rho(\mathbf{I}_{D_k})), \tag{6.20}$$

where ρ can be either ρ^{ex} or ρ^{im}. $S(x)$ is a saturation function with maximum at $x = 0$:

$$S(x) = \tau \left(\frac{2}{1 + e^{-2(\eta - x)/\tau}} - 1 \right), \tag{6.21}$$

with shape parameters τ and η. We set $\tau = 6$ and η is locally adaptive: $\eta = \|\mathbf{I}_{D_k}\|^2$, where $\mathbf{I}_{D_k}$ denotes the local image patch. We call $r(\mathbf{I}_{D_k})$ the response of the feature (prototype B_k or $\mathbf{h}_k$) on patch $\mathbf{I}_{D_k}$.

Projecting Image Patches to 1D Responses

Though the image patches are from heterogeneous subspaces of varying dimensions with different metrics, we project them into the one-dimensional feature response $r(\mathbf{I}_{D_k})$, on which we can calculate the statistics (expectation) of $r(\mathbf{I}_{D_k})$ over the training set regardless of the types of patches. This way it is easy to integrate them into a probabilistic model.

In the following, we discuss the details of computing the responses for the four different image subspaces.

Given an input color image $\mathbf{I}$ on lattice D, we first transform it into an HSV space with HS being the chromatic information and V the gray level image. We apply a common set of filters Δ to the gray level image. The dictionary Δ includes Gabor filters (sine and cosine) at 3 scales and 16 orientations. The Gabor filter of the canonical scale and orientation is of the form: $F(x, y) \propto \exp\{-(x/\sigma_1)^2 - (y/\sigma_2)^2\}e^{ix}$ with $\sigma_1 = 5, \sigma_2 = 10$.

1. *Calculating responses on primitives.* When a patch $\mathbf{I}_{D_k}$ contains a prominent primitive, such as an edge or a bar, it is dominated by a filter that inhibits all the other filters. Thus the whole patch is represented by a single filter, which is called a basis function $B_k \in \Delta$. The response is calculated as the local maximum over the activity δ_k,

$$r^{\text{skt}}(\mathbf{I}_{D_k}) = \max_{\delta x, \delta y, \delta \theta} S(\|\mathbf{I} - c B_{x+\delta x, y+\delta y, o+\delta o}\|^2). \tag{6.22}$$

The local maximum pooling is proposed by [214] as a possible function of complex cells in V1.

2. *Calculating responses on texture.* In contrast to the primitives, a texture patch usually contains many small elements, such as the patch on the hedgehog body in Fig. 6.6. As a result, many filters have medium responses on the image patch. Thus we pool a histogram of these filters collectively over the local patch to form a histogram descriptor $H(\mathbf{I})$.

The texture response is calculated by

$$r^{\text{txt}}(\mathbf{I}_{D_k}) = S(\|H(\mathbf{I}_{D_k}) - \mathbf{h}\|^2),$$ (6.23)

where $\mathbf{h}$ is a pre-computed histogram prototype (one may consider it as a cluster center of similar texture patches). More specifically, $\mathbf{h}$ is obtained by averaging the histograms at the same position of roughly aligned positive example images. For texture, we are only interested in the medium to strong strengths along certain directions. So we replace the indicator function, which is often used in histogram binning, by a continuous function $a(x) = \frac{12}{1+e^{-x/3}} - 6$. The histogram is then weighted into one bin for each filter,

$$H_o(\mathbf{I}_{D_k}) = \frac{1}{|D_k|} \sum_{(x,y)\in D_k} a(|F_o * \mathbf{I}_{D_k}|^2).$$ (6.24)

Thus, we obtain the oriented histogram for all filters as a vector,

$$H(\mathbf{I}_{D_k}) = (H_1, \dots, H_{|\mathcal{O}|}).$$ (6.25)

It measures the strengths in all orientations.

3. *Calculating responses on flat patch.* By flat patch, we mean image areas that are void of structures, especially edges. Thus filters have near-zero responses. They are helpful for suppressing false alarms in cluttered areas. As a textureless measure, we choose a few small filters $\Delta^{\text{flt}} = \{\nabla_x, \nabla_y, LoG\}$ and further compress the texture histogram into a single scalar,

$$H(\mathbf{I}_{D_k}) = \sum_{F\in\Delta^{\text{flt}}} \sum_{(x,y)\in D_k} b(|F_o * \mathbf{I}_{D_k}|^2).$$ (6.26)

$b()$ is a function that measures the featureless responses. It takes the form of a sigmoid function like $S()$ but with different shape parameters. In Fig. 6.8, we plot the four functions $a()$, $b()$, $\mathbf{1}()$, and $S()$ for comparison. Then the flatness response is defined as

$$r^{\text{flt}}(\mathbf{I}_{D_k}) = S(H(\mathbf{I}_{D_k}) - h).$$ (6.27)

In the above $h = 0$ is a scalar for the flatness prototype.

4. *Calculating responses on color.* The chromatic descriptors are informative for certain object categories. Similar to orientation histogram, we calculate a histogram $H^{\text{clr}}(\mathbf{I}_{D_k})$ on the color space (we use the 2D HS space in the HSV format). Then the color patch response is defined as the saturated distance between the color histogram of the observed image and the prototype histogram $\mathbf{h}$,

$$r^{\text{clr}}(\mathbf{I}_{D_k}) = S(\|H^{\text{clr}}(\mathbf{I}_{D_k}) - \mathbf{h}\|^2).$$ (6.28)

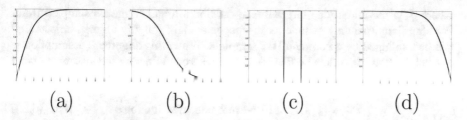

Fig. 6.8 Reprinted with permission from [217]. Plotting the four functions. (**a**) $a(x)$. (**b**) $b(x)$. (**c**) $\mathbf{1}(x)$. (**d**) $S(x)$

In summary, a HIT template consists of K prototypes $\{B_k \text{ or } \mathbf{h}_k, k = 1, \cdots, K\}$ for sketch, texture/gradient, flatness, and color patches, respectively, which define K-subspaces (or ϵ-balls) $\Omega(B_k)$ or $\Omega(\mathbf{h}_k)$ of varying dimensions. These ϵ-balls quantize the image space with different metrics. An input image $\mathbf{I}$ on lattice D is then projected to the HIT and is represented by a vector of responses:

$$\mathbf{I} \to (r_1, r_2, \ldots, r_K), \tag{6.29}$$

where r_k is a soft measure for whether the image patch $\mathbf{I}_{D_k}$ belongs to the subspace defined by the corresponding prototype. In the next section, we will define a probability model on image $\mathbf{I}$ based on these responses.

Template Pursuit by Information Projection

We present an algorithm for learning the hybrid image templates automatically from a set of image examples. It pursues the image patches, calculates their prototypes, and derives a probability model sequentially until the information gain is within the statistical fluctuation.

Let $f(\mathbf{I})$ be the underlying probability distribution for an image category, and our objective is to learn a series of models that approach f from an initial or reference model q,

$$q = p_0 \to p_1 \to p_2 \to \cdots \to p_K \approx f. \tag{6.30}$$

These models sequentially match the observed marginal statistics collected from the samples of f. With more marginal statistics matched between the model p and f, p will approach f in terms of reducing the Kullback–Leibler divergence $\mathrm{KL}(f \| p)$ monotonically.

The main input to the learning algorithm is a set of positive examples

$$\mathcal{I}^+ = \{\mathbf{I}_1, \ldots, \mathbf{I}_n\} \sim f, \tag{6.31}$$

where f is the underlying target image distribution and $\sim$ means sampled from. For simplicity, we may assume these images contain roughly aligned objects that can be explained by a common HIT template. When this alignment assumption is not satisfied, we can adopt an EM-like iterative procedure with the unknown object localization as missing data. We are also given a set of negative examples

$$\mathcal{I}^- = \{\mathbf{J}_1, \ldots, \mathbf{J}_N\} \sim \text{ reference distribution } q. \tag{6.32}$$

The negative examples are only used for pooling marginal histograms of one-dimensional feature responses in a pre-computation step.

The image lattice D is divided into overlapping patches for multiple scales by a scanning window with a step size about 10% of the window size. Then we calculate their corresponding prototypes and responses for all images in $\mathcal{I}^+$. The sketch prototypes B_i are specified by the Gabor dictionary Δ, and the histogram prototypes $\mathbf{h}_k$ are obtained by computing the histograms for positive examples in the same region of the template lattice and then taking the average. As a result, we obtain an excessive number of candidate patches.

$$\Omega_{\text{cand}} = \{D_j, \ell_j, \{B_j \text{ or } \mathbf{h}_j\} : j = 1, 2, \ldots, M\}. \tag{6.33}$$

From Ω_{cand}, we will select the most informative patches and their corresponding prototypes for HIT.

By induction, at the k-th step, we have a HIT with $k - 1$ patches and a model $p = p_{k-1}$:

$$\text{HIT}_{k-1} = (\{D_j, \ell_j, B_j \text{ or } \mathbf{h}_j, \delta_j, j = 1, \ldots, k-1\}, \Theta_{k-1}). \tag{6.34}$$

Consider a new candidate patch D_k in Ω_{cand} and its responses on n positive examples and N negative examples:

$$\{r_{k,i}^+, i = 1, \ldots, n\} \quad \{r_{k,i}^-, i = 1, \ldots, N\}. \tag{6.35}$$

And let $\bar{r}_k^+$ and $\bar{r}_k^-$ be the sample means on the two sets.

The gain of adding this patch to the template is measured by the KL-divergence between the target marginal distribution $f(r_k)$ and the current model $p_{k-1}(r_k)$, as this represents the new information in the training data that is not yet captured in the model. Among all the candidate patches, the one with the largest gain is selected.

To estimate this gain, we use Monte Carlo methods with samples from $f(r_k)$ and $p_{k-1}(r_k)$. Obviously $\{r_{k,i}^+\}$ is a fair sample from $f(r_k)$. While to sample from $p_{k-1}(r_k)$, one may use importance sampling on $\{r_{k,i}^-\}$, i.e., re-weighting the examples by $\frac{p_{k-1}(r_k)}{q(r_k)}$. Here we simplify the problem by a conditional independence assumption as stated in the previous section. A feature response $r_1(\mathbf{I}_{D_1})$ is roughly uncorrelated with $r_2(\mathbf{I}_{D_2})$ if one of the following holds: (i) the two patches D_1 and D_2 have little overlap and (ii) D_1 and D_2 are from different scales. If at the k-th step we have removed from Ω_{cand} all the candidate patches that overlap with selected patches, then r_k is roughly uncorrelated with all the previously selected responses $r_1, \ldots, r_{k-1}$. As a result, $p_{k-1}(r_k) = q(r_k)$ and $\{r_{k,i}^-\}$ can be used as a sample of $p_{k-1}(r_k)$. The exact formula for estimating the gain (i.e., KL-divergence between $p(r_k)$ and $p_{k-1}(r_k)$) is given, once we have derived the parametric form of p in the following.

For a selected patch D_k, the new model $p = p_k$ is required to match certain observed statistics (e.g., first moment), while it should be also close to the learned model p_{k-1} to preserve the previous constraints.

$$p_k^* = \arg\min \quad \text{KL}(p_k \| p_{k-1}) \qquad (6.36)$$

$$s.t. \quad \text{E}_{p_k}[r_k] = \text{E}_f[r_k]. \qquad (6.37)$$

By solving the Euler–Lagrange equation with Lagrange multipliers $\{\lambda_j\}$ and γ,

$$\frac{\partial}{\partial p_k} \left[\sum_{\mathbf{I}} p_k(\mathbf{I}) \log \frac{p_k(\mathbf{I})}{p_{k-1}(\mathbf{I})} + \lambda_k (\text{E}_{p_k}[r_j] - \text{E}_f[r_j]) \right.$$

$$\left. + \gamma \left(\sum_{\mathbf{I}} p_k(\mathbf{I}) - 1 \right) \right] = 0, \qquad (6.38)$$

we have

$$p_k(\mathbf{I}) = p_{k-1}(\mathbf{I}) \frac{1}{z_k} \exp\{-\lambda_k r_k(\mathbf{I})\}. \qquad (6.39)$$

$z_k = \text{E}_q \left[\exp\{\lambda_k r_k(\mathbf{I}_{D_k})\} \right]$ is a normalizing constant. This can be estimated by the negative samples,

$$z_k \approx \frac{1}{N} \sum_{i=1}^{N} e^{\lambda_k r(\mathbf{J}_i, D_k)}. \qquad (6.40)$$

λ_k is the parameter (Lagrange multiplier) to satisfy the constraint in Eq. (6.37),

$$\text{E}_p[r_k] \approx \frac{1}{N} \sum_{i=1}^{N} \left[r(\mathbf{J}_i, D_k) e^{\lambda_k r(\mathbf{J}_i, D_k)} \right] \frac{1}{z_k} = \bar{r}_k^+. \qquad (6.41)$$

In computation, we can look up $\bar{r}_k^+$ in the table to find the best λ_k. The importance sampling is a good estimation in calculating λ_k and z_k because in our model r is one-dimensional.

By recursion, we have a factorized log-linear form,

$$p_K(\mathbf{I}) = q(\mathbf{I}) \prod_{j=1}^{K} \left[\frac{1}{z_j} \exp\{\lambda_j r_j(\mathbf{I}_{D_j})\} \right].$$
(6.42)

Example: Vector Fields for Human Hair Analysis and Synthesis

In this subsection, we demonstrate an example of applying the primal sketch model to human hair analysis and synthesis [25]. Specifically, the hair images can be treated as 2D piecewise smooth vector (flow) fields, and thus the representation is view-based in contrast to the physically based 3D hair models in graphics. The primal sketch model has three levels. The bottom level is the high-frequency band of the hair image. The middle level is a piecewise smooth vector field for the hair orientation, gradient strength, and growth directions. The top level is an attribute sketch graph for representing the discontinuities in the vector field. Besides the three-level representation, the shading effects, i.e., the low-frequency band of the hair image, are modeled by a linear superposition of some Gaussian image basis functions, and the hair color is encoded by a color map. Figure 6.9 shows an example.

Let $\mathbf{I}^{obs}$ denote an observed color hair image. By a Luv transform, we obtain an intensity image $\mathbf{I}_Y^{obs}$ and a color channel image $\mathbf{I}_{UV}^{obs}$. The color channel $\mathbf{I}_{UV}^{obs}$ is discretized into a small number of colors and represented by a color map from the intensity $[0, 255]$ of $\mathbf{I}_Y^{obs}$ to a color. The intensity image $\mathbf{I}_Y^{obs}$ is further decomposed into a low-frequency band $\mathbf{I}_L^{obs}$ for illumination and shading with a low-pass Gaussian filter and the remaining. The high-frequency band is the texture for the hair pattern $\mathbf{I}_H^{obs}$. The low-frequency band is simply represented by a linear superposition of Gaussian image basis functions plus a mean intensity μ,

$$\mathbf{I}_L^{obs}(x, y) = \mu + \sum_{i=1}^{K_L} \alpha_i G(x - x_i, y - y_i; \theta_i, \sigma_{xi}, \sigma_{yi}) + \text{noise}.$$
(6.43)

Usually $K_L = O(10)$. Each Gaussian basis function is represented symbolically by an ellipse for editing and it has five parameters for the center, orientation, and standard deviation along the two axes. The coefficients $\{\alpha_i\}$ can be positive or negative for highlights and shadows, respectively. The matching pursuit algorithm is used to automatically extract the coefficients from the input image.

Our study is focused on the texture appearance $\mathbf{I}_H^{obs}$ with a three-level primal sketch model. A hair texture $\mathbf{I}_H$ on a lattice D is generated by a hidden layer $\mathbf{V}$—the vector field for hair growth flow, and $\mathbf{V}$ is in turn generated by an attribute hair sketch $\mathbf{S}$ which is a number of sketch curves representing the boundaries of hair strands and wisps with direction $\mathbf{d}_S$.

$$\text{Sketch } (\mathbf{S}, \mathbf{d}_S) \xrightarrow{\Delta_{sk}} \text{Vector field } \mathbf{V} \longrightarrow \text{hair image } \mathbf{I}_H,$$
(6.44)

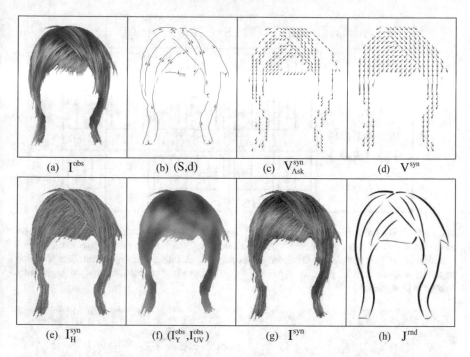

Fig. 6.9 Reprinted with permission from [25]. Example of hair model and inference. (**a**) is an input color image $\mathbf{I}^{obs}$. (**b**) is the computed sketch $\mathbf{S}$ with directions $\mathbf{d}_S$. (**c**) is the sketchable vector field $\mathbf{V}_{D_{sk}}$ generated from $(\mathbf{S}, \mathbf{d}_S)$. (**d**) is the overall vector field $\mathbf{V}$ after filling in non-sketchable part. (**e**) is the high-frequency hair texture image $\mathbf{I}_H^{syn}$ generated from the vector field. (**f**) is the shading and lighting image. (**g**) is the synthesized color image $\mathbf{I}^{syn}$ after adding the shading and color. We render an artistic sketch $\mathbf{J}^{rnd}$ in (**h**)

where Δ_{sk} is a dictionary of sketch primitives shown in Fig. 6.10. Each primitive is a rectangular window (say 5×7 pixels) and some examples are shown in Fig. 6.11.

For synthesizing hair images, we assume that $(\mathbf{S}, \mathbf{d}_S)$ is either inferred from a hair image or edited manually through a simple user interface. From $(\mathbf{S}, \mathbf{d}_S)$, we synthesize a hair image $\mathbf{I}_H^{syn}$ in three steps according to the generative model.

1. Synthesizing the vector field from the sketch $\mathbf{V}^{syn} \sim p(\mathbf{V}|\mathbf{S}, \mathbf{d})$.
2. Synthesizing the hair texture from the vector field $\mathbf{I}_H^{syn} \sim p(\mathbf{I}_H|\mathbf{V}^{syn})$.
3. Synthesizing color image $\mathbf{I}^{syn}$ by adding a shading image $\mathbf{I}_L^{syn}$ to $\mathbf{I}_H^{syn}$ and then transferring the grey image to color by the color map. Figure 6.12 shows three examples of hair synthesis.

The inference algorithm is divided into two stages: (i) compute the undirected orientation field and sketch graph from an input image and (ii) compute the hair growing direction for the sketch curves and the orientation field using a Swendsen–Wang cut algorithm. Both steps maximize a joint Bayesian posterior probability. This generative primal sketch model provides a straightforward way for synthesizing realistic hair images and stylistic drawings (rendering) from a sketch graph and a few Gaussian basis functions.

Primitives	(a)	(b)	(c)		(d)	(e)
orientation fields	$\|$	$-\|$	$\|-$		$-\|-$	$\|\|\|$
vector fields with (d_l, d_r)	↑ (1,1)	→ (1,1)	↑→ (1,1)	↑← (1,-1)	←→ (1,1)	↑↑ (1,1)
	↓ (1,-1)	← (1,-1)	↓→ (-1,1)	↓← (-1,-1)		↓↓ (-1,-1)

Fig. 6.10 Reprinted with permission from [25]. Five primitives for the orientation field and eleven primitives for the directed vector field **V** in a dictionary Δ_{sk}. (**a**) Side boundary. (**b**) Source (origin) or sink (end) of hair strands. (**c**) Occluding boundary. (**d**) Dividing line. (**e**) Stream line. The line segments and arrows in the primitive windows show the canonical orientations, and the angles may change in $[-\pi/6, \pi/6]$

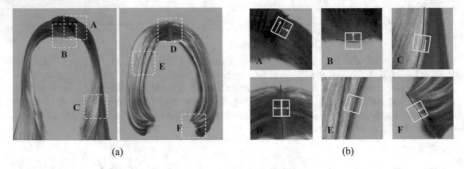

(a)   (b)

Fig. 6.11 Reprinted with permission from [25]. (**a**) Windows A-F are 6 primitive examples. (**b**) Zoomed-in views of the six windows

6.4   HoG and SIFT Representations

In this section, we will draw some connections between the primal sketch and Histogram of Oriented Gradients (HOGs) [34] and Scale-Invariant Feature Transform (SIFT) [158, 159] representations, which were widely adopted in computer vision as generic image features and object templates before deep neural networks.

Histogram of Oriented Gradient (HOG) [34] is based on evaluating normalized local histograms of image gradient orientations in a dense grid. The basic idea is that local object appearance and shape can often be characterized rather well by the distribution of local intensity gradients or edge directions, even without precise knowledge of the corresponding gradient or edge positions. In practice, this is implemented by dividing the image window into small spatial regions ("cells"),

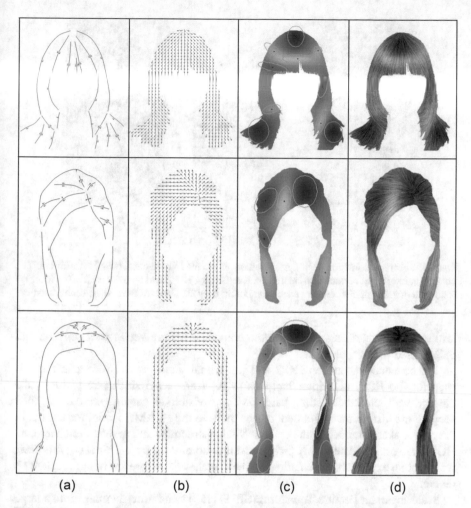

| (a) | (b) | (c) | (d) |

Fig. 6.12 Reprinted with permission from [25]. Three examples of hair drawing and synthesis. (**a**) Manually input hair sketch **S** with directions $\mathbf{d}_S$. (**b**) Synthesized vector field $\mathbf{V}^{\mathrm{syn}}$ given $(\mathbf{S}, \mathbf{d}_S)$. (**c**) Edited shading maps with a small number of ellipses. (**d**) Synthesized color images $\mathbf{I}^{\mathrm{syn}}$

for each cell accumulating a local 1D histogram of gradient directions or edge orientations over the pixels of the cell (Fig. 6.13). The combined histogram entries form the representation. For better invariance to illumination, shadowing, etc., it is also useful to contrast-normalize the local responses before using them. This can be done by accumulating a measure of local histogram "energy" over somewhat larger spatial regions ("blocks") and using the results to normalize all of the cells in the block. The normalized descriptor blocks are usually referred to as Histogram of Oriented Gradient (HOG) descriptors. A typical application of HOG descriptors is human detection, which is achieved by tilting the detection window with a dense

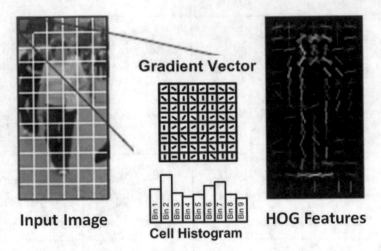

Fig. 6.13 Feature extraction with the histogram of oriented gradients (HOGs) descriptors. The image window is first divided into small spatial regions ("*cells*"), and then for each cell a local 1D histogram of gradient directions or edge orientations over the pixels of the cell is accumulated

grid of HOG descriptors and using the combined feature vector in a conventional SVM-based window classifier.

The connection between HOG and the primal sketch is as follows. If the cell used for the HOG descriptor happens to be a fine-scale stochastic texture (like the sky, wall, clothes, shading patch, with no visible sketchable structures), then keeping the histogram of gradient corresponds to the FRAME model for texture in the primal sketch model. If the cell has visible structures and is thus sketchable, its HOG histogram will be highly picked and thus dominated by 1-2 bins. For this case, we adopt an explicit representation by the sketches (structure) in the primal sketch model.

Scale-Invariant Feature Transform (SIFT) [159] transforms an image into a large collection of local feature vectors, each of which is invariant to image translation, scaling, and rotation, and partially invariant to illumination changes and affine or 3D projection. It consists of four major stages: (1) scale-space peak selection, (2) keypoint localization, (3) orientation assignment, and (4) keypoint descriptor. In the first stage, potential interest points are identified by scanning the image over location and scale. This is implemented efficiently by constructing a Gaussian pyramid and searching for local peaks (termed keypoints) in a series of difference-of-Gaussian (DoG) images. In the second stage, candidate keypoints are localized to sub-pixel accuracy and eliminated if found to be unstable. The third stage identifies the dominant orientations for each keypoint based on its local image patch. The assigned orientation(s), scale, and location for each keypoint enable SIFT to construct a canonical view for the keypoint that is invariant to similarity transforms. The final stage builds a local image descriptor for each keypoint, based on a patch of pixels in its local neighborhood.

Fig. 6.14 Reprinted with permission from [159]. The training images for two objects are shown on the left. These can be recognized in a cluttered image with extensive occlusion, shown in the middle. The results of recognition are shown on the right. A parallelogram is drawn around each recognized object showing the boundaries of the original training image under the affine transformation solved during recognition. Smaller squares indicate the keypoints that were used for recognition

SIFT can be applied to object recognition, which is performed by first matching each keypoint independently to the database of keypoints extracted from training images. Many of these initial matches will be incorrect due to ambiguous features or features that arise from background clutter. Therefore, clusters of at least 3 features are first identified that agree on an object and its pose, as these clusters have a much higher probability of being correct than individual feature matches. Then, each cluster is checked by performing a detailed geometric fit to the model, and the result is used to accept or reject the interpretation. Figure 6.14 shows an example of object recognition using SIFT for a cluttered and occluded image containing 3D objects. The training images of a toy train and a frog are shown on the left. The middle image contains instances of these objects hidden behind others and with extensive background clutter so that detection of the objects may not be immediate even for human vision. The image on the right shows the final correct identification superimposed on a reduced contrast version of the image. The keypoints that were used for recognition are shown as squares with an extra line to indicate orientation. The sizes of the squares correspond to the image regions used to construct the descriptor. An outer parallelogram is also drawn around each instance of recognition, with its sides corresponding to the boundaries of the training images projected under the final affine transformation determined during recognition.

The SIFT keys can be considered an extension to the keypoints in images, like color or complex object patterns. It corresponds to a local sub-graph in the primal sketch, with some scale invariance.

Chapter 7
2.1D Sketch and Layered Representation

Primal sketch is a generic two-layer 2D representation in terms of how image content is explained away with respect to either explicit basis functions or feature statistics. Primal sketch seeks to decompose an image domain into the texton structural domain and the remaining texture domain. In this chapter, we shall study how to decompose an input image into multiple layers with partial occluding order relation inferred and the occluded contour completed if possible (Fig. 7.1). In the pioneering work, the resulting representation is called *the 2.1D sketch*, by Nitzberg and Mumford [185], to bridge low-level image primitives (including textons such as edges and junctions and texture atomic regions) and middle-level Marr's 2.5D sketch (to be presented in the next chapter), and termed *layered image representation*, by Adelson and Wang [2, 245, 246], in image coding and motion analysis.

In practice, the idea of layered representation has been widely used for image manipulation in image editing software such as Adobe's Photoshop (e.g., adding text to an image, or adding vector graphic shapes, or applying a layer style to add a special effect such as a drop shadow or a glow). In the literature of computer vision, the 2.1D sketch or layered image representation has also been studied from other perspectives including line drawing interpretations [162, 243], segmentation [53, 232], occlusion recovery [211, 251], contour illusory and completion [50, 51, 106, 200, 244], and figure–ground separation [201, 225].

The 2.1D sketch stems from the figure–ground separation–organization problem, a type of perceptual grouping that contributes to visual object recognition. In Gestalt psychology, the figure–ground separation problem is usually posed as identifying a *figure* (such as the two girls in Fig. 7.1) from the *background* (such as the dancing practice room in Fig. 7.1). Computing the figure–ground organization of an input image can help resolve perceptual ambiguities in, for example, the face-vase and martini-bikini drawings. The 2.1D sketch captures the partial occluding order between multiple object surfaces/generic regions in a scene, representing the rank information of relative depth among them, and thus providing a critical

S.-C. Zhu, Y. N. Wu, *Computer Vision*, https://doi.org/10.1007/978-3-030-96530-3_7

(a) input image   (b) curve completion at layer 2   (c) curve completion at layer 3

(d) layer 1   (e) layer 2 after fill-in   (f) layer 3 after fill-in

Fig. 7.1 Illustration of the 2.1D sketch and layered image representation. From a 2D sketch to a 2.1D layered representation by reconfiguring the bond relations between regions in different layers. (**a**) is an input image from which a 2D primal sketch is computed. This is transferred to a 2.1D sketch representation with three layers shown in (**d**), (**e**), and (**f**), respectively. The inference process reconfigures the bonds of the image primitives shown in red in (**b**) and (**c**)

representation scheme for addressing the multi-stable perception phenomenon in vision.

In this chapter, we focus on the problem of computing the 2.1D sketch from a monocular image under the variational formulation framework, the energy minimization framework, and the Bayesian inference framework, respectively.

7.1 Problem Formulation

In this section, we introduce the notation and present a high-level formulation. We will elaborate on different components in the sections followed.

Denote by $\mathcal{D}$ an image domain and $\mathbf{I}$ an image defined on the domain $\mathcal{D}$. To infer the 2.1D sketch, we will build up a three-layer model from the input image $\mathbf{I}$, to its 2D representation, denoted by W_{2D}, and to the 2.1D sketch, denoted by $W_{2.1D}$. We have

$$\mathbf{I} \Rightarrow \quad W_{2D} \Rightarrow \quad W_{2.1D}. \tag{7.1}$$

W_{2D} represents a set of 2D elements to be layered, which can be 2D atomic regions or 2D curves and curve groups, as well as the associated attributes if needed in inference. For example, if we only consider 2D regions, we have

Fig. 7.2 Reprinted with permission from [236]. A Hasse diagram for a partial order relation

$$W_{2D} = \left\{ n, (R_i, l_i, \theta_i)_{i=1}^{n} \right\}, \tag{7.2}$$

which consists of n regions, R_i, each of which is represented by a region model indexed by l_i in a predefined model family with parameters θ_i.

$W_{2.1D}$ represents a set of surfaces, their partial occluding order, and contours used to complete regions in different surfaces. A surface consists of one 2D region or more than one 2D region with completed contours. Any 2D region in W_{2D} belongs to one and only one surface in the layered representation. We have

$$W_{2.1D} = \{m, \{S_i\}_{i=1}^{m}, \mathcal{PR}\}, \tag{7.3}$$

which consists of m surfaces ($m \leq n$) and $\cup_i S_i = \mathcal{D}$. $\mathcal{PR}$ represents the partial occluding order between the set of m surfaces. Consider a set $A = \{a, b, c, d, e, f\}$, and we define a partially ordered set, *poset* [220], $\mathcal{PR} = \langle A, \preceq \rangle$. $b \preceq c$ means that surface b occludes surface c or b is on top of c. $\mathcal{PR}$ is represented by a directed acyclic graph (DAG) called a Hasse diagram. Figure 7.2 shows an example of the Hasse diagram for $\mathcal{PR} = \{\langle a, b \rangle, \langle b, d \rangle, \langle a, d \rangle, \langle a, c \rangle, \langle c, d \rangle, \langle e, f \rangle\}$ on the set A.

Denote the "visible" portion of a surface S_i by

$$S_i' = S_i \setminus \cup_{S_j \preceq S_i} S_j. \tag{7.4}$$

Then, the recovered curve(s) in the surface S_i can be defined by

$$C_i = \partial S_i \setminus \partial S_i'. \tag{7.5}$$

Some C_i's may be empty. For example, the front-most surface does not have occluded parts, and the occluded parts in the background are usually left open.

To infer the 2.1D sketch, we will study two different formulations that cast the inference under the energy minimization framework and under the Bayesian framework, respectively. In either case, there are two different assumptions:

- Assume the 2D representation W_{2D} has been computed already and will be fixed in the inference of the 2.1D sketch. This is typically adopted in the literature.
- Compute the 2D representation W_{2D} and the 2.1D sketch $W_{2.1D}$ jointly. This leads to a much larger search space and usually requires more powerful search algorithms such as DDMCMC [235].

7.2 Variational Formulation by Nitzberg and Mumford

In their seminal work [185], Nitzberg and Mumford proposed a variational formulation for inferring the 2.1D sketch extended from the Mumford–Shah energy functional [177]. This belongs to the energy minimization framework, and we briefly introduce the formulation in this section.

The Energy Functional

We first overview the Mumford–Shah energy functional for low-level image segmentation. It is a piecewise smooth model that aims to segment an image into as few and simple regions as possible while keeping the color of each region as smooth and/or slowly varying as possible. The functional is defined such that it takes its minimum at an optimal piecewise smooth approximation to a given image $\mathbf{I}$ defined on the domain $\mathcal{D}$. An approximation function, denoted by f, is smooth except at a finite set Γ of piecewise contours that meet the boundary of the image domain, $\partial \mathcal{D}$, and meet each other only at their endpoints. So, the contours of Γ segment the image domain into a finite set of disjoint regions, denoted by $R_1, \cdots, R_n$, i.e., the connected components of $\mathcal{D} \setminus \Gamma$. The Mumford–Shah functional is defined to measure the match between an image $\mathbf{I}$ and a segmentation f, Γ:

$$E_{M-S}(f, \Gamma | \mathbf{I}) = \mu^2 \int_{\mathcal{D}} (f - \mathbf{I})^2 dx + \int_{\mathcal{D} \setminus \Gamma} \|\nabla f\|^2 dx + v \int_{\Gamma} ds, \qquad (7.6)$$

where on the right-hand side the first term measures how good f approximates $\mathbf{I}$, the second asks that f varies slowly except at boundaries, and the third asks the set of contours to be as short, and hence as simple and straight as possible.

Similar in spirit to the Mumford–Shah functional, Nitzberg and Mumford proposed an energy functional that achieves a minimum at the optimal overlapping layering of surfaces. For simplicity, assume S_i is a closed subset of $\mathcal{D}$ with piecewise smooth boundary and connected interior. The energy functional is defined by

$$E_{2.1D}(\{S_i\}, \mathcal{PR} | \mathbf{I}) = \sum_{i=1}^{n} \left\{ \mu^2 \int_{S_i'} (\mathbf{I} - m_i) dx + \epsilon \int_{S_i} dx + \int_{\partial S_i \setminus \partial \mathcal{D}} \phi(\kappa) ds \right\},$$
$$(7.7)$$

where m_i is the mean of the image $\mathbf{I}$ on S_i' and κ is the curvature of the boundary ∂S_i. The function $\phi : \mathbf{R} \to \mathbf{R}$ is defined by

$$\phi(\kappa) = \begin{cases} v + \alpha \kappa^2, & \text{for } |\kappa| < \beta/\alpha \\ v + \beta|\kappa|, & \text{for } |\kappa| \geq \beta/\alpha. \end{cases} \qquad (7.8)$$

The functional above can only address inferring the 2.1D sketch from simple input images, which is a reasonable assumption in the early day of computer vision. First, the model does not allow self-overlapping "woven" surfaces such as what a garden hose would project, nor folded surfaces such as that produced by an image of a sheet which also disappears around the back of an arm. Second, the model is piecewise constant with a constant mean being used for the surface model. Third, transparency and shadows are not handled. We will study more expressive models with more powerful inference algorithms for handling those issues frequently observed in modern real images.

The Euler Elastica for Completing Occluded Curves

To recover the occluded curves, Nitzberg and Mumford adopted the Euler Elastica method to interpolate them. Suppose a surface S_i disappears behind occluding objects at a point $P_0 \in \partial S_i$ and reappears at P_1. Let t_0 and t_1 be the unit tangent vectors to ∂S_i at P_0 and P_1, respectively. Then, computing the occluded curve C_i between P_0 and P_1 is posed as a minimization problem where the energy functional is

$$E(C_i | S_i, P_0, t_0, P_1, t_1) = \int_{C_i} (v + \alpha \kappa^2) ds. \tag{7.9}$$

Computationally, the simplest way to solve the Elastica seems to be hill-climbing. It starts with a convenient chain $x_0 = P_0, x_1, \cdots, x_N = P_1$ of points for which $x_1 - x_0 \propto t_0$ and $x_{N-1} - x_N \propto t_1$. X_i's $(i = 1, \cdots, N-1)$ are computed by letting them evolve to decrease the Elastica functional.

7.3   Mixed Markov Random Field Formulation

In this section, we address the 2.1D sketch problem with the mixed Markov random field (MRF) representation and under the Bayesian inference framework. We consider 2D image regions only as layering primitives. As illustrated in Fig. 7.3, given the set of segmented regions (in different colors) of an input image, our objective consists of two components: *region layering or coloring* to divide the set of regions into an unknown number of layers, and *contour completion* to recover the occluded contours between regions in the same layer. In addition, we will also fit the probability models of regions in different layers and preserve multiple distinct interpretations accounting for the intrinsic ambiguities if needed. Two key observations in modeling and computing the 2.1D sketch problem are as follows:

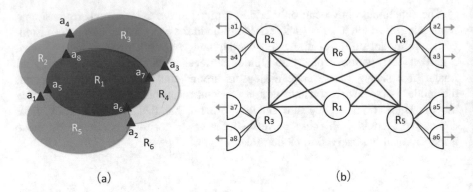

Fig. 7.3 A simple example of the 2.1D sketch and the mixed MRF modeling. (**a**) is a set of 2D regions to be layered; (**b**) is the initial reconfigurable graph with mixed MRF including two types of nodes, region nodes and terminator nodes. See text for details

(i) *Long-range interactions* between 2D regions. In order to determine which 2D regions are in the same layer and to complete the contour of occluded parts, long-range interactions are entailed, which imposes different requirements in modeling than traditional local relation modeling using Markov random field (MRF) models. Consider regions 3 and 5 in Fig. 7.1. Whether they should merge into a single surface or should remain independent in the same layer depends on their compatibility across long-range interactions.

(ii) *Dynamic neighborhood system.* In the contour completion process, given a current layering assignment, the neighborhood of any endpoint on an occluded contour looking for its corresponding point is constrained to match among the set of points in the same layer. This means the neighborhood system changes according to different layering assignments at each step in the inference, which is different from the fixed neighborhood system in the traditional MRF.

The two properties stated above lead traditional MRFs to fail to model the 2.1D sketch problem and call for new methods, such as the mixed Markov random field [61, 63] to be used in this chapter. Both MRF and mixed MRF can be presented by graphical models. A mixed MRF differs from traditional MRFs in its definition of its neighborhood system, which is static in the latter but dynamic in the former due to the introduction of a new type of nodes in the graph. Concretely speaking, a mixed MRF has the following two characteristics:

(i) Nodes are inhomogeneous with different degrees of connections, which are inferred from images on the fly.

(ii) The neighborhood of each node is no longer fixed but inferred as open bonds or address variables, which reconfigures the graph structure.

Following Eq. (7.1), we adopt a three-layer generative image model consisting of the input image, the 2D representation, and the 2.1D sketch. We use 3-degree junctions such as T-junctions, Y-junctions, or arrow junctions as cues for occlusion.

The 2D representation consists of a set of 2D regions (see $R_1, R_2, \cdots, R_6$ in Fig. 7.3a) and a set of terminators (see $a_1, a_2, \cdots, a_8$) broken from detected T-, Y-, or arrow junctions during the process of layering and, if possible, to be completed in the inferred layered representation. So, in the graphical representation (see Fig. 7.3b), there are two types of nodes: one consists of the region nodes and the other terminator nodes. Region nodes constitute the region adjacency graph to be partitioned during the process of layering. The terminator nodes make the graph reconfigurable since the neighborhood system of terminator nodes is not static and depends on the current assignment of layering. For such reconfigurable graphical models, it has been shown that a probability model defined on them still observes a suitable form of the Hammersley–Clifford theorem and can be simulated by Gibbs sampling [61].

Definition of W_{2D} and $W_{2.1D}$

For the clarity of the presentation, we mainly focus on inferring the 2.1D sketch with the 2D representation given and fixed.

The 2D Representation W_{2D} consists of a set of 2D regions, V_R, and a set of 3-degree junctions, V_T, such as T-, Y-, or arrow junctions:

$$W_{2D} = (V_R, V_T) \tag{7.10}$$

$$V_R = (R_1, R_2, \ldots, R_N) \tag{7.11}$$

$$V_T = (T_1, T_2, \ldots, T_M). \tag{7.12}$$

V_R is the regions set, which can be many atomic regions or composed of a few image patches. V_T consists of those junctions selected from the detected 3-degree junctions [252] that are assigned to the corresponding regions.

Open Bonds or Address Variables 3-degree junctions will be broken into a set of terminators, illustrated in Fig. 7.4a and b for a T-junction, as the open bonds, or address variables, V_B, of corresponding regions during the inference. The open bonds are like pointers to regions, initially kept open but to be assigned an address variable and completed during inference. In practice, each bond has a set of attributes, A_B, including those belonging to itself and those inherited from the region. These attributes often include geometric transformation features, such as location, orientation, and length, and some appearance features such as features computed from region models. They are used to test the compatibility of any two bonds, deciding whether or not to link together. Often, besides the foreground and background regions, each region has two or more (assumed no more than $m \leq M$) open bonds with their ownerships defined, denoted as $B(R_i), \forall i \leq N$.

Fig. 7.4 The open bounds or address variable. A T-junction is shown in (**a**); it is broken into a terminator in (**b**) as an open bound described as an address variable including location (x, y), the orientation of the terminator, and appearance information inherited from the region it belongs to; (**c**) is the Elastica computed using two terminators in the process of contour completion

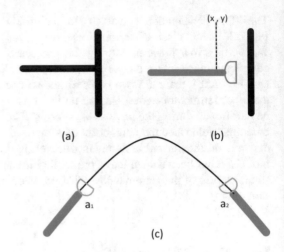

$$V_B = (a_1, a_2, \ldots, a_M) \tag{7.13}$$

$$A_B = (A_{a1}, A_{a2}, \ldots, A_{aM}), a_i \in V_B, \forall i \leq M \tag{7.14}$$

$$B(R_i) = (a_{i1}, a_{i2}, \ldots, a_{im}), a_{ij} \in V_B, \forall i \leq N, \forall j \leq m. \tag{7.15}$$

Generative Models of Regions   We adopt generative models for each region. Let $R \subset \Lambda$ denote a 2D region and $\mathbf{I}_R$ the intensities in R or color values in (r, g, b) color space. The model assumes constant intensity or color value with additive noise η modeled by a non-parametric histogram H. The model can be learned offline, and more sophisticated models can easily be added to the algorithm.

$$\mathbf{J}(x, y, \theta) = \mu, \mathbf{I}_R(x, y) = \mathbf{J}(x, y, ; \theta) + \eta, \eta \sim H. \tag{7.16}$$

With a slight abuse of notation, we denote the parameters used in a region by $\theta = (\mu, H)$. Here μ is the mean of a region R or a connected component CP. The likelihood is

$$p(\mathbf{I}_R | R, \theta) \propto \prod_{(x, y \in R)} H(\mathbf{I}_R(x, y) - \mathbf{J}(x, y, \theta)). \tag{7.17}$$

The prior for a region R assumes a short boundary length ∂R. This is to encourage smoothness, and a compact area $|R|$,

$$p(R) \propto \exp \left\{ -\gamma_r |R|^\rho - \frac{1}{2}\lambda |\partial R| \right\}, \tag{7.18}$$

where ρ and λ are fixed constants and γ is a fixed scale factor for regions.

Terminator Representation and the Elastica When a terminator a_i is broken from a junction, it is represented by its attributes $A_{ai} = (x, y; ori, \kappa, p_f)$, where (x, y) is the location, ori is the orientation of the terminator, κ is the curvature, and p_f is the profile or some attributes inherited from the region.

Given two terminators, a_i and a_j, the common to be completed between them denoted by Γ^*, is decided by minimizing the Elastica cost function in a contour space Ω_Γ [110, 129, 177, 185]

$$\Gamma^* = \arg\min_{\Gamma \in \Omega_\Gamma} E(\Gamma; a_i, a_j)$$

$$= \arg\min_{\Gamma \in \Omega_\Gamma} \int_\Gamma \left[\left(v_1 + \alpha_1 \kappa_1^2\right) + \left(v_2 + \alpha_2 \kappa_2^2\right) \right] ds, \tag{7.19}$$

where κ_1 is the curvature of a_1, κ_2 is the curvature of a_2, v_1 and v_2 are constants, and α_1 and α_2 are scalable coefficients. The parameters v_1, v_2, α_1, and α_2 are learned from an image dataset [273].

The 2.1D Sketch $W_{2.1D}$ is represented by a set of labels, X, for the layer information of regions and a set of variables, Y, describing address variable assignments. Both of them are inferred from the image. In addition, according to the layering labels and assignments of address variables, a set of surfaces, S_f, is formed, which consists of one or more regions merged through a set of recovered contours, C_t.

$$W_{2.1D} = (X, Y; S_f, C_t) \tag{7.20}$$

$$X = (x_{R_1}, x_{R_2}, \ldots, x_{R_N}) \tag{7.21}$$

$$Y = (y_{a_1}, y_{a_2}, \ldots, y_{a_M}) \tag{7.22}$$

where $x_{Ri} \leq K, \forall i \in [1, N]$, $y_{aj} \in V_B$, $a_j \in V_B, \forall j \in [1, M]$, K is an unknown number of layers to be inferred.

X represents the partition of V_R into K layers with the partial occlusion relations represented in a Hasse diagram along with Y. The assignments among address variables indicate to whom the open bonds a_i are connected or assigned. Each terminator has the same label for its layer information as the region to which it is assigned.

Because the surface is merged from regions so that its generative model includes two constraints: one is appearance defined as the region model and the other shape constrained by some generic shape priors (Elastica in this chapter) or some specific object templates. Recovered contours are computed using Elastica based on the results of assignments.

The Mixed MRF and Its Graphical Representation

Given the 2D representation defined above, there are two kinds of nodes in its graphical representation: region nodes (the nodes found in a traditional MRF) and terminator nodes (a newly introduced set of address nodes). Terminator nodes are dynamically broken from 3-degree junctions in the process of region layering, which means that their neighborhood systems are determined on the fly. According to the discussion stated above, we know that this will build up a reconfigurable graph with a mixed MRF as illustrated in Fig. 7.3b. We clarify some definitions in this section.

Let $G = \langle V, E \rangle$ be the graph for the 2D representation, where $V = V_R \cup V_B$ is the set of vertices and $E = E_R \cup E_B$ the set of edges. E_R is the set of edges linking regions into a region adjacency graph, and E_B is the dynamic set of edges linking open bonds.

The edges decide the topology of the graph and can link, generally speaking, any two vertices. In a mixed MRF, the introduction of address variables makes the neighborhood system dynamic and different from traditional graphical models.

A mixed MRF can be defined in the same way that an MRF is defined as a probability distribution that is factorized into a product of local terms, the only difference being in what "local" means after introducing address variables. The idea of "locality" in a mixed MRF can handle long-range interactions through open bonds, meaning that a clique denoted by C in a mixed MRF may contain both standard region nodes and address nodes.

Definition 1 (Mixed Neighbor Potential) Let C denote the set of cliques in G. A family of nonnegative functions λ_C is called a mixed neighbor potential if for any pair of configurations x and y, $x_C = y_C$ and $x_{x_a} = y_{y_a}, \forall a \in C \cap V_B$. Thus a mixed potential function λ_C depends on both the values of the standard nodes in C and that of those pointed to by open bonds in C.

Definition 2 (Mixed Markov Random Field) A probability distribution P defined on G is a mixed MRF if P can be factorized into a product of mixed potential functions over the cliques: $P(\mathbf{I}) \propto \prod_C \lambda_C(\mathbf{I}_C, \mathbf{I}_{\mathbf{I}_C})$, where $\mathbf{I}_{\mathbf{I}_C}$ is the vector of states of those standard variables pointed to from within the clique C.

Equivalence Between the Mixed MRF and the Gibbs Distribution It was shown that the originally established equivalence between MRFs and the Gibbs distribution by Hammersley–Clifford theorem is also applicable for mixed MRFs so that the probability models defined on a mixed MRF can be simulated by Gibbs sampling [61] or SW cuts [9] according to Friedman's proof in [61].

Given a region R_i, its neighborhood is

$$N(R_i) = N_{\text{adjacency}}(R_i) \cup N_{\text{pointer}}(R_I), \tag{7.23}$$

where $N_{adjacency}(R_i) = \{R_j : \forall R_j \in V_R \text{ adjacent to } R_i\}$ represents the usual definition of a neighborhood found in MRF models and defines an adjacency graph. At the same time, open bonds of R_i will cause it to link to those regions that are not locally adjacent, that is, $N_{pointer}(R_i) = \{R_j : B(R_j) \text{ and } B(R_i) \text{ are connected}\}$. Then R_j is

$$E_R = \{\langle R_i, R_j \rangle, \forall R_i \in V_R \text{ and } R_j \in N(R_i)\}. \tag{7.24}$$

Given an open bond $a_i \in V_B$, its neighborhood is

$$N(a_i) = \{a_j : a_j \text{ and } a_i \text{ are in the same layer}\}. \tag{7.25}$$

Hence, $E_B = \langle a_i, a_j \rangle, \forall a_i \in V_T$ and $a_j \in N(a_i)$. At the initial stage, all the open bonds are open.

Bayesian Formulation

In the Bayesian framework, the three-layer model in Eq. (7.1) is described as

$$p(\mathbf{I}, W_{2D}, W_{2.1D}) = p(\mathbf{I}|W_{2D}, W_{2.1D})p(W_{2D}|W_{2.1D})p(W_{2.1D}), \tag{7.26}$$

where $p(\mathbf{I}|W_{2D}, W_{2.1D})$ is the likelihood model. We want to maximize the posterior joint probability of $W_{2.1D}$ given W_{2D} in a solution space $\Omega_{W2.1D}$

$$\begin{aligned} W_{2.1D}^* &= \arg\max_{\Omega_{W_{2.1D}}} p(W_{2.1D}|W_{2D}; \mathbf{I}) \\ &= \arg\max_{\Omega_{W_{2.1D}}} p(W_{2D}|W_{2.1D})p(W_{2.1D}). \end{aligned} \tag{7.27}$$

Graph Partition Perspective Given the graphical representation G defined with a mixed MRF, we are interested in a partitioning, or coloring, of the vertex, i.e., $V = V_R \cup V_B$, in G. An n-partition is denoted by

$$\pi_n = (V_1, V_2, \ldots, V_n), \bigcup_{i=1}^{n} V_i = V, V_i \bigcap V_j =, \forall i \neq j. \tag{7.28}$$

Each subset $V_i, i = 1, 2, \ldots, n$ (that is a surface in 2.1D representation), is assigned a color c_i that represents its model. For region nodes, this model consists of layering information, and for open bonds, the model consists of the connected contours. Let $\Omega_{\pi_n}(\pi_n \in \Omega_{\pi_n})$ be the space of all possible n-partitions of V, Ω_{lr} the set of types of region models, $\Omega_{\theta r}$ the model parameter family, Ω_{lc} the set of types of Elastica, and $\Omega_{\theta c}$ the Elastica parameter space. Thus, the solution space for $W_{2.1D}$ is

$$\Omega = \bigcup_{n=1}^{N} \left\{ \Omega_{\pi_n} \times \Omega_{l_r}^n \times \Omega_{\theta_{r1}} \times \ldots \times \Omega_{\theta_{rn}} \times \Omega_{l_c}^n \times \Omega_{\theta_{c1}} \times \ldots \times \Omega_{\theta_{cn}} \right\}. \qquad (7.29)$$

This leads us to extend the Swendsen–Wang Cuts algorithm [9] to perform the inference. In the formulation, the prior model is

$$p(W_{2.1D}) = p(K)p(X)p(Y) \prod_{i=1}^{N} p(R_i), \qquad (7.30)$$

where $p(K)$ is an exponential model $p(K) \propto \exp(-\lambda_0 K)$, $p(X) = \prod_{i=1}^{N} p(X_{R_i})$ and $p(Y) \propto \exp -\beta_0 \sum_{\forall a_i, a_j, i \neq j} 1(y_{a_i} = y_{a_j})$ penalize the situation that more than one terminator is assigned to the same junction and $p(R_i)$ is defined in Eq. (7.18).

The likelihood is

$$p(W_{2D}|W_{2.1D}) \propto \prod_{i=1}^{K} \prod_{j=1}^{K(i)} \prod_{(x,y) \in s_j} H(\mathbf{I}(x, y) - \mathbf{J}(x, y; \theta_j))$$

$$\times \prod_{i=1}^{M} \exp\{-\lambda_0 E\}, \qquad (7.31)$$

where $\mathbf{J}(x, y; \theta_j)$ is defined in Eq. (7.16) and λ_0 is a scalable factor and also learned from the dataset [273]. The first term in the likelihood handles the region coloring/layering problem, and the second term handles the contour completion problem. They are solved jointly.

Inference Algorithm   Based on the graphical representation with a mixed MRF, the presented inference algorithm proceeds in two ways: (1) region coloring/layering based on the Swendsen–Wang Cuts algorithm [9] for the partitioning of the region adjacency graph to obtain partial occluding order relations; (2) address variable assignments based on Gibbs sampling for the completion of open bonds. The basic goal is to realize a reversible jump between any two successive states π and π_0 in the Metropolis–Hastings method.

Experiments   We first demonstrate the inference algorithm using the Kanizsa image. Figure 7.5 shows the 2D representation and the graphical representation with mixed MRF. Figure 7.6 shows the inference procedure.

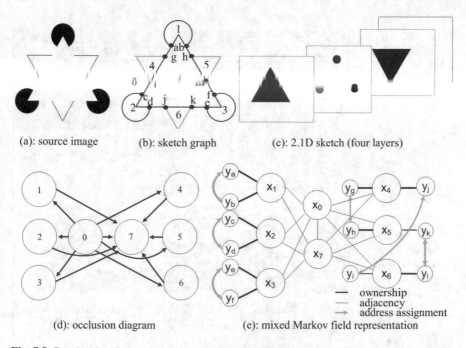

(a): source image   (b): sketch graph   (c): 2.1D sketch (four layers)

(d): occlusion diagram   (e): mixed Markov field representation

Fig. 7.5 Reprinted with permission from [73]. A running example of inferring the 2.1D sketch. (**a**) is the original image, Kanizsa figure. (**b**) is the sketch graph computed by the primal sketch model after interactively labeling. There are 8 atomic regions and 12 terminators broken from T-junctions. (**c**) is the 2.1D sketch. There are 4 layers, and the contour completion is performed using Elastica rules. (**d**) is the Hasse diagram for partial order relation, such as $\langle 7, 1 \rangle$ means region 7 occludes region 1. (**e**) is the graph representation with a mixed Markov random field. Each big circle denotes a vertex of the atomic region, each red bar denotes one terminator, and each little circle denotes a vertex of open bound described as address variable. Each region may have two or more terminators. The blue line segments connect the two neighboring regions. The green two-way arrows connect two terminators, and each terminator is assigned another terminator's address variable

7.4   2.1D Sketch with Layered Regions and Curves

In this section, we handle images with both regions and curves that are frequently observed in natural images as illustrated in Fig. 7.7. Given an input image, our objective is to infer an unknown number of regions, free curves, parallel groups, and trees, with recovered occlusion relation and their probability models selected and fitted—all in the process of maximizing (or simulating) a Bayesian posterior probability. This algorithm searches for optimal solutions in a complex state space that contains a large number of subspaces of varying dimensions for the possible combinations of regions, curves, and curve groups.

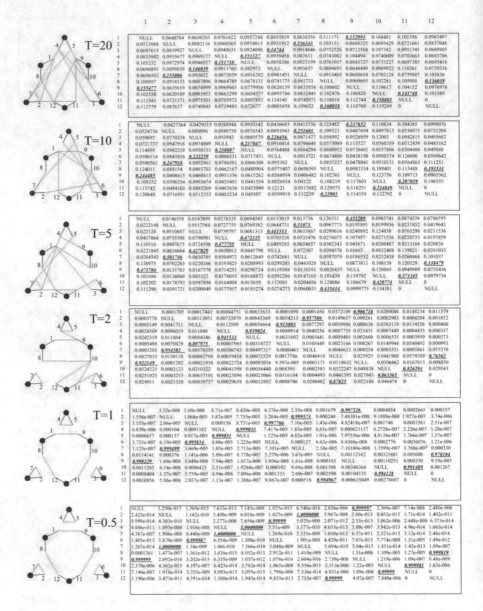

Fig. 7.6 Reprinted with permission from [73]. Illustration of the Gibbs inference procedure. Each red node denotes a region, the black line segment denotes the terminator, and the green line segment shows the inferred connection or assignment of address variables. Inference starts from an initial temperature $T = 20$, and $(a) \sim (h)$ are the results in different temperatures. After $T = 1$, the result is right as in the figure

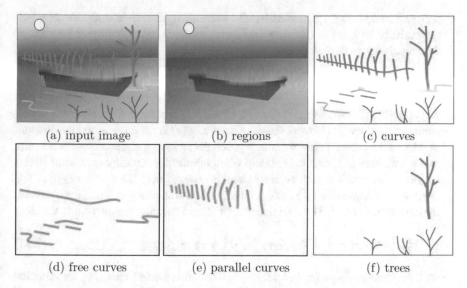

Fig. 7.7 Reprinted with permission from [236]. Illustration of the 2.1D sketch with a layered representation of regions and curves. (**a**) is an input image that is decomposed into two layers—(**b**) a layer of regions and (**c**) a layer of curves. These curves are further divided into (**d**) free curves, (**e**) a parallel curve group for the fence, and (**f**) trees. Curves observe a partial order occlusion relation

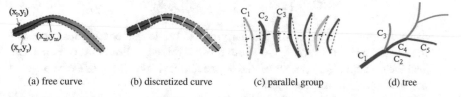

Fig. 7.8 Reprinted with permission from [236]. Representations of curves and curve groups. (**a**) A free curve in continuous representation. (**b**) A free curve is discretized into a chain of "bars." (**c**) Curves for a parallel group. (**d**) Curves for a Markov tree

Generative Models and Bayesian Formulation

In this section, we present generative models for both regions and curve structures.

Generative Models of Curves

We consider three types of curve models that are illustrated in Fig. 7.8 and described as follows.

1. **Free curves**. A free curve, denoted by C, is represented by its medial axis $\mathbf{c}_m(s) = (x_m(s), y_m(s))$ and its width $2w(s)$ for $s = [0, L]$. L is the curve length. In a continuous representation, a free curve C occupies an elongated

area or domain $\mathcal{D}(C)$ bounded by the left and right side boundaries denoted, respectively, by $\mathbf{c}_l(s) = (x_l(s), y_l(s))$ and $\mathbf{c}_r = (x_r(s), y_r(s))$. Figure 7.8a shows the boundaries in dashed lines.

$$\mathbf{c}_l(s) = \mathbf{c}_m(s) - w(s)\mathbf{n}(s), \mathbf{c}_r(s) = \mathbf{c}_m(s) + w(s)\mathbf{n}(s), \tag{7.32}$$

where $\mathbf{n}(s)$ is the unit normal of $\mathbf{c}_m(s)$. Intuitively, a curve is a degenerated region parameterized by its 1D medial axis. Usually, $w(s)$ is only $1 - 3$ pixels wide and $w \ll L$. This causes major topology problems in image segmentation where the two boundaries $\mathbf{c}_l(s)$ and $\mathbf{c}_r(s)$ could often intersect generating numerous trivial regions. This problem will be resolved with the explicit 1D representation. The intensities of a curve often exhibit globally smooth shading patterns, for example, the curves in Fig. 7.8. Thus we adopt a quadratic function for curve intensities

$$\mathbf{J}(x, y; \theta_0) = ax^2 + bxy + cy^2 + dx + ey + f, (x, y) \in \mathcal{D}(C), \tag{7.33}$$

with parameters $\theta_0 = (a, b, c, d, e, f)$. The validation of choosing an inhomogeneous model to capture the smoothly changing intensity patterns can be found in [235]. Therefore, a free curve is described by the following variables in the continuous representation

$$C = \left(L, \mathbf{c}_m(s)_{s=0}^L, w(s)_{s=0}^L, \theta_0, \sigma \right), \tag{7.34}$$

where σ is the variance of the intensity noise. While this continuous representation is a convenient model, we should also work on a discrete representation. Then the domain $\mathcal{D}(C)$ is a set of pixels in a lattice, and C is a chain of elongated bars as Fig. 7.8b illustrates.

The prior model for $p(C)$ prefers smooth medial axes, narrow and uniform width, and it also has a term for the area of the curve in order to match with the region prior.

$$p(C) \propto p(\mathcal{D}(C))p(\mathbf{c}(s))p(w(s)) \propto e^{-E(C)}. \tag{7.35}$$

The energy $E(C)$ is the sum of three terms

$$E(C) = \gamma_c|\mathcal{D}(C)|^\rho + \lambda L + E_o(w), \tag{7.36}$$

where ρ, λ are the constants and are fixed in our experiments, and γ_c is a scale factor that can be adjusted to control the number of curves. $E_o(w)$ is a term that constrains width $w(s)$ to be small. We denote the intensities inside the curve domain by $\mathbf{I}_{\mathcal{D}(C)}$ and assume the reconstruction residue follows i.i.d. Gaussian $\mathcal{N}(0; \sigma^2)$. The image likelihood therefore is

$$p(\mathbf{I}_{\mathcal{D}(C)}|C) = \prod_{(x,y)\in\mathcal{D}(C)} \mathcal{N}(\mathbf{I}(x, y) - \mathbf{J}(x, y; \theta_0); \sigma^2). \tag{7.37}$$

2. **Parallel curve groups**. A parallel curve group consists of a number of nearly parallel curves as Fig. 7.8c shows. Each curve C_i, $i = 1, 2, \ldots, n$, is summarized by a short line segment connecting its endpoints. They represent curve structures, such as zebra stripes, grids, and railings shown in the experiments. Grouping curves into a parallel group is encouraged in the model as it reduces the coding length, and it is useful for perceiving an object, for example, a zebra. We denote a parallel curve group by

$$pg = (n, \{C_1, C_2, \ldots, C_n\}, \{\alpha_1, \alpha_2, \ldots, \alpha_n\}), \tag{7.38}$$

where $\alpha_i \in \{1, \ldots, n\}$ is the index to the curve preceding C_i in the chain.

The prior model for a pg is a first-order Markov model in a Gibbs form with singleton energy on an individual curve and pairwise energy for two consecutive curves as

$$p(pg) \propto \exp\left\{-\lambda_0 n - \sum_{i=1}^{n} E(C_i) - \sum_{i=2}^{n} E_{pg}(C_i, C_{\alpha_i})\right\}. \tag{7.39}$$

The singleton $E(C_i)$ is inherited from the free curve model. For the pair energy, we summarize each curve C_i by five attributes: center (x_i, y_i), orientation θ_i of its associate line segment, length L_i of the line segment, curve average width (thickness) $\bar{w}_i$, and average intensity μ_i. $E_{pg}(C_i, C_{\alpha_i})$ measures the differences between these attributes.

3. **Markov trees**. Figure 7.8d shows a number of curves in a Markov tree structure. We denote it by

$$T = (n, \{C_1, C_2, \ldots, C_n\}, \{\beta_1, \beta_2, \ldots, \beta_n\}). \tag{7.40}$$

$\beta_i \in \{1, \ldots, n\}$ is the index to the parent curve of C_i. Thus the prior probability is

$$p(T) \propto \exp\left\{-\lambda_0 n - \sum_{i=1}^{n} E(C_i) - \sum_{\alpha_i \neq \emptyset} E_T(C_i, C_{\beta_i})\right\}. \tag{7.41}$$

Again, $E(C_i)$ is inherited from the free curve. The term for C_i and its parent C_{α_i}, $E_T(C_i, C_{\alpha_i})$, measure the compatibilities such as endpoint gap, orientation continuity, thickness, and intensity between the parent and child curves.

The parallel group pg and tree T inherit the areas from the free curve, thus

$$\mathcal{D}(pg) = \cup_{i=1}^{n} \mathcal{D}(C_i), \text{ and } \mathcal{D}(T) = \cup_{i=1}^{n} \mathcal{D}(C_i). \tag{7.42}$$

It also inherits the intensity function $\mathbf{J}(x, y; \theta_i)$ from each free curve $C_i, i = 1, 2, \ldots, n$. In summary, the intensity models for C, pg, T are all generative for image $\mathbf{I}$ as

$$\mathbf{I}(x, y) = \mathbf{J}(x, y; \theta) + \mathcal{N}(0; \sigma^2), (x, y) \in \mathcal{D}(C_i), \mathcal{D}(pg), \text{ or } \mathcal{D}(T). \qquad (7.43)$$

Generative Models of Regions

Once the curves explain away the elongated patterns, what is left within each image are the regions in the background. We adopt two simple region models in comparison to the four models in [235]. We denote a 2D region by $R \subset D$ and $\mathbf{I}_R$ the intensities inside R.

The first model assumes constant intensity with additive noise modeled by a non-parametric histogram $\mathcal{H}$,

$$\mathbf{J}(x, y; 1, \theta) = \mu, \mathbf{I}(x, y) = \mathbf{J}(x, y) + \eta, \eta \sim \mathcal{H}, (x, y) \in R. \qquad (7.44)$$

With a slight abuse of notation, we denote by $\theta = (\mu, \mathcal{H})$ the parameters used in a region.

The second model assumes a 2D Bezier spline function with additive noise. The spline accounts for global smooth shadings,

$$\mathbf{J}(x, y; 2, \theta) = B'(x)MB(y), \mathbf{I}(x, y) = \mathbf{J}(x, y; \theta_2) + \eta, \eta \sim \mathcal{H}, (x, y) \in R, \qquad (7.45)$$

where $B(x) = ((1 - x)^3, 3x(1 - x)^2, 3x^2(1 - x), x^3)$ is the basis and M is a 4×4 control matrix. This is to impose an inhomogeneous model for capturing the gradually changing intensity patterns, e.g., the sky. This model is important since regions with shading effects will be segmented into separate pieces with homogeneous models. The parameters are $\theta = (M, \mathcal{H})$, and more details with a validation can be found in [235], where we compare different models for different types of images.

The likelihood is

$$p(\mathbf{I}_R | R, \theta) \propto \prod_{(x,y) \in \mathcal{D}(R)} \mathcal{H}(\mathbf{I}(x, y) - \mathbf{J}(x, y; \ell, \theta)), \quad \ell \in \{1, 2\}. \qquad (7.46)$$

The prior for a region R assumes short boundary length ∂R (smoothness) and compact area $|\mathcal{D}(R)|$,

$$p(R) \propto \exp\left\{ -\gamma_r |\mathcal{D}(R)|^\rho - \frac{1}{2}\lambda |\partial R| \right\}, \qquad (7.47)$$

where ρ and λ are constants that are fixed for all the experiments in this section, and γ_r is a scale factor that can be adjusted to control the number of regions in the segmentation.

Bayesian Formulation for Probabilistic Inference

Given an image $\mathbf{I}$, our objective is to compute a representation of the scene (world W) in terms of a number of regions W^r, free curves W^c, parallel curve groups W^{pg}, trees W^t, and a partial order $\mathcal{PR}$. We denote the representation by variables

$$W = (W^r, W^c, W^{pg}, W^t, \mathcal{PR}). \tag{7.48}$$

The region representation W^r includes the number of regions K^r, and each region R_i has a label $\ell_i \in \{1, 2\}$ and parameter θ_i for its intensity model

$$W^r = (K^r, \{(R_i, \ell_i, \theta_i) : i = 1, 2, \ldots, K^r\}). \tag{7.49}$$

Similarly, we have $W^c = (K^c, C_1, \ldots, C_{K^c})$, $W^{pg} = (K^{pg}, pg_1, pg_2, \ldots, pg_{K^{pg}})$, and $W^t = (K^t, T_1, T_2, \ldots, T_{K^t})$. In this model, there is no need to define the background since each pixel either belongs to a region or is explained by a curve/curve group.

The problem is posed as Bayesian inference in a solution space Ω,

$$W^* = \arg \max_{W \in \Omega} p(\mathbf{I}|W)p(W). \tag{7.50}$$

By assuming mutual independence between W^r, W^c, W^{pg}, W^t, we have the prior model

$$p(W) = \left(p(K^r) \prod_{i=1}^{K^r} p(R_i) \right) \left(p(K^c) \prod_{i=1}^{K^c} p(C_i) \right) \left(p(K^{pg}) \prod_{i=1}^{K^{pg}} p(pg_i) \right) \left(p(K^t) \prod_{i=1}^{K^t} p(T_i) \right). \tag{7.51}$$

The priors for individuals $p(R)$, $p(C)$, $p(pg)$, $p(T)$ are given in the previous subsections.

As there are N curves in total including the free curves, and curves in the parallel groups and trees, then the likelihood follows the lattice partition and Eqs. (7.37) and (7.46).

$$p(\mathbf{I}|W) = \prod_{i=1}^{K^r} \prod_{(x,y) \in D_{R_i}} \mathcal{H}((\mathbf{I}(x, y) - \mathbf{J}(x, y; \ell_i, \theta_i))$$

$$\cdot \prod_{j=1}^{N} \prod_{(x,y)\in D_{C_j}} \mathcal{N}((\mathbf{I}(x, y) - \mathbf{J}(x, y; \theta_j); \sigma_j^2). \qquad (7.52)$$

Since all objects use generative models for reconstructing $\mathbf{I}$, these models are directly comparable and they compete to explain the image. This property is crucial for the integration of region segmentation and curve grouping.

Inference Algorithm Here we briefly summarize the design of the algorithm. The goal is to design an algorithm to make inference of the W^* that maximizes the posterior $p(W|\mathbf{I})$ by sampling W in the solution space with a fast simulated annealing procedure. Since W^* is usually highly peaked, we hope that it will most likely be sampled if the algorithm converges to the target distribution. This poses rather serious challenges even though we have simplified the image models above. The main difficulty is dealing with objects with different structures and exploring a large number of possible combinations of regions, curves, and curve groups in an image. Especially our objective is to achieve automatic and nearly globally optimal solutions. The following are three basic considerations in our MCMC design.

First, the Markov chain should be irreducible so that it can traverse the entire solution space. This is done by designing a number of pairs of jumps to form an ergodic Markov chain. The resulting Markov chain can reach any state from an arbitrary initialization.

Second, each jump operates on $1 - 2$ curves or curve elements. We study the scopes of the jumps within which the algorithm proposes the next state according to a conditional probability. This is like a Gibbs sampler. The proposal is then accepted in a Metropolis–Hastings step, hence its name the Metropolized Gibbs Sampler (MGS [157]).

Third, the computational cost at each jump step should be small. The proposal probability ratios in our design are factorized and computed by discriminative probability ratios. These discriminative probabilities are computed in bottom-up processes that are then used to activate the generative models in a top-down process. As Fig. 7.9 illustrates, each jump maintains a list of "particles" that are weighted hypotheses with the weights expressing the discriminative probability ratios. Then a particle is proposed at a probability proportional to its weight within the list (scope). The higher the weight is, the more likely a particle will be chosen.

Experiments

The proposed algorithm searches for the optimal solution W^* by sampling $p(W|\mathbf{I})$. It starts from a segmentation with regions obtained at a coarse level by the Canny edge detector. Our method does not rely much on the initial solution due to the use of various MCMC dynamics guided by bottom-up proposals, which help the algorithm to jump out of local minimums. However, we do use an annealing strategy to allow

S_{1r}: birth row S_{1l}: death row

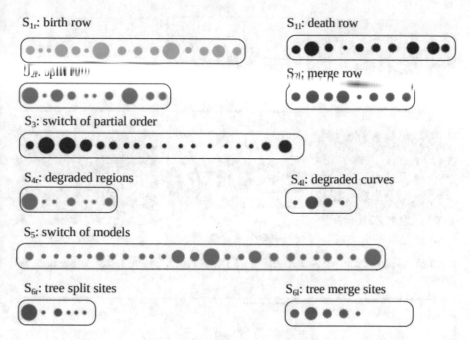

S_{2r}: split row S_{2l}: merge row

S_3: switch of partial order

S_{4r}: degraded regions S_{4l}: degraded curves

S_5: switch of models

S_{6r}: tree split sites S_{6l}: tree merge sites

Fig. 7.9 Reprinted with permission from [236]. The 6 simple jumps maintain 10 sets of "particles" whose sizes illustrate their weights. The sets are updated and re-weighted in each jump step, and they encode the proposal probabilities in a non-parametric representation

large change of W at high temperatures and to focus more on local modes with the temperature gradually cooling down. The optimal solution W^* is found when the algorithm converges since $p(W^*|I)$ is in general highly peaked for many vision problems, especially at a low temperature.

Experiment A: Computing Regions and Free Curves

An example is shown in Fig. 7.10. For each example, the first row displays the input image $\mathbf{I}_{obs}$, the computed free curves W^c, and the region segmentations W^r in the background. The second row shows the synthesized image according to the generative models for the regions $\mathbf{I}^r_{syn} \sim p(\mathbf{I}|W^r)$, the curves $\mathbf{I}^c_{syn} \sim p(\mathbf{I}|W^c)$, and the overall synthesis $\mathbf{I}_{syn}$ by occluding $\mathbf{I}^c_{syn}$ on $\mathbf{I}^r_{syn}$.

We construct a synthesis image to verify how an input image is represented in W^*. In the experiment, two parameters in the prior models are adjustable: (1) γ_r in Eq. (7.47) and (2) γ_c in Eq. (7.36). The two parameters control the extent of the segmentation, i.e., the number of regions and curves. Therefore, they decide how detailed we like the parsing to be. Usually, we set $\gamma_r = 5.0$ and $\gamma_c = 3.5$, and other parameters are fixed.

In the second experiment, we further compute the parallel groups and trees by turning on the two composite jumps $\mathbf{J}_7, \mathbf{J}_8$. Figure 7.11 shows an example: The

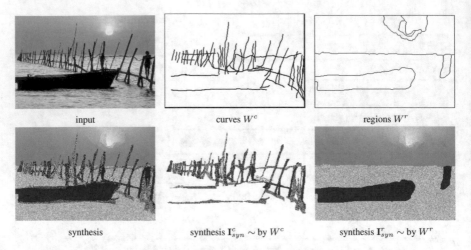

Fig. 7.10 Reprinted with permission from [236]. Experiment A: parsing images into regions and free curves

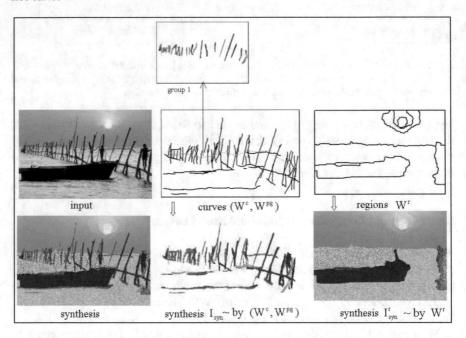

Fig. 7.11 Reprinted with permission from [236]. Experiment B: parsing an image into regions, curves, and parallel curve groups

top row shows the parallel groups or trees grouped from the simple curves. The second and third rows are displayed as before. From the results, we can see that the algorithm successfully segments, detects, and groups regions, curves, and curve groups, respectively.

Chapter 8
2.5D Sketch and Depth Maps

In previous chapters, we discuss the first stage of early visual processing, i.e., representing the changes and structures in the image with the primal sketch and 2.1D sketch. In general, the primal sketch is a generic two-layer 2D representation describing image features such as intensity changes, local geometrical structures, and illumination effects such as light sources, highlights, and transparency. Based on the primal sketch, the 2.1D sketch, a layered representation, is analyzed to describe the surfaces with occluding relations, defining the visibility of surfaces and contours in the given image. However, such rough descriptions of the spatial relations in images are not sufficient for our overall goal to thoroughly understand the vision. To understand how to obtain descriptions of the world efficiently and reliably from images, we need to go one step further. In this chapter, we introduce the 2.5D sketch, which aims to analyze the depth information that describes the relative positions of surfaces in a more precise way.

The idea of the 2.5D sketch first appeared in Marr and Nishihara's research [170], whose original goal is to provide a viewer-centered representation of the visible surfaces. As shown in Fig. 8.1, the 2.5D sketch construction is considered to be a significant part of the mid-level vision, the last step before surface interpretation, and the end, perhaps, of pure perception. We start with Marr's definition of 2.5D sketch and the viewer-centered representation, to analyze the vision process from the primal sketch to 2.5D sketch from a relatively intuitive perspective. Besides, we introduce some methods to reconstruct the 2.5D presentation, i.e., the depth information, for image analysis, including the shape from stereo, shape from shading, and direct estimation.

8.1 Marr's Definition

As introduced by Marr in [169], the distances from the observed objects to the human's two eyes are different, making the two eyes form different images of the

© Springer Nature Switzerland AG 2023
S.-C. Zhu, Y. N. Wu, *Computer Vision*, https://doi.org/10.1007/978-3-030-96530-3_8

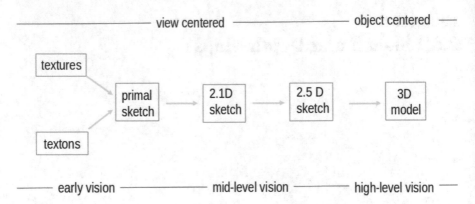

Fig. 8.1 Representation in Marr's paradigm. The 2.1D sketch was proposed by Nitzberg and Mumford [185]

world. The disparities in the two images help human brains estimate the relative distances of the objects from the observer. Marr divides this account into two parts, the first is concerned with measuring disparity and the second is concerned with using it. Marr separates both parts into three levels, i.e., computational theory, representation and algorithm, and hardware implementation. More specifically, the three steps involved in measuring stereo disparities are: (1) selecting a particular location on a surface in the scene from one image; (2) identifying the same location in the other image, and (3) measuring the disparity between the two corresponding image points. Before making use of the measured disparity, we need to solve another fundamental problem in binocular fusion, i.e., the elimination or avoidance of false targets. The abundance of matchable features and the disparities between these matches lead to difficulties in deciding which pair is false and useless. With the validated measured disparities, we can compute the distance from the viewer to surfaces in the images and also analyze the surface orientations from the disparity changes.

According to Marr, constructing the orientation-and-depth map of the visible surfaces around a viewer is one of the most critical goals of early visual processing. In this map, information is combined from several different and probably independent processes to interpret disparity, motion, shading, texture, and contour information. These ideas are called 2.5D sketch by Marr and Nishihara [170]. The full 2.5D sketch would include rough distances to the surfaces as well as their orientations. Besides, the places where surface orientations change sharply and where depth is discontinuous can be labeled as contours, which are informative for further analysis.

Many types of information can be extracted from images by different kinds of early visual processes. For example, the stereopsis outputs disparity, hence the continuous or small local changes in relative depth. At the same time, the optical flow generates the relative depth and the local surface orientation. Although in principle processes such as stereopsis and motion can directly deliver depth-related information, they are more likely to deliver local changes of depth by measuring

Fig. 8.2 Reprinted with permission from [170]. Example of a 2.5D Sketch described in [169]. The surface orientation is represented by arrows, as explained in the text. Occluding contours are shown with full lines, and surface orientation discontinuities with dotted lines. Depth is not shown in the figure, though it is thought that rough depth is available in the representation

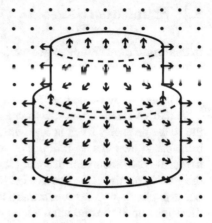

local disparities in practice. However, the main function of the visual representation we seek should not be limited to making explicit information about depth, local surface orientations, and discontinuities in these quantities. We want to create and maintain a global representation of depth that is consistent with the local cues obtained from these sources. Such a representation is called the 2.5D sketch.

To prepare for a more thorough discussion, we first describe the original definition for a viewer-centered representation that uses surface primitives of small size as described in [169]. Figure 8.2 illustrates such a representation, which is like having a gradient space at each point in the visual field. It includes a representation of contours of surface discontinuity. It has enough internal computational structure to maintain its descriptions of depth, surface orientation, and surface discontinuity in a consistent state. Depth may be represented by a scalar quantity r, the distance from the viewer of a point on a surface, and the surface discontinuities may be represented by oriented line elements. Surface orientation may be represented as a vector (p, q) in two-dimensional space, equivalent to covering the image with needles. The length of each needle defines the surface slant (or dip) at the point. Here, zero length corresponds to a surface that is perpendicular to the vector from the viewer to that point. The length of the needle increases as the surface slants away from the viewer. The needle orientation defines the tilt, that is, the direction of the surface's slant. In principle, the relationship between depth and surface orientation is straightforward—one is simply the integral of the other, taking over regions bounded by surface discontinuities. Therefore, it is possible to devise a representation with intrinsic computational facilities that can maintain the two variables of depth and surface orientation in a consistent state.

We will discuss the approaches to generate the 2.5D sketch in the following sections, including two generative methods (in applications of shape from stereo and shape from shading) that generate the 2.5D sketch from the primal sketch and a recent discriminative approach in deep learning.

8.2   Shape from Stereo

In this section, we introduce two works related to shape from stereo. We first briefly introduce the stereopsis work proposed by Peter Belhumeur in [14]. After that, we introduce the visual knowledge representation applied in the stereo reconstruction work proposed by Barbu and Zhu [4].

A pair of stereo images have been known to encode the information detailing the scene geometry since at least the time of Leonardo da Vinci [31]. The animal brain has known this for millions of years and has developed yet barely understood neuronal mechanisms for decoding it. Everywhere animals gaze, they are aware of the relative depths of the observed objects. Even though stereo vision is not the only cue to depth, monocular cues are still less exact and often ambiguous in determining the depth and 3D spatial relations.

Binocular stereopsis algorithms use the data in a pair of images taken from slightly different viewpoints to construct a depth map of the 3D surfaces captured within the images [14]. The 3D surfaces are estimated by first matching pixels in the images that correspond to the same point on a 3D surface and then computing the point's depth as a function of its displacement (or disparity) in the two images. The task of matching points between the two images is known as the correspondence problem.

In the last 50 years, researchers have tried to reconstruct the scene geometry from a pair of stereo images. Unfortunately, like most computer vision problems, the stereo problem has proven to be more difficult than initially anticipated. Argued by Belhumeur [14], the solutions for properly handling occluded regions and salient features in the scene geometry have been largely overlooked. Generally speaking, there are the following four major problems in matching the two images in binocular stereopsis: (1) Handling the feature noise in the left/right images caused by quantization error, imperfect optics, imaging system, lighting variation between images, specular reflection, etc. (2) Handling the indistinct image features near pixels where the matching is ambiguous. (3) Preserving the salient features, such as depth discontinuities at object boundaries, in the 3D scene geometry to produce an accurate reconstruction. (4) Correctly handling the half-occlusion regions by first matching half-occluded points to mutually visible points and then estimating the depth at these points.

For years, people have offered solutions to the correspondence problem without adequately addressing all of these complications. In the early area-based algorithms, the disparity was assumed to be constant within a fixed-sized window. Other methods integrate a type of smoothness (flatness) constraint into the matching process, again biasing toward reconstructions with persistent disparity. Although algorithms using smoothness constraints show effectiveness in handling the first two problems, their performance deteriorated at salient features in the scene geometry. These algorithms generally over-smooth the depth discontinuities at object boundaries ("breaks") and in surface orientation ("creases"). Sometimes erratic results would be produced. Line processes were introduced to solve the image segmentation problem. While this method helps preserve the boundaries of objects, it overlooks three critical complications. First, the introduction of line processes to model object boundaries gave rise to highly nonlinear optimization problems. Second, no prescription was given for preserving the other salient features in the scene geometry. Third, it is challenging to identify whole regions of half occlusion caused by discontinuity.

Facing the previous challenges, [14] developed a computational model for stereopsis by making explicit all of the assumptions about the nature of image coding and the structure of the world. In designing computer vision models, researchers often skip this step and, consequently, have no way of testing whether the underlying assumptions are valid. The developed model in [14] is designed within a Bayesian framework and attempts to explain the process by which the information detailing the 3D geometry of object surfaces is encoded in a pair of stereo images. Belhumeur [14] starts by deriving the model for image formation, introducing a definition of half-occluded regions, and deriving simple equations relating these regions to the disparity function. In this chapter, the author shows that the disparity function alone contains enough information to determine the half-occluded regions. These relations are utilized to derive a model for image formation in which the half-occluded regions are explicitly represented and computed. The prior model is presented in a series of three stages, or "worlds," where each world considers an additional complication to the prior, constructed from all of the local quantities in the scene geometry, i.e., depth, surface orientation, object boundaries, and surface creases.

For computer vision problems, the Bayesian paradigm seeks to extract scene information from an image or a sequence of images by balancing the content of the observed image with prior expectations about the observed scene's content. This method is general and can be applied to a wide range of vision problems, including binocular stereopsis. Let $\mathbf{S}$ indicate the scene geometry given the left and the right images by $\mathbf{I}_l$ and $\mathbf{I}_r$. Within the Bayesian paradigm, one infers $\mathbf{S}$ by considering $P(\mathbf{S}|\mathbf{I}_l, \mathbf{I}_r)$, the posterior probability of the state of the world given the measurement. From Bayes theorem, we have

$$P(\mathbf{S}|\mathbf{I}_l, \mathbf{I}_r) = \frac{P(\mathbf{I}_l, \mathbf{I}_r|\mathbf{S})P(\mathbf{S})}{P(\mathbf{I}_l, \mathbf{I}_r)}. \tag{8.1}$$

Sometimes $P(\mathbf{I}_l, \mathbf{I}_r|\mathbf{S})$ is referred to as the "image formation model," which measures how well $\mathbf{S}$ matches the observed images. $P(\mathbf{S})$ is usually referred to as the "prior model" that measures how probable a particular $\mathbf{S}$ is a priori before the images are observed. We denote the maximum a posteriori (MAP) estimate $\hat{\mathbf{S}}$. For notational convenience, we define the energy functional as below

$$E(\mathbf{S}) = -\log(P(\mathbf{I}_l, \mathbf{I}_r|\mathbf{S})P(\mathbf{S})) = E_D + E_P, \tag{8.2}$$

where $E_D = -\log(P(\mathbf{I}_l, \mathbf{I}_r|\mathbf{S}))$ is referred as the "data term," and $E_P = -\log(P(\mathbf{S})$ is referred as the "prior term." To employ these energy functionals with the underlying assumptions taken into consideration, one must be careful in choosing random variables to be estimated and in the assumed relations between these random variables in developing a Bayesian formulation of a vision problem.

The Image Formation Model

In this subsection, we introduce the image formation model proposed by Belhumeur [14]. Besides, the definition of the epipolar line will be mentioned.

In deriving the model for image formation, we choose the simplest possible geometry: pinhole cameras with parallel optical axes. Assume the cameras are calibrated and the epipolar geometry is known, we define disparity relative to an imaginary cyclopean image plane placed halfway between the left and right cameras. Here we derive explicit relations between disparity and depth, as well as disparity and half-occlusion, showing that the disparity function correctly determines the half-occluded regions in the left and right image planes. We then use these relations to derive our image formation model.

Assume that we have two pinhole cameras whose optical axes are parallel and separated by a distance of w. The cameras each have focal length l, with f_l the focal point of the left image, and f_r the right. Then we create an imaginary cyclopean camera in the same manner, placing its focal point f halfway along the baseline, i.e., the line connecting the left and right focal points. Then we restrict the cameras' placement so that the baseline is parallel to the image planes and perpendicular to the optical axes, as shown in Fig. 8.3. A point p on the surface of an object in 3D space that is visible to all three cameras is projected through the focal points onto the image planes. Each image plane has a 2D coordinate system with its origin determined by the intersection of its optical axis with the image plane. The brightness of each point projected onto the image planes creates image luminance functions $\mathbf{I}_l$, $\mathbf{I}_r$, and $\mathbf{I}$ in the left, right, and cyclopean planes, respectively.

A horizontal plane through the baseline intersects the three image planes in what are called *epipolar lines*, which we denote by X_l, X_r, and X, with coordinates $x_l \in X_l, x_r \in X_r$, and $x \in X$, respectively. The coordinates of the epipolar lines run from right to left, so that when a point in the world moves from left to right, its coordinates in the image planes increase. When the same point is visible from all three eyes, it is easy to check that $x = (x_l + x_r)/2$. Thus, we can relate the coordinates of points projected onto all three image planes by a positive disparity function $d(x)$ via $x_l = x + d(x)$ and a negative one via $x_r = x - d(x)$. Thus we have $d(x) = \frac{x_l - x_r}{2}$. Suppose $z(x)$ represents the perpendicular distance from a line connecting the focal points to the point p on the surface of the object, then the disparity $d(x)$ can be related to the distance $z(x)$ by $d(x) = \frac{lw}{2z(x)}$.

Some early work assumed that none of the points in either of the left or right images is half-occluded (visible in one camera, but not in the other). However, the vast majority of the millions of images we view every day contains large regions of half-occluded points, and half-occlusion is also a positive cue for the human visual

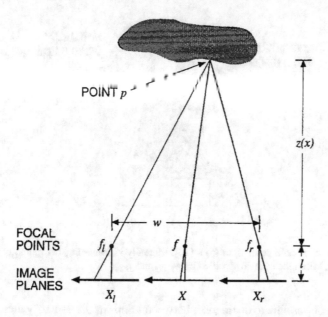

Fig. 8.3 Reprinted with permission from [14]. Camera geometries: The figure shows the left and right image planes, plus an imaginary cyclopean image plane. Both the disparity and distance functions are defined relative to the cyclopean image plane

system to determine depth. Thus, computer vision systems must take advantage of the cues provided by half-occlusion as human visual system [14].

Here we give the definition of the mutually visible point as proposed in [14]. A point p is mutually visible to both eyes if the triangle formed by p, f_i, and f_r is free of obstructing objects, as shown in Fig. 8.4. Note that according to this definition, if any object is contained within the triangle formed by p, f_i, and f_r, then the point p is not considered mutually visible, even though the point may be visible to all three eyes. To determine from the disparity function when a point is mutually visible, it is convenient to introduce a morphologically filtered version $d^*(x)$ of $d(x)$, $d^*(x) = \max_a(d(x+a) - |a|)$. $d^*(x) = d(x)$ if and only if the point p visible to the cyclopean eye in direction x is mutually visible to the left and right eyes. Thus, the function $d^*(x)$ tracks the mutually visible points. Then we can define the half-occluded points as the points that are not mutually visible as in [14]. The half-occluded points $O \in X$ are the closure of the set of points x such that $d^*(x) > d(x)$.

Before further derivations, we must define the previous quantities in a discrete manner. Take the fixed interval $[-a, a]$ of the cyclopean epipolar line X and sample it at n evenly spaced points represented by $X = x_1 \ldots . . X_N$ such that $x_1 = -a$, $x_{i+1} - x_i = \delta$ and $x_N = a$. Let the disparities at the sampled points be represented by $D = d(x_1) \ldots d(x_N) = d_1 \ldots d_N$. We define the half-occluded points $O \in X$ as $O = \{x_i | d_j - d_i| > |j - i|$ for some $x_j\}$. Finally, we discretize the range of possible disparity values with sub-pixel fineness, so that $d_i \in 0, \frac{1}{k}, \frac{2}{k}, \ldots, 1, 1 + \frac{1}{k}, \ldots, d_{max}$ for some k specifying the disparity resolution.

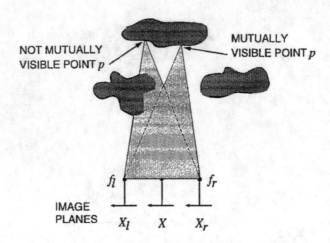

Fig. 8.4 Reprinted with permission from [14]. Mutually visible points: A mutually visible point has no object within the triangle specified by p, f_l, and f_r

Since it is not possible to distinguish between jumps in the disparity along a sloping surface and jumps in disparity at the boundaries of objects unless using sub-pixel resolution, these methods will falsely assume that the jumps along sloping surfaces produce half-occluded points. Yet, all surfaces visible to both the left and right eyes have $[d_{i+1} - d_i] < 1$ as pointed out in [14].

Keeping within the Bayesian framework, a probabilistic model needs to be developed for the joint distribution $P(\mathbf{I}_l, \mathbf{I}_r | \mathbf{S})$. To do this, assume we are given a scene of objects in 3D space with Lambertian illumination (i.e., an object's brightness is independent of the viewing angle). Label points on the surfaces of objects by elements of a set Ψ. To each point $p \in \Psi$, there is a brightness $\gamma(p)$. Define $\prod_l$ and $\prod_r$ to be the maps that take points in the image planes to the point on the surface of the closest visible object, i.e., $\prod_l : X_l \rightarrow \Psi$ and $\prod_r : X_r \rightarrow \Psi$. The brightness of a visible point once projected into the image plane is corrupted by noise. Assuming additive Gaussian white noise, image functions can be written as $\mathbf{I}_l(x_l) = \gamma(\prod_l(x_l)) + \eta_l(x_l)$ and $\mathbf{I}_r(x_r) = y(\prod_r(x_r)) + \eta_r(x_r)$, where η_l and η_r are independent and identically distributed (i.i.d.) Gaussian noise processes with mean zero and variance v^2. For notational convenience, we only consider the image functions $\mathbf{I}_l$ and $\mathbf{I}_r$ along corresponding epipolar lines. The joint density for any set of N samples, which we denote by $x_{l1}, \ldots, x_{lN}$, from the left image function $\mathbf{I}_l$, given γ, is

$$P(\mathbf{I}_l(x_{l1}), \ldots, \mathbf{I}_l(x_{lN}) | \gamma) = P(\mathbf{I}_l | \gamma) = \frac{1}{(2\pi v^2)^{N/2}} \prod_{i=1}^{N} e^{-\frac{\eta_l^2(x_{li})}{2v^2}}, \quad (8.3)$$

where $\eta_l(x_{li}) = I_l(x_{li}) - \gamma(\prod_l(x_{li}))$. Likewise, we can get the joint density from the right image function $\mathbf{I}_r$. Using the fact that η_l and η_r are independent, we can write the combined joint density as

$$P(\mathbf{I}_l, \mathbf{I}_r|\gamma) = \frac{1}{(\sqrt{2\pi}v)^N} \prod_{i}^{N} e^{-\frac{\eta_l^2(x_{li})+\eta_r^2(x_{ri})}{2v^2}}. \tag{8.4}$$

Choose the N samples from the left and right epipolar lines that correspond to the evenly spaced points $x_1 \ldots .x_N$ on the cyclopean epipolar line. So we choose $x_{li} = x_i + d_i$ and $x_{ri} = x_i - d_i$. Because the brightness function γ is unknown, we approximate γ, with its maximum likelihood estimator $\hat{\gamma}(\prod_l(x_{li})) = \hat{\gamma}(\prod_r(x_{ri})) = \frac{\mathbf{I}_l(x_{li})+\mathbf{I}_r(x_{ri})}{2}$. This approximation yields $\eta_l^2(x_{li}) + \eta_r^2(x_{ri}) \approx \frac{(\mathbf{I}_l(x_{li})-\mathbf{I}_r(x_{ri}))^2}{2}$; then we will have the joint density $P(\mathbf{I}_l, \mathbf{I}_r|\hat{\gamma}) = \frac{1}{(2\pi v^2)^N} \prod_{i=1}^{N} e^{-\frac{(\mathbf{I}_l(x_{li})-\mathbf{I}_r(x_{ri}))^2}{4v^2}}$. In this way, we can compute this quantity from the data if the point x_i is mutually visible. However, if the x_i is half-occluded, such formulation is not feasible.

To solve for the half-occluded points, we approximate the squared difference $(\mathbf{I}_l(x_{li}) - \mathbf{I}_r(x_{ri}))^2/2$ by its expected value v^2. Then we have the combined joint density as below

$$P(\mathbf{I}_l, \mathbf{I}_r|\hat{\gamma}) = \frac{1}{(2\pi v^2)^N} \prod_{i=1, x_i \notin O}^{N} e^{-\frac{(\mathbf{I}_l(x_{li})-\mathbf{I}_r(x_{ri}))^2}{4v^2}} \prod_{i=1, x_i \in O}^{N} e^{-\frac{1}{2}}. \tag{8.5}$$

Or equivalently, the combined joint distribution can be rewritten in terms of the cyclopean epipolar points X and the corresponding disparities D as

$$P(\mathbf{I}_l, \mathbf{I}_r|\hat{\gamma}, D) = \frac{1}{(2\pi v^2)^N} e^{-E_D}, E_D = \frac{1}{4v^2} \sum_{X-O} (\mathbf{I}_l(x_i+d_i) - \mathbf{I}_r(x_i-d_i))^2 + \sum_{O} \frac{1}{2}, \tag{8.6}$$

with E_D being the data term in the model.

We have introduced the data model under the assumption of Lambertian illumination so far. Let us generalize the above equation so that the data term considers, as simply opposed to image intensity, other, possibly more viewpoint invariant, features (e.g., edges, texture, filtered intensity, etc.). In doing this, we rewrite the above equation by replacing the intensity functions $\mathbf{I}_l$ and $\mathbf{I}_r$ with general feature functions F_l and F_r. Thus, the data term becomes $E_D = \frac{1}{4v^2} \sum_{X-O} (F_l(x_i + d_i) - F_r(x_i - d_i))^2 + \sum_{O} \frac{1}{2}$.

Furthermore, [14] derives the prior model for the Bayesian estimator, arguing that to capture the quantities in the scene geometry, namely depth, surface orientation, object boundaries, and surface creases, one should explicitly represent these quantities as random variables or continuous time random processes in the prior model. The derivation is broken up into three stages, or worlds, with each succeeding world considering additional complications in the scene geometry.

The first world, i.e., *World I*, is *surface smoothness*. As shown in [14], we assume a simple world in which the scenes captured in a stereo pair contain only one object.

On the surface of this object, we further assume that the 2D distance function of the cyclopean coordinate system is everywhere continuous, so a particular epipolar line has both a distance function D and a disparity function d that are also everywhere continuous. Because the relation between disparity and depth is known, we do not explicitly represent the depth, but rather the disparity in the derivation of the prior model. In the second world, i.e., *World II—Object Boundaries*, we assume a slightly more complicated world than *World I*: here we consider the possibility of more than one object in a scene. For this world, the assumption is that the disparity function d is a sample path of the sum of a scaled Brownian motion process and a compound Poisson process with i.i.d., uniform random variables. In this way, we were able to consider multiple objects in a scene by introducing random variables that explicitly represent the discontinuities in disparity at the boundaries of objects. The third world, *World III*, is a more complicated one, which is the world for *Surface Slope and Creases*. In this world, not only do we consider more than one object in a scene, but we also consider that the surfaces of objects may be steeply sloping and may have creases. We recommend [14] for further implementation details about the proposed model.

In the rest of this section, we present a two-level generative model that incorporates generic visual knowledge for dense stereo reconstruction introduced by Barbu and Zhu [4]. In this chapter, the visual knowledge is represented by a dictionary of surface primitives including various categories of boundary discontinuities and junctions in parametric form. The work takes advantage of the fact that depth discontinuities usually happen at object boundaries and that depth maps have much less complexity than color images. In stereo, textured areas contain a trove of depth information, while textureless areas are the most difficult and their depth must be propagated from far away via priors. So here the 2D sketch comes into play to capture the potential depth boundaries, and the inference algorithm propagates the depth information along the sketch to obtain a 3D consistent representation, which is a 3D sketch.

Given a stereo pair, we first compute a primal sketch representation that decomposes the image into a structural part for object boundaries and high-intensity contrast represented by a 2D sketch graph, and a structureless part represented by Markov random field (MRF) on pixels. Then we label the sketch graph and compute the 3D sketch (like a wireframe) by fitting the dictionary of primitives to the sketch graph. The surfaces between the 3D sketches are filled in by computing the depth of the MRF model on the structureless part using conventional stereo methods. These two levels interact closely with the primitives acting as boundary conditions for MRF and the MRF propagating information between the primitives. The two processes maximize a Bayesian posterior probability jointly. Barbu and Zhu [4] propose an MCMC algorithm that simultaneously infers the 3D primitive types and parameters and estimate the depth (2.5D sketch) of the scene. The experiments show that this representation can infer the depth map with sharp boundaries and junctions for textureless images, curve objects, and free-form shapes.

The overall data flow of the algorithm is illustrated in Fig. 8.5. Given a stereo pair of images, [4] first compute a primal sketch representation [90] that decomposes the

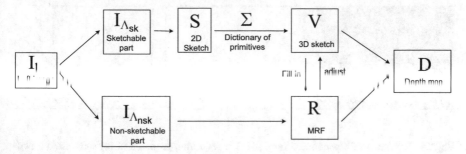

Fig. 8.5 Reprinted with permission from [4]. The flow diagram of the algorithm

image into two layers: (i) A structural layer for object boundaries and high-intensity contrast represented by a 2D sketch graph and (ii) a structureless layer represented by Markov random field on pixels. The sketch graph in the structural layer consists of a number of isolated points, line segments, and junctions that are considered vertices of different degrees of connection.

Barbu and Zhu [4] then study the 3D structures for these points, line segments, and junctions and develop a dictionary for different configurations. The boundary primitives correspond to places where the depth map is not smooth, namely the boundaries between objects in the scene (first-order discontinuities) and the places where the surface normally experiences large changes in direction (second-order discontinuities). The curve primitives describe thin curves of different intensities from the background and usually represent wire-like 3D objects such as thin branches of a tree or electric cables, etc. The point primitives represent feature points in the image that have reliable depth information. The valid combinations of these 3D primitives are summarized in a dictionary of junctions. Figures 8.8 and 8.9 show the dictionaries of line segments and junctions, respectively. Each is a 3D surface primitive specified by a number of variables. The number of variables is reduced for degenerated (accidental) cases.

Barbu and Zhu [4] adopt a probability model in a Bayesian framework, where the likelihood is described in terms of the matching cost of the primitives to images, while the prior has terms for continuity and consistency between the primitives, and a Markov random field (MRF) that is used to fill in the depth information in the structureless areas. This Markov random field together with the labeling of the edges can be thought of as a mixed Markov model [63], in which the neighborhood structure of the MRF depends upon the types of the primitives, and changes dynamically during the computation.

The inference algorithm simultaneously finds the types of the 3D primitives, their parameters, and the full depth map (2.5D sketch). To make up for the slow-down given by the long-range interactions between the primitives through the MRF, the algorithm makes use of data- driven techniques to propose local changes (updates) in the structureless areas.

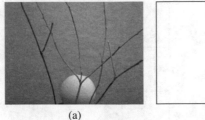 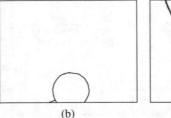 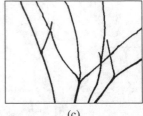

(a)   (b)   (c)

Fig. 8.6 Reprinted with permission from [4]. The algorithm proposed in [4] starts from a two-layer sketch representation. (**a**) Input image, (**b**) region layer, (**c**) curve layer

The model proposed in [4] is different from other models existent in the literature in the close relationship between the MRF and the boundary primitives. The non-horizontal boundary primitives serve as control points for the MRF, while the horizontal primitives and the occluded sides of the primitives have their disparity determined by the MRF.

In the following subsections, we introduce more details about the work in [4].

Two-Layer Representation

Given a stereo pair $\mathbf{I}_l$, $\mathbf{I}_r$ of images, we would like to find the depth of all pixels in $\mathbf{I}_l$. Assuming that the camera parameters are known, this is equivalent to finding for each pixel, the horizontal disparity that matches it to a corresponding pixel in $\mathbf{I}_r$. Let D be the disparity map that needs to be inferred and Λ be the pixel lattice.

We assume the disparity map D is generally continuous and differentiable, with the exception of a number of curves Λ_{sk} where the continuity or differentiability assumption does not hold. These curves are augmented with disparity values and are considered to form a 3D sketch D_s that acts as boundary conditions for the MRF modeling the disparity on $\Lambda_{\text{nsk}} = \Lambda \setminus \Lambda_{\text{sk}}$.

A. The Sketch Layer—from 2D to 3D Assume that the places where the disparity is discontinuous or non-differentiable are among the places of intensity discontinuity. The intensity discontinuities are given in the form of a sketch S consisting of a region layer S_R and a curve layer S_C, as illustrated in Fig. 8.6. The curve layer is assumed to occlude the region layer. These sketches can be obtained as in [90, 236]. The sketch edges are approximated with line segments $S = \{s_i, i = 1, .., n_e\}$. The segments that originated from the region layer $s_i \in S_R$ will be named *edge segments*, while the segments originating from the curve layer $s_i \in S_C$ will be named *curve segments*.

Each edge segment $s_i \in S_R$ from the region layer is assigned two 5-pixel-wide *edge regions* l_i, r_i, on the left and on the right of s_i, respectively, as shown in Fig. 8.7, left. Each curve segment $s_j \in S_C$ is assigned a *curve region* r_j along the segment,

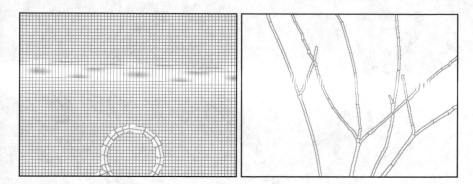

Fig. 8.7 Reprinted with permission from [4]. Division of the image in Fig. 8.6 into sketch primitives and 6×6 pixel square regions. Region layer (left) and curve layer (right)

of width equal to the width of the curve, as shown in Fig. 8.7, right. Denote the pixels covered by the edge and curve regions by Λ_R, Λ_C, respectively.

Because away from the places of discontinuity, the surfaces are in general very smooth, and to reduce the dimensionality of the problem, the pixels of $\Lambda \setminus \Lambda_R$ are grouped into *square regions* of size 6×6 pixels, by intersecting $\Lambda \setminus \Lambda_R$ with a 6×6 rectangular grid. Small regions at the boundary between the edge regions and the rectangular grid are merged into the edge regions. This way, the entire lattice Λ is divided into atomic regions that either are along the sketch S_C or are on the rectangular grid, as shown in Fig. 8.7. This structure allows the use of the thin plate spline model for the MRF and also enables the implementation of good boundary conditions by the 3D primitives.

Then all line segments $s_i \in S$ are augmented with parameters to become 3D sketch primitives, as shown in Fig. 8.8. Depending on the type of segments they originated from, there are *boundary primitives* and *curve primitives*.

B. A Dictionary of Primitives Let

$$V_1 = \{\pi_i = (s_i, [l_i, o_i^l], r_i, o_i^r, t_i, p_i, d_i, w_i[, f_i]), i = 1, .., n_e\} \tag{8.7}$$

be the set of all primitives, where the parameters in brackets might be missing, depending on the primitive type. The variables of each primitive are:

1. The edge segment $s_i \in S_R$ or curve segment $s_i \in S_C$.
2. The left and right regions (wings) l_i, r_i in case of an edge segment, or the curve as a region r_i in case of a curve segment.
3. An occlusion label o_i^l, o_i^r for each of the regions l_i, r_i, representing whether the region is occluded (value 0) or not (value 1).
4. The label $t_i = t(\pi_i) \in \{1, .., 8\}$ indexing the type of the primitive from the *primitive dictionary* with the restriction that edge segments $s_i \in S_R$ can only be

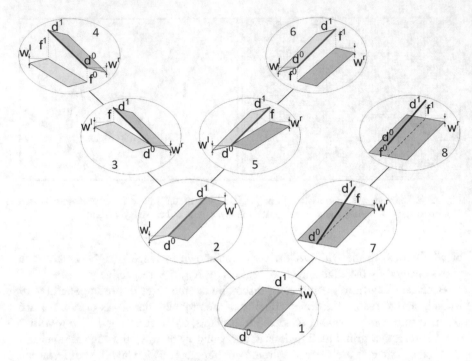

Fig. 8.8 Reprinted with permission from [4]. Each sketch segment is augmented to a primitive from the dictionary, ordered by generality

assigned types from $\{1, .., 6\}$, while curve segments $s_i \in S_C$ can only be assigned types from $\{1, 7, 8\}$. These types are illustrated in Fig. 8.8:

- Type 1 represents edges or curves that are on the surface of the objects.
- Type 2 represents first-order discontinuities, i.e., places where the surface is continuous, but the normal is discontinuous.
- Types 3, 4, 5, 6 represent occluding edges where the occluded surface is on the left (types 3, 4) or on the right (types 5, 6) of the edge.
- Types 7, 8 represent 3D curves, either connected with one end to the surface behind or totally disconnected.

5. A label p_i specifying whether this primitive is a control point (value 1) of the thin plate spline or not (value 0). All horizontal edges have $p_i = 0$ at all times.
6. The disparities $d_i = d(\pi_i) = (d_i^0, d_i^1)$ at the endpoints of the segment or the disparity $d_i = d(\pi_i)$ at the center of the segment if the segment is short (less than 6 pixels long).
7. The left and right control arms $w_i = w(\pi_i) = (w_i^l, w_i^r)$ representing the slope of the disparity map D in the direction perpendicular to the line segment.
8. For types 3–6, the disparity $f_i = f(\pi_i) = (f_i^0, f_i^1)$ of the occluded surface at the ends of the segment, or the disparity $f_i = f(\pi_i)$ at the center of the edge segment if the segment is short (less than 6 pixels long).

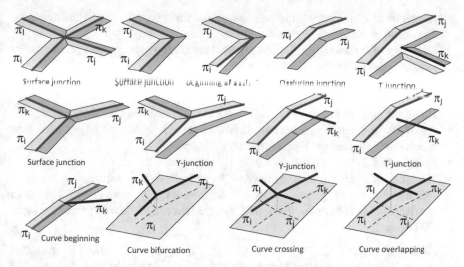

Fig. 8.9 Reprinted with permission from [4]. These are the main types of junctions between boundary and curve primitives

Each of the regions l_i, r_i of the primitive $\pi_i = (s_i, [l_i, o_i^l], r_i, o_i^r, t_i, p_i, d_i, w_i[, f_i])$ is assigned a matching cost where for each pixel $v \in r_i$, the disparity $d_v(\pi_i)$ is the linear interpolation based on the parameter d representing the disparity at the ends of the region, in the assumption that $w = 0$.

$$c(r_i, d) = \begin{cases} 0 & \text{if } r_i \text{ intersects the curve sketch } S_C \\ \sum_{v \in r_i} |\mathbf{I}_l(v) - \mathbf{I}_r(v - d_v(\pi_i))| & \text{if } o_i^r = 1 \\ \alpha & \text{else} \end{cases}.$$
(8.8)

Then the matching cost of the primitive π_i is

$$c(\pi_i) = c(r_i, [l_i], t_i, d_i, [f_i]) = \begin{cases} c(r_i, d_i) & \text{if } t_i = 7, 8, 1 \text{(curve)} \\ c(l_i, d_i) + c(r_i, d_i) & \text{if } t_i = 2, 1 \text{(region)} \\ c(l_i, f_i) + c(r_i, d_i) & \text{if } t_i = 3, 4 \\ c(l_i, d_i) + c(r_i, f_i) & \text{if } t_i = 5, 6. \end{cases}$$
(8.9)

The primitives form a graph by the natural adjacency relation between the underlying edge segments.

C. Modeling Junctions Between the Primitives To increase the model accuracy, the junction points of two or more primitives are modeled. Similar to [212], certain types of possible junctions depending on the degree (the number of primitives) of the junction are introduced below and illustrated in Fig. 8.9. These junctions include:

- Junctions of 2 boundary primitives have three main types: surface junctions, the beginning of occlusion, and occlusion junctions.
- Junctions of 3 boundary primitives have three main types: surface junctions, Y-junctions, and T-junctions.
- Junctions of 4 or more boundary primitives are accidental and are assumed to be all surface junctions.
- No junctions between one or two curve primitives and one boundary primitive.
- Junctions of 1 curve primitive with two boundary primitives have three main types: curve beginning, Y-junctions, and T-junctions.
- Junctions of 2 curve primitives have only one type.
- Junctions of 3 curve primitives have only one type, namely bifurcation.
- Junctions of 4 curve primitives have two types, namely curve crossing or curve overlapping. In both cases, the opposite primitives can be seen as part of the same 3D curve.

Let $J = \{\phi_i = (t, k, \pi_{i_1}, \ldots, \pi_{i_k}), \pi_{i_1}, \ldots, \pi_{i_k} \in V_1, i = 1, \ldots, n_J\}$ be the set of junctions, each containing the list of primitives that are adjacent to it. The variable t is the junction type and restricts the types of the primitives $\pi_{i_1}, \ldots, \pi_{i_k}$ to be compatible with it.

Each junction $\phi_i \in J$ imposes a prior model that depends on the junction type, and the types and directions of the 3D primitives $\pi_{i_1}, \ldots, \pi_{i_k}$ meeting in this junction. This prior is composed of a 3D geometric prior on the primitives and a 2D occurrence prior on each particular junction type.

Thus

$$P(\phi) \propto P(\pi_{i_1}, \ldots, \pi_{i_k} | t, \phi^{2D}) P(\phi^{2D}, t) = P(\phi^{3D} | t, \phi^{2D}) P(t | \phi^{2D}) \qquad (8.10)$$

since the 2d geometry ϕ^{2D} of the junction is fixed.

We will now discuss $P(\phi^{3D} | t, \phi^{2D})$ for each junction type. To simplify the notation, we define two continuity priors:

$$p_c(\pi_i, \pi_j) = \frac{1}{Z_c} \exp\left\{-\beta_c |d_i^\phi - d_j^\phi|^2\right\},$$

$$p_s(\pi_i, \pi_j) = \frac{1}{Z_s} \exp\left\{-\beta_c |d_i^\phi - d_j^\phi|^2 - \beta_s \left(|d_i^{\overline{\phi}} - 2d_i^\phi + d_j^{\overline{\phi}}|^2 - |d_i^{\overline{\phi}} - 2d_j^\phi + d_j^{\overline{\phi}}|^2\right)\right\},$$
$$(8.11)$$

where $d_i = (d_i^\phi, d_i^{\overline{\phi}})$ is the disparity of the primitive π_i, with d_j^ϕ being the disparity at the junction ϕ endpoint:

(1) All the surface junctions of 3 or more boundary primitives and the curve bifurcation or crossing have a prior that prefers the same disparity for all primitives meeting at this junction.

$$P(\phi^{3D} | t, \phi^{2D}) = \frac{1}{Z_1} \prod_{\pi_j, \pi_k \in \phi} p_c(\pi_j, \pi_k). \qquad (8.12)$$

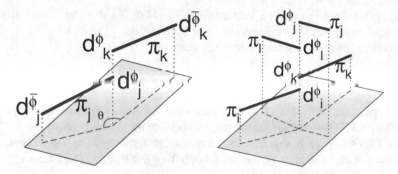

Fig. 8.10 Reprinted with permission from [4]. Left: The prior of the junction between two boundary or curve primitives depends on the angle θ between the primitives. Right: The prior of the curve overlapping junction encourages continuity of each pair of opposite curves

(2) The prior of junctions of two boundary or two curve primitives depends on the angle θ between the primitives at the junction.

$$P(\phi^{3D}|t, \phi^{2D}) = \frac{1}{Z_2} \begin{cases} p_c(\pi_j, \pi_k) & \text{if } |\theta - \pi| > \pi/6 \\ p_s(\pi_j, \pi_k) & \text{else} \end{cases}, \tag{8.13}$$

as shown in Fig. 8.10, left.

(3) For the curve overlapping junction involving four curve primitives, the prior is defined in terms of the continuity of each pair of opposite curves.

$$P(\phi^{3D}|t, \phi^{2D}) = p_s(\pi_i, \pi_k) p_s(\pi_j, \pi_l), \tag{8.14}$$

as shown in Fig. 8.10, right.

(4) For the Y-junctions of 3 boundary primitives and for the curve beginning, the prior encourages all three primitives to be adjacent, and the primitives π_i, π_j (refer to Fig. 8.9) to have a good continuation as in case (2).

$$P(\phi^{3D}|t, \phi^{2D}) = \frac{1}{Z_4} p_s(\pi_i, \pi_j) \prod_{\pi_u, \pi_v \in \phi, \{u,v\} \neq \{i,j\}} p_c(\pi_u, \pi_v). \tag{8.15}$$

(5) For the T-junctions, the prior encourages continuity of the occluding edge.

$$P(\phi^{3D}|t, \phi^{2D}) = p_s(\pi_i, \pi_j). \tag{8.16}$$

Since the disparity space of each primitive is discretized, the normalizing constant for each junction can be computed effectively.

The prior $P(t|\phi^{2D})$ can be learned from hand-labeled data, independently for each degree (the number of primitives) k of the junction.

Based on the matching cost, a saliency map

$$\psi_{\pi_i}(d, [f]) = \exp(-c(r_i, [l_i], t_i, d, [f])/10) \qquad (8.17)$$

toward all possible values of d, f is computed for each primitive $\pi_i \in V_1$. This information will be used to find the disparities d_i of the sketch primitives.

We also compute a saliency map toward the three main types of boundary primitives, namely surface (types $1, 2$), occluding left (types $3, 4$), occluding right (types $5, 6$), based on the feature

$$\xi(\pi_i) = \frac{\min_d c(l_i, d)}{|l_i|} - \frac{\min_d[c(l_i, d) + c(r_i, d)]}{|l_i| + |r_i|}, \qquad (8.18)$$

which measures how well both wings of the primitive fit the same disparity, as compared to the left wing alone.

From hand-labeled data, we obtained histograms H_{12}, H_{34}, H_{56} of the values of ξ for each of the three main types of boundary primitives. We fit these histograms with Gaussians to even out the small amount of training data and eliminate the need for histogram bins. From here, we obtain a likelihood $L_\pi(t)$ toward the three main types of boundary primitives.

$$L_\pi(t) = \begin{cases} 60e^{-\xi^2/2} & \text{if } t = 1, 2 \\ 4.4e^{-(\xi+1.18)^2/1.42} + 3.67e^{-(\xi+8.21)^2/6} & \text{if } t = 3, 4 \\ 9.18e^{-(\xi-0.58)^2/0.87} + 3.06e^{-(\xi-7.36)^2/7.5} & \text{if } t = 5, 6. \end{cases} \qquad (8.19)$$

Using the intensity-driven likelihood for the boundary primitives, we construct a likelihood, driven simultaneously by the image intensity and the geometry (relative position of primitives), for each junction $\phi = \{\pi_1, \ldots, \pi_k\}$:

$$L_\phi(t) = P(\phi)L_{\pi_1}(t_1) \ldots L_{\pi_k}(t_k). \qquad (8.20)$$

D. The Free-Form Layer The primitives $\pi \in V_1$ discussed in the previous section are elongated primitives corresponding to line segments, so they can be considered of dimension 1. Other sketch primitives that are involved in the free-form layer are the zero-dimensional primitives corresponding to feature points with reliable disparity information, i.e., *point primitives*. These primitives are a subset of the rectangular atomic regions and together with the one-dimensional boundary primitives are the control points of the thin plate spline. The curve primitives are not involved in the MRF computation.

Let R be the set of all rectangular atomic regions. For each region $r \in R$, we compute a saliency map

$$\rho_r(d) \propto \exp\left\{-\sum_{v\in r}|\mathbf{I}_l(v) - \mathbf{I}_r(v-d))|/10\right\} \tag{8.21}$$

to all possible disparities $d \in [d_{\min}, d_{\max}]$. Then the square regions

$$R = \left\{r_i = \left(d_i, o_i, p_i, \mu_i, \sigma_i^2\right), i = 1, .., n_r\right\} \tag{8.22}$$

have the following parameters:

1. The disparity $d_i = d(r_i)$ of the center of the region
2. A label o_i specifying whether this region is occluded (value 0) or not (value 1)
3. A label $p_i = p(r_i) \in \{0, 1\}$ representing whether the region is a point primitive (i.e., control point for the thin plate spline) or not
4. The mean μ_i and variance σ_i^2 of the saliency map ρ_{r_i}

Following [14], all regions (edge regions, curve regions, and square regions) will have their occlusion labels deterministically assigned based on the disparities of the boundary and curve primitives. For example, for an occlusion primitive π_i of type 4, the left region l_i and other regions horizontally to the left of the edge at a horizontal distance less than the disparity difference between the right and left wings of π_i will be labeled as occluded.

The matching cost for each region $r_i \in R$ is

$$c(r_i) = \begin{cases} \alpha & \text{if } o_i = 0 \\ \sum_{v\in r_i}|\mathbf{I}_l(v) - \mathbf{I}_r(v-d_i))| & \text{if } o_i = 1. \end{cases} \tag{8.23}$$

The set of point primitives is denoted by

$$V_0 = \{r_i \in R, s_i = 1\}. \tag{8.24}$$

Figure 8.11 shows the labeled graph, i.e., primitive types (middle), and the point and boundary primitives that act as control points for the Λ_{nsk} part (right). The

Fig. 8.11 Reprinted with permission from [4]. Left image of a stereo sequence, the graph labeling, and the control points (point and boundary primitives) of the thin plate spline

depth and disparity maps obtained this way are shown in Fig. 8.15. Observe that the horizontal edges are not control points.

The dense disparity map D is obtained from V_1 and R by interpolation. By using the boundary primitives to model the places of discontinuity, the obtained disparity map has crisp discontinuities at the object boundaries and is smooth everywhere else, as shown in Fig. 8.15.

E. Bayesian Formulation We formulate our model using the Bayes rule:

$$P(V_1, R|\mathbf{I}_l, \mathbf{I}_r) \propto P(\mathbf{I}_l|\mathbf{I}_r, V_1, R)P(R - V_0|V_0, V_1)P(V_0, V_1). \qquad (8.25)$$

The likelihood $P(\mathbf{I}_l|\mathbf{I}_r, V_1, R)$ is expressed in terms of the likelihood $L_{\pi_i}(t_i)$ and matching cost $c(r_j)$ of the sketch primitives.

$$P(\mathbf{I}_l|\mathbf{I}_r, V_1, R) \propto \prod_{i=1}^{n_e} L_{\pi_i}(t_i) \exp[-\sum_{r_j \in R} c(r_j)]. \qquad (8.26)$$

The prior

$$P(R - V_0|V_0, V_1) \propto \exp\{-E_c(R) - \beta_b E_b(R, V_1)\} \qquad (8.27)$$

is defined in terms of the energy of the soft control points:

$$E_c(R) = \sum_{r_j \in V_0} (d_j - \mu_j)/2\sigma_j^2, \qquad (8.28)$$

and the thin plate bending energy:

$$E_b(R, V_1) = \sum_{(x,y) \in G} \left[d_{xx}^2(x, y) + 2d_{xy}(x, y)^2 + d_{yy}^2(x, y) \right], \qquad (8.29)$$

which is computed on a 6×6 grid G containing the centers of all the square regions and neighboring grid points on the boundary primitives. For example, if the point $(x, y) \in G$ is the center of $r_j \in R$ and $r_N, r_{NW}, r_W, r_{SW}, r_S, r_{SE}, r_E, r_{NE}$ are the 8 neighbors of r_j, then

$$d_{xx}(x, y) = d_W - 2d_j + d_E$$
$$d_{yy}(x, y) = d_N - 2d_j + d_S$$
$$d_{xy}(x, y) = (d_{NE} + d_{SW} - d_{NW} - d_{SE})/4.$$

Similar terms in the bending energy $E_b(R, V_1)$ can be written for cases where one or many of the neighbors are boundary primitives. However, there are no terms involving the left and right atomic regions $l_i, r_i \in \pi_i$ belonging to the same edge primitive π_i.

The prior $P(V_0, V_1) = P(V_0)P(V_1)$ assumes a uniform prior on V_0, while $P(V_1)$ is defined in terms of the junction priors $P(\phi_i)$ defined above, $P(V_1) = \prod_{\phi_i \in J} P(\phi_i)$.

The Inference Algorithm

In our problem formulation, there are two types of variables, discrete and continuous. The discrete variables are

$$\Delta = V_1^d \cup R^d \tag{8.30}$$

consisting of $V_1^d = \{(t(\pi), o^l(\pi), o^r(\pi), p(\pi)), \forall \pi \in V_1\}$ and $R^d = \{(s(r), o(r), p(r)), \forall r \in R\}$. All other variables are continuous variables, namely $V_1^c = V_1 \setminus V_1^d$ and $R^c = R - R^d$, and can be divided into the boundary conditions

$$\Gamma = V_0^c \cup \{d(\pi), \forall \pi \in V_1, p(\pi) = 1\}, \tag{8.31}$$

and the fill-in variables

$$\Psi = \{([w(\pi)], [f(\pi)]), \forall \pi \in V_1\} \cup \{d(\pi), \forall \pi \in V_1, p(\pi) = 0\} \cup R^c - V_0^c. \tag{8.32}$$

The posterior probability can then be written as

$$p(V_1, R|\mathbf{I}_l, \mathbf{I}_r) = p(\Delta, \Gamma, \Psi|\mathbf{I}_l, \mathbf{I}_r). \tag{8.33}$$

In a MAP formulation, our algorithm needs to perform the following three tasks:

1. Reconstruct the 3D sketch to infer the parameters Γ of the primitives.
2. Label the primitive graph to infer the discrete parameters Δ, i.e., associate the primitives with the appropriate types. This represents the detection of surface boundaries and of the feature points of the image.
3. Perform "fill in" of the remaining parts of the image, using the MRF and Γ, Δ as boundary conditions, to infer Ψ and obtain a dense disparity map D.

The algorithm will proceed as follows. In an initialization phase, the first two steps will be performed to compute an approximate initial solution. Then steps (2) and (3) will be performed to obtain the final result.

A. Initialization Initializing the system purely based on the local depth ψ_π and likelihood $L_\pi(t)$ information existent at the primitives $\pi \in V_1$ results in an

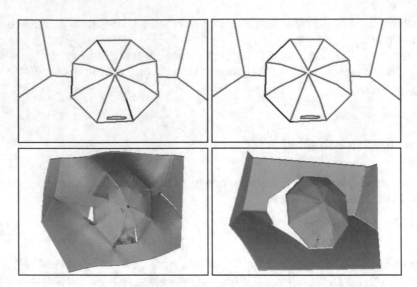

Fig. 8.12 Reprinted with permission from [4]. Left: An initialization purely based on local information is not satisfactory. Right: By propagating the junction priors along the sketch, a much better initialization can be quickly obtained

inconsistent initial solution that is valid only at places with reliable local depth information, as shown in Fig. 8.12left.

A major improvement can be achieved by using the junction prior $P(\phi)$ that has been defined in Sect. 8.2C, which provides a way to propagate depth information quickly along the edges of the sketch, from the places where it is available. This is why we use an approximation of the posterior probability that only takes into account the matching cost of the edge regions $\pi_i \in V_1$ and the junction prior.

$$P(V_1|\mathbf{I}_l, \mathbf{I}_r) \propto \prod_{i=1}^{n_e} L_{\pi_i}(t_i) \prod_{\phi_i \in J} P(\phi_i). \tag{8.34}$$

At this stage, the variables that highly depend on the thin plate spline prior will be assigned some default values. Thus, the wing parameters $w_i, \forall \pi_i \in V_1$ will be assigned value 0 (i.e., all wings will be horizontal), while the occlusion labels o_i will be assigned value 1 (not occluded).

The initialization algorithm alternates the following MCMC steps:

- A single node move that changes one variable d_i at a time.
- A move that simultaneously shifts all d_i at the same junction ϕ by the same value. This move is capable of adjusting the disparity of primitives at a junction at times when changing the disparity of only one primitive will be rejected because of the continuity prior.

- A labeling move as described in the MCMC algorithm below, which proposes a new labeling for a set of primitives and junctions. The move is accepted using the Metropolis–Hastings method based on the posterior probability from Eq. (8.34).

The algorithm is run for many steps and obtains the initialization result shown in Fig. 8.12, right in about 10 s. The initialization algorithm is very fast because the fill-in of the interior pixels is not performed, eliminating the expensive MRF computation.

The 3D reconstruction of the curve primitives is performed separately in a similar manner. The labeling move is much simpler since the curve primitives basically accept two labels, surface/non-surface. The rectangular regions with low matching cost and small variance (less than 1) will initially be labeled as control points for the thin plate spline.

B. Updating the Fill-in Variables Ψ Observe that in our formulation of the energy, if Δ, Γ are fixed, the conditional $-\log(P(\Psi|\Delta, \Gamma))$ is a quadratic function in all the variables Ψ, so it can be minimized analytically. This implies that Ψ can be regarded as a function on Δ, Γ, $\Psi = \Psi(\Delta, \Gamma)$. This restricts the problem to maximizing the probability $P(\Delta, \Gamma, \Psi(\Delta, \Gamma)|\mathbf{I}_l, \mathbf{I}_r)$, of much smaller dimensionality.

Inside each of the regions C bounded by the control point sketch primitives, the variables depend only on the control points inside and on the boundary of this region. So the computation can be localized to each of these regions independently, as shown in Fig. 8.13. Additional speedups can be obtained following the approximate thin plate spline methods from [45].

Observe that the update can affect some non-control point edges, such as the horizontal edges from Fig. 8.13.

For each such region C, we define relative labels l_C of the edges adjacent to C that only take into account the side of the edge that belongs to C. For example, an occluding edge of type 4 and an edge of type 1 will have the same label relative to the region C containing the right wing of the edge. Using these relative labels, we reduce the computation expense by defining the energy of the region

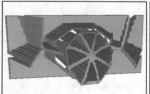

Fig. 8.13 Reprinted with permission from [4]. The fill-in can be restricted to the connected components bounded by control point boundary primitives. In a few steps, the initial 3D reconstruction before graph labeling is obtained. Shown are the 3D reconstructions after 0, 1, and 5 connected components have been updated. The horizontal edges change the disparity at the same time as the interior because they are not control points

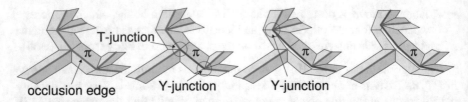

Fig. 8.14 Reprinted with permission from [4]. Each graph labeling move changes the types of a set of primitives in a consistent manner. First, a primitive π is chosen and its type is sampled from the likelihood $L_\pi(t)$; then the adjacent junctions change their type conditional on the chosen type of π, which in turn determines the types of the other primitives of the junctions, etc. The labeling move is accepted based on the Metropolis–Hastings method. Illustrated is the left side of the umbrella image

$$E(C, l_C) = E_c(C) + \mu_b E_b(C) + \sum_{r \in C \cap R} c(r). \tag{8.35}$$

The full posterior probability can be recovered from the energy of the regions and the junction prior:

$$P(V_1, R|\mathbf{I}_l, \mathbf{I}_r) \propto \prod_{i=1}^{n_e} L_{\pi_i}(t_i) \exp\left[-\sum_C E(C, l_C) \right] \prod_{\phi \in J} P(\phi). \tag{8.36}$$

C. The MCMC Optimization Algorithm After the initialization, the 3D sketch variables $\Gamma = \Gamma_0$ will be fixed. The algorithm will only update the primitive types Δ and the fill-in variables Ψ.

To maximize $P(\Delta, \Gamma_0, \Psi(\Delta, \Gamma_0)|\mathbf{I}_l, \mathbf{I}_r)$, we will use a Markov chain Monte Carlo algorithm that will sample $P(\Delta, \Gamma_0, \Psi(\Delta, \Gamma_0)|\mathbf{I}_l, \mathbf{I}_r)$ and obtain the most probable solutions.

At each step, the algorithm proposes, as shown in Fig. 8.14, new types for a set of primitives N and junctions J in one move, as follows:

1. Grow a set N of primitives as follows:

   1. Choose a random non-horizontal primitive π.
   2. Initialize $N = \{\pi\}$ and $J = \{\phi_1, \phi_2\}$ where ϕ_1, ϕ_2 are the two junctions adjacent to π.
   3. Sample the primitive type $t(\pi)$ from the local likelihood $L_\pi(t)$.
   4. Sample the type of $\phi \in J$ from $L_\phi(t)$, conditional on the primitive type $t(\pi)$. This determines the types of all primitives of $N_n = \{\pi' \notin N, \pi' \sim \phi \text{ for some } \phi \in J\}$,
      where $\pi \sim \phi$ means π is adjacent to ϕ.
   5. Set $N \leftarrow N \cup N_n$.
   6. Initialize $J_n = \emptyset$.

7. For each $\pi \in N_n$, pick the adjacent junction $\phi \notin J$. If π changed its type at step 4, set
   $J_n \leftarrow J_n \cup \{\phi\}$, else set $J_n \leftarrow J_n \cup \{\phi\}$ with probability 0.5.
8. Set $I \leftarrow I \cup J_n$.
9. Repeat steps 4–8 for each $\pi \in N_n$ and each $\psi \in J_n$, $\pi \quad \phi$.

2. Update the fill-in variables $\Psi(\Delta, \Gamma)$ for the connected components C where it is necessary.
3. Accept the labeling move based on the full posterior probability, computed using Eq. (8.36).

Example Results

Experiments are presented in Fig. 8.15 where four typical images for stereo matching are shown. The first two have textureless surfaces, and the most information is from the surface boundaries. The fourth image has curves (twigs). For these images,

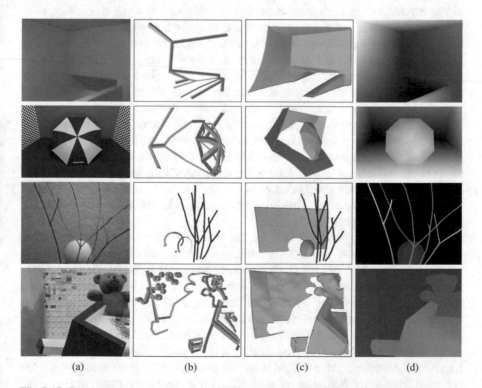

(a) (b) (c) (d)

Fig. 8.15 Reprinted with permission from [4]. Results obtained using our method. (**a**) Left image of the stereo pair, (**b**) 3D sketch using the primitives, (**c**) 3D depth map, (**d**) disparity map

it is not a surprise to see that the graph cut method with simple MRF models on pixels produces unsatisfactory results. The second and fourth images have free-form surfaces with or without textures from [153]) and [213]. On the teddy-bear sequence, the percentage of pixels with an error of at least 1, as compared to the ground truth, is 3.3%. In comparison, the graph cuts result observes a 7.3% error rate. We have also shown the interactions of the two layers in Fig. 8.13 and the effects of sketch labeling in Fig. 8.14.

8.3   Shape from Shading

In this section, we first briefly introduce the Lambertian model and the classic Horn's shape-from-shading, or in short SFS, work. Then we briefly present a two-level generative model for representing the images and surface depth maps of drapery and clothes.

In computer vision, the techniques to recover shape are called shape-from-X techniques, where X can be shading, stereo, motion, texture, etc. Shape from shading (SFS) deals with the recovery of shape from a gradual variation of shading in the image. Artists have long exploited lighting and shading to convey vivid illusions of depth in paintings. It is essential to study how the images are formed in order to solve the SFS problem. A simple model of image formation is the Lambertian model, in which the gray level at a pixel in the image depends on the light source direction and the surface normal. In SFS, given a gray level image, the aim is to recover the light source and the surface shape at each pixel in the image. However, real images do not always follow the Lambertian model. Even if we assume Lambertian reflectance and known light source direction, and if the brightness can be described as a function of surface shape and light source direction, it is still not straightforward. The reason is that if the surface shape is described in terms of the surface normal, we have a linear equation with three unknowns, and if the surface shape is described in terms of the surface gradient, we have a nonlinear equation with two unknowns. Therefore, finding a unique solution to SFS is difficult; it requires additional constraints.

Shading plays an essential role in the human perception of surface shape. Researchers in human vision have attempted to understand and simulate the mechanisms by which our eyes and brains use the shading information to recover the 3D shapes. The extraction of SFS by the visual system is also strongly affected by stereoscopic processing. Barrow and Tenenbaum discovered that it is the line drawing of the shading pattern that seems to play a central role in interpreting shaded patterns [12]. Mingolla and Todd's study of the human visual system based on the perception of solid shape [173] indicated that the traditional assumptions in SFS—Lambertian reflectance, known light source direction, and local shape recovery—are not valid from a psychological point of view. From the above discussion, one can observe that the human visual system uses SFS differently than computer vision does typically.

In the 1990s, Horn et al. [112] discovered that some impossibly shaded images exist, which could not be shading images of any smooth surface under the assumption of uniform reflectance properties and lighting. For this kind of image, SFS will not provide a correct solution, so it is necessary to detect impossibly shaded ⅢⅢⅢⅢ.

SFS techniques can be divided into four groups: minimization approaches, propagation approaches, local approaches, and linear approaches. Minimization approaches obtain the solution by minimizing an energy function. Propagation approaches propagate the shape information from a set of surface points (e.g., singular points) to the whole image. Local approaches derive shape based on the assumption of surface type. Linear approaches compute the solution based on the linearization of the reflectance map.

One of the earlier minimization approaches, which recovered the surface gradients, was by Ikeuchi and Horn [117]. Since each surface point has two unknowns for the surface gradient, and each pixel in the image provides one gray value, we have an underdetermined system. To overcome this, they introduced two constraints: the brightness constraint and the smoothness constraint. The brightness constraint requires that the reconstructed shape produces the same brightness as the input image at each surface point, while the smoothness constraint ensures a smooth surface reconstruction. The shape was computed by minimizing an energy function that consists of the above two constraints. In general, the shape at the occluding boundary was given for the initialization to ensure a correct convergence. Since the gradient at the occluding boundary has at least one infinite component, the stereographic projection was used to transform the error function to a different space. Additionally, Brooks and Horn [23] minimized the same energy function in terms of the surface normal using these two constraints. For further improvements, Frankot and Chellappa [58] enforced integrability in Brooks and Horn's algorithm to recover integrable surfaces (surfaces for which $z_{xy} = z_{yx}$). Surface slope estimates from the iterative scheme were expressed in terms of a linear combination of a finite set of orthogonal Fourier basis functions. The enforcement of integrability was done by projecting the non-integrable surface slope estimates onto the nearest (in terms of distance) integrable surface slopes. This projection was fulfilled by finding the closest set of coefficients that satisfy integrability in the linear combination. Their results showed improvements in both accuracy and efficiency over Brooks and Horn's algorithm [23]. Later, Horn [111] replaced the smoothness constraint in his approach with an integrability constraint. The major problem with Horn's method is its slow convergence. Szeliski [228] sped it up using a hierarchical basis pre-conditioned conjugate gradient descent algorithm. Based on the geometrical interpretation of Brooks and Horn's algorithm, Vega and Yang [240] applied heuristics to the variational approach in an attempt to improve the stability of Brooks and Horn's algorithm.

As for traditional propagation approaches, Horn proposed the characteristic strip method [109]. A characteristic strip is a line in the image along which the surface depth and orientation can be computed if these quantities are known at the starting point of the line. Horn's method constructs initial surface curves around the neighborhoods of singular points (singular points are the points with maximum intensity) using a spherical approximation. The shape information is propagated simultaneously along the characteristic strips outward, assuming no crossover of adjacent strips. The direction of characteristic strips is identified as the direction of intensity gradients. To get a dense shape map, new strips have to be interpolated when neighboring strips are not close to each other.

Examples for local approaches are [194] and [147]. Pentland's local approach [194] recovered shape information from the intensity and its first and second derivatives. He used the assumption that the surface is locally spherical at each point. Under the same spherical assumption, Lee and Rosenfeld [147] computed the slant and tilt of the surface in the light source coordinate system using the first derivative of the intensity.

The approaches by Pentland and Tsai and Shah are linear approaches that linearize the reflectance map and solve for shape. Pentland [147] used the linear approximation of the reflectance function in terms of the surface gradient and applied a Fourier transform to the linear function to get a closed-form solution for the depth at each point. Tsai and Shah [196] applied the gradient's discrete approximation first and then employed the linear approximation of the reflectance function in terms of the depth directly. Their algorithm recovered the depth at each point using a Jacobi iterative scheme.

Except for the traditional SFS work, we show in the following sections a two-level generative model for representing the images and surface depth maps of drapery and clothes. The upper level consists of a number of folds that will generate the high-contrast (ridge) areas with a dictionary of shading primitives (for 2D images) and fold primitives (for 3D depth maps). These primitives are represented in parametric forms and are learned in a supervised learning phase using 3D surfaces of clothes acquired through the photometric stereo. The lower level consists of the remaining flat areas that fill between the folds with a smoothness prior (Markov random field). We show that the classical ill-posed problem—shape from shading (SFS)—can be much improved by this two-level model for its reduced dimensionality and incorporation of middle-level visual knowledge, i.e., the dictionary of primitives. Given an input image, we first infer the folds and compute a sketch graph using a sketch pursuit algorithm as in the primal sketch [90, 91]. The 3D folds are estimated by parameter fitting using the fold dictionary, and they form the "skeleton" of the drapery/cloth surfaces. Then the lower level is computed by the conventional SFS method using the fold areas as boundary conditions. The two levels interact at the final stage by optimizing a joint Bayesian posterior probability on the depth map. We show a number of experiments that demonstrate robust results. In a broader scope, our representation can be viewed as a two-level inhomogeneous MRF model that is applicable to general shape-from-X problems. Our study is an attempt to revisit Marr's idea [169] of computing the 2.5D sketch from primal sketch.

Overview of Two-Layer Generation Model

The dataflow of our method is illustrated in Fig. 8.16, and a running example is shown in Fig. 8.17. The problem is formulated in a Bayesian framework, and we adopt a stepwise greedy algorithm by minimizing various energy terms sequentially. Given an input image $\mathbf{I}$ on a lattice Λ, we first compute a sketch graph G for the folds by a greedy sketch pursuit algorithm. Figure 8.17b is an exemplary graph G. The graph G has attributes for the shading and fold primitives. G decomposes the image domain into two disjoint parts: the fold part $\mathbf{I}_{fd}$ for pixels along the sketch and non-fold part $\mathbf{I}_{nfd}$ for the remaining flat areas. We estimate the 3D surface $\hat{\mathbf{S}}_{fd}$ for the fold part by fitting the 3D fold primitives in a fold dictionary Δ_{fd}. Figure 8.17c shows an example of $\mathbf{S}_{fd}$. This will yield gradient maps (p_{fd}, q_{fd}) for the fold surface. Then we compute the gradient maps (p_{nfd}, q_{nfd}) for the non-fold part by the traditional shape-from-shading method on the lower-level pixels, using gradient maps in the fold area as boundary conditions. Then we compute the joint surface $\mathbf{S} = (\mathbf{S}_{fd}, \mathbf{S}_{nfd})$ from the gradient maps (p, q) of both fold part and non-fold part. Therefore, the computation of the upper-level fold surfaces $\mathbf{S}_{fd}$ and the lower-level flat surface $\mathbf{S}_{nfd}$ is coupled. Intuitively, the folds provide the global "skeleton" and therefore boundary conditions for non-fold areas, and the non-fold areas propagate information to infer the relative depth of the folds and to achieve a seamless surface $\mathbf{S}$. The two-level generative model reduces to the traditional smoothness MRF model when the graph G is null. Since the two-layer generative model is similar to the one described in the previous section, we skip the formulation part and show some qualitative results.

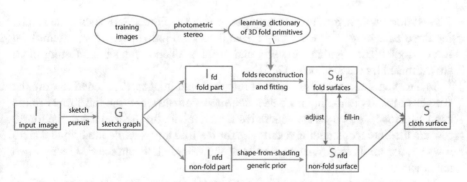

Fig. 8.16 Reprinted with permission from [93]. The data flow of our method for computing the 3D surface $\mathbf{S}$ of drapery/cloth from a single image $\mathbf{I}$ using the two-layer generative model. See text for interpretation

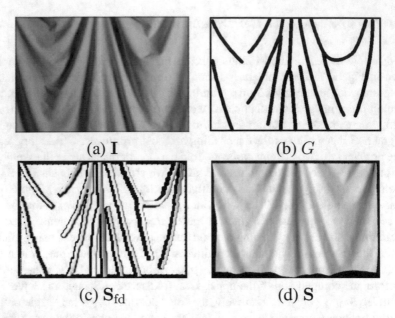

(a) $\mathbf{I}$ (b) G

(c) $\mathbf{S}_{\mathrm{fd}}$ (d) $\mathbf{S}$

Fig. 8.17 Reprinted with permission from [93]. (**a**) A drapery image under approximately parallel light. (**b**) The sketch graph for the computed folds. (**c**) The reconstructed surface for the folds. (**d**) The drapery surface after filling in the non-fold part. It is viewed at a slightly different angle and lighting direction

Results

We test our whole algorithm on a number of images. Figure 8.18 shows the results for three images of drapery hung on wall and a cloth image (last column) on some people. The lighting direction and surface albedos for all the testing cloth are estimated by the method in [275].

In the experimental results, the first row are input images, the second row are the sketches of folds in the input images and their domains, and the third row are the syntheses for $\mathbf{I}_{\mathrm{fd}}$ based on the generative sketch model for the fold areas, the fourth row are the 3D reconstruction results $\mathbf{S}_{\mathrm{fd}}$ for the fold areas, while the fifth and sixth rows are the final reconstruction results of the whole cloth surface $\mathbf{S}$ shown in two novel views.

In these results, the folds in row (d) have captured most of the perceptually salient information in the input images, and they can reconstruct the surface without too much skewing effect. It makes sense to compute them before the non-fold part. We observe that the SFS for the non-fold parts indeed provides useful information for the 3D positions of the folds.

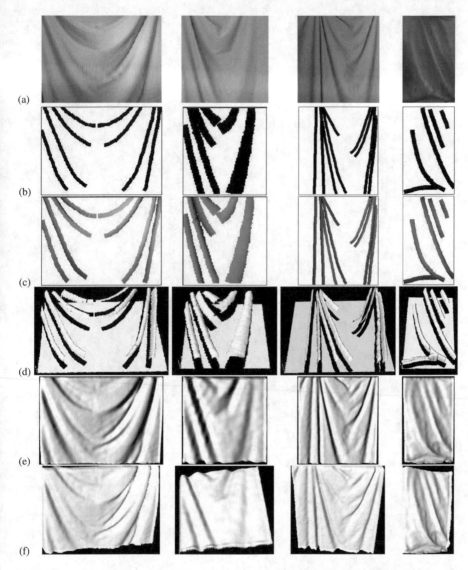

Fig. 8.18 Reprinted with permission from [93]. (**a**) Input cloth image. (**b**) 2D folds and their image domains. (**c**) Synthesis for 2D fold sketches $\mathbf{I}_{fd}$. (**d**) 3D reconstruction $\mathbf{S}_{fd}$ for fold areas. (**e**)–(**f**) Final reconstructed surface $\mathbf{S}$ in novel views

Chapter 9
Learning by Information Projection

In this chapter, a general framework [37, 281] for learning a statistical model as an approximation to the true distribution that generates images is considered.

9.1 Information Projection

Suppose training images $\{\mathbf{I}_m, m = 1, \ldots, M\} \sim f(\mathbf{I})$ are observed. The goal is to find a good approximation to the unknown true distribution f that generates the training examples. Suppose M is large so that it is feasible to estimate the expectations with respect to f accurately from the training examples.

Consider starting from a reference distribution $q(\mathbf{I})$, e.g., the white noise distribution. Suppose there is a set of features $H(\mathbf{I}) = (H_k(\mathbf{I}), k = 1, \ldots, K)$. The following may be estimated:

$$\mathrm{E}_f[H(\mathbf{I})] \approx \frac{1}{M} \sum_{m=1}^{M} H(\mathbf{I}_m). \tag{9.1}$$

$\mathrm{E}_f[H(\mathbf{I})]$ is all that is known about the unknown f as far as the feature H is concerned.

Again, the goal is to find a distribution p to be an approximation to the unknown distribution f. Such a distribution should better reproduce the feature statistics, i.e.,

$$\mathrm{E}_p[H(\mathbf{I})] = \mathrm{E}_f[H(\mathbf{I})]. \tag{9.2}$$

Call p such an eligible distribution. Let

$$\Omega = \{p : \mathrm{E}_p[H(\mathbf{I})] = \mathrm{E}_f[H(\mathbf{I})]\} \tag{9.3}$$

© Springer Nature Switzerland AG 2023
S.-C. Zhu, Y. N. Wu, *Computer Vision*, https://doi.org/10.1007/978-3-030-96530-3_9

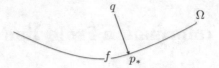

Fig. 9.1 In the above illustration, each point is a probability distribution. f is the true distribution that generates the training examples. q is the reference distribution or the null model. The curve Ω consists of all the distributions that reproduce the feature statistics of f. p_* is the projection of q onto Ω

be the family of all the eligible distributions. Clearly, $f \in \Omega$. See Fig. 9.1 for an illustration, in which each point is a probability distribution. f is the true distribution that generates the training examples. q is the reference distribution or the null model. The curve Ω consists of all the distributions that reproduce the feature statistics of f.

So starting from q, the updated goal is to find a distribution in Ω so that it has the minimum distance from q, i.e., an eligible distribution that can be obtained from a minimum modification of q, so that artificial features beyond H are not introduced. Let p_* be such a distribution. p_* may be thought of as the projection of q onto the family Ω, and it is hence called the information projection. More specifically, the objective is to find

$$p_* = \arg \min_{p \in \Omega} \mathrm{KL}(p \| q). \tag{9.4}$$

Orthogonality and Duality

In order to solve for $p_* = \arg \min_{p \in \Omega} \mathrm{KL}(p \| q)$, the Langevin multiplier may be used as it was for the FRAME model. In this chapter, an approach is adopted that is less direct but is more geometrically meaningful. A dual minimization problem may be found by introducing another family of distributions that is called the model family. Specifically, the following exponential family models are defined:

$$p(\mathbf{I}; \lambda) = \frac{1}{Z(\lambda)} \exp\{\langle \lambda, H(\mathbf{I}) \rangle\} q(\mathbf{I}), \tag{9.5}$$

in which $\lambda = (\lambda_k, k = 1, \ldots, K)$, $\langle \lambda, H(\mathbf{I}) \rangle = \sum_{k=1}^{K} \lambda_k H_k(\mathbf{I})$, and

$$Z(\lambda) = \int \exp\{\langle \lambda, H(\mathbf{I}) \rangle\} q(\mathbf{I}) = \mathrm{E}_q[\exp\{\langle \lambda, H(\mathbf{I}) \rangle\}] \tag{9.6}$$

is the normalizing constant. For simplicity, $p(\mathbf{I}; \lambda)$ as p_λ is written. Let

$$\Lambda = \{p_\lambda, \forall \lambda\} \tag{9.7}$$

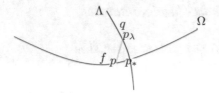

Fig. 9.2 The eligible family Ω and the model family Λ are orthogonal, $\Omega \perp \Lambda$, because for any $p \in \Omega$, $p_\lambda \in \Lambda$, $\mathrm{KL}(p\|p_\lambda) = \mathrm{KL}(p\|p_*) + \mathrm{KL}(p_*\|p_\lambda)$, in which $p_* = \Omega \cap \Lambda$ is the intersection

be the model family. Clearly, $q \in \Lambda$ with $\lambda = 0$. See Fig. 9.2 for an illustration, in which the model family Λ is illustrated by the vertical curve.

Let $p_* = p(\mathbf{I}; \lambda_*) = \Omega \cap \Lambda$ be the intersection between the eligible family and the model family. It shall be shown that $p_* = \arg\min_{p \in \Omega} \mathrm{KL}(p\|q)$, which is the projection that is sought after. The key is that $\Omega \perp \Lambda$; that is, Ω is orthogonal to Λ, in the sense of the following Pythagorean theorem [37]:

Theorem 1 *For any $p_\lambda \in \Lambda$ and any $p \in \Omega$:*

$$\mathrm{KL}(p\|p_\lambda) = \mathrm{KL}(p\|p_\star) + \mathrm{KL}(p_\star\|p_\lambda). \tag{9.8}$$

Proof

$$\mathrm{KL}(p\|p_\lambda) = \mathrm{E}_p[\log p(\mathbf{I})] - \mathrm{E}_p[\log p(\mathbf{I}; \lambda)]. \tag{9.9}$$

$$\mathrm{KL}(p\|p_\star) = \mathrm{E}_p[\log p(\mathbf{I})] - \mathrm{E}_p[\log p(\mathbf{I}; \lambda_\star)]. \tag{9.10}$$

$$\mathrm{KL}(p_\star\|p_\lambda) = \mathrm{E}_{p_\star}[\log p(\mathbf{I}; \lambda_\star]] - \mathrm{E}_{p_\star}[\log p(\mathbf{I}; \lambda)]. \tag{9.11}$$

Meanwhile, $\mathrm{E}_p[\log p(\mathbf{I}; \lambda_\star)] = \mathrm{E}_{p_\star}[\log p(\mathbf{I}; \lambda_\star]]$, and $\mathrm{E}_p[\log p(\mathbf{I}; \lambda)] = \mathrm{E}_{p_\star}[\log p(\mathbf{I}; \lambda)]$, because $\mathrm{E}_p[H(\mathbf{I})] = \mathrm{E}_{p_\star}[H(\mathbf{I})]$, as both p and $p_\star$ belong to Ω. Thus, the result follows. $\qquad\square$

The above result leads to the following duality result:

$$p_* = \arg\min_{p \in \Omega} \mathrm{KL}(p\|q) = \arg\min_{p_\lambda \in \Lambda} \mathrm{KL}(f\|p_\lambda). \tag{9.12}$$

Thus, it is seen that $p_* = \arg\min_{p \in \Omega} \mathrm{KL}(p\|q)$ by finding $p_* = \arg\min_{p_\lambda \in \Lambda} \mathrm{KL}(f\|p_\lambda)$. Since $\mathrm{KL}(f\|p_\lambda) = \mathrm{E}_f[\log f(\mathbf{I})] - \mathrm{E}_f[\log p(\mathbf{I}; \lambda)]$,

$$p_* = \arg\min_{p_\lambda \in \Lambda} \mathrm{KL}(f\|p_\lambda) = \arg\max_{p_\lambda \in \Lambda} \mathrm{E}_f[\log p(\mathbf{I}; \lambda)]. \tag{9.13}$$

$\mathrm{E}_f[\log p(\mathbf{I}; \lambda)]$ is actually the log-likelihood in the limit.

Maximum Likelihood Implementation

If $\{\mathbf{I}_m, m = 1, \ldots, M\} \sim f(\mathbf{I})$ is observed, then

$$\mathrm{E}_f[\log p(\mathbf{I}; \lambda)] \approx \frac{1}{M} \sum_{m=1}^{M} \log p(\mathbf{I}_m; \lambda), \qquad (9.14)$$

so λ_* can be approximated by the maximum likelihood estimate $\hat{\lambda} = \arg \max_\lambda L(\lambda)$, in which

$$L(\lambda) = \frac{1}{M} \sum_{m=1}^{M} \log p(\mathbf{I}_m; \lambda) = \frac{1}{M} \sum_{m=1}^{M} \langle \lambda, H(\mathbf{I}_m) \rangle - \log Z(\lambda) \qquad (9.15)$$

is the log-likelihood function of the exponential family model (9.5).

It can be shown that

$$\frac{\partial}{\partial \lambda} \log Z(\lambda) = \mathrm{E}_\lambda[H(\mathbf{I})], \qquad (9.16)$$

in which E_λ denotes the expectation with respect to $p(\mathbf{I}; \lambda)$.

$$\frac{\partial^2}{\partial \lambda^2} \log Z(\lambda) = \mathrm{Var}_\lambda[H(\mathbf{I})], \qquad (9.17)$$

in which Var_λ denotes the variance with respect to $p(\mathbf{I}; \lambda)$. Thus,

$$\frac{\partial}{\partial \lambda} L(\lambda) = \frac{1}{M} \sum_{m=1}^{M} H(\mathbf{I}_m) - \mathrm{E}_\lambda[H(\mathbf{I})], \qquad (9.18)$$

and

$$\frac{\partial^2}{\partial \lambda^2} L(\lambda) = \mathrm{Var}_\lambda[H(\mathbf{I})]. \qquad (9.19)$$

That is, $L(\lambda)$ is a concave function with a unique maximum, provided that $\mathrm{Var}[H(\mathbf{I})]$ is positive definite, which is the case if the components of $H(\mathbf{I})$ are linearly independent. At the maximum $\hat{\lambda}$,

$$\mathrm{E}_{\hat{\lambda}}[H(\mathbf{I})] = \frac{1}{M} \sum_{m=1}^{M} H(\mathbf{I}_m), \qquad (9.20)$$

in which E_λ denotes the expectation with respect to $p(\mathbf{I}; \lambda)$. Thus, at the maximum likelihood estimate, the model reproduces the observed feature statistics.

If $M \to \infty$, $L(\lambda) \to E_f[\log p(\mathbf{I}; \lambda)]$, and at the maximum, $E_f[H(\mathbf{I})] = E_{\lambda_*}[H(\mathbf{I})]$, i.e., $p_{\hat{\lambda}} \to p_* = \Lambda \cap \Omega$.

Crucially, the information projection viewpoint is deeper than the maximum likelihood estimation of the exponential family model (9.5). The former provides ⷧ ɉⷧⷧⷧⷧⷧⷧⷧⷧⷧⷧⷧ Ꮳⷧⷧ Ɩⷧⷧ 1ⷧⷧⷧ ⷧ ⷧⷧⷧⷧ Ɩⷧⷧ Ɩⷧⷧⷧ ⷧⷧ Ɩⷧ ⷧⷧⷧⷧⷧⷧⷧⷧⷧ ⷧⷧⷧ ᖴⷧ�ⷧⷧⷧ⥂ⷧ

9.2 Minimax Learning Framework

Suppose there are two different sets of features $H(\mathbf{I})$ and $\tilde{H}(\mathbf{I})$; then, there are two different eligible families Ω and $\tilde{\Omega}$. If the same reference distribution q is projected onto Ω and $\tilde{\Omega}$, respectively, p_* and $\tilde{p}_*$ will be the results, respectively. In Fig. 9.3, the solid curve illustrates Ω, while the dotted curve illustrates $\tilde{\Omega}$. Due to the Pythagorean theorem,

$$\text{KL}(f\|q) = \text{KL}(f\|p_*) + \text{KL}(p_*\|q) = \text{KL}(f\|\tilde{p}_*) + \text{KL}(\tilde{p}_*\|q). \quad (9.21)$$

Thus, if the desire is to make $\text{KL}(f\|p_*)$ small, $\text{KL}(p_*\|q)$ needs to be made large. So if there are many different choices of H, then the one that maximizes $\text{KL}(p_*\|q)$ should be chosen. Recall that $p_* = \arg\min_{p\in\Omega} \text{KL}(p\|q)$; thus, the goal is to solve the following max–min problem:

$$\max_{H} \min_{p\in\Omega(H)} \text{KL}(p\|q), \quad (9.22)$$

in which $\Omega(H)$ is the eligible family defined by the set of features H. Because of the duality, the above problem is equivalent to the maximum likelihood problem,

$$\max_{H} \min_{\lambda} \text{KL}(f\|p_{H,\lambda}), \quad (9.23)$$

in which $p_{H,\lambda}$ is the exponential family model defined by H in Eq. (9.5). Here H is made explicit in $p_{H,\lambda}$ because different sets of features are being considered. Thus, the log-likelihood L in (9.15) can be maximized over both λ and H.

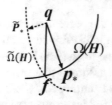

Fig. 9.3 The solid curve and the dotted curve illustrate two eligible families defined by two different sets of feature statistics. q should be projected onto the solid curve instead of the dotted curve in order to get closer to the target distribution f

Fig. 9.4 Learning a sequence
of distributions p_k to
approach the target
distribution f. Each time, the
current distribution p_{k-1} is
projected onto the eligible
family defined by H_k to
obtain p_k

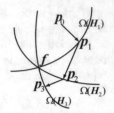

Intuitively, for a given set of features H that defines an eligible family $\Omega(H)$, the goal is to choose p that is closest to q to avoid adding artificial features that are not in H. Meanwhile, for different sets of features, the set of features so that the change from q to the corresponding p_* is the largest should be chosen.

The minimax entropy learning [281] is a special case of the above learning scheme, in which the reference distribution q is the uniform measure.

Model Pursuit Strategies

$H(\mathbf{I}) = (H_k(\mathbf{I}), k = 1, \ldots, K)$ may be obtained by selecting each H_k sequentially to pursue a sequence of models p_k that get closer and closer to the target distribution f. Figure 9.4 illustrates the idea of sequential projection. The starting point is $p_0 = q$, the reference distribution or the null model. After selecting the first feature H_1, the eligible family $\Omega(H_1)$ that consists of all the distributions that reproduce $\mathrm{E}_f[H_1(\mathbf{I})]$ is obtained. Then, p_0 is projected onto $\Omega(H_1)$ to obtain p_1. Then, the second feature H_2 is selected, and p_1 is projected onto $\Omega(H_2)$ to obtain p_2, and so forth. Because of the Pythagorean theorem, each iteration approaches the target f.

Sequential projection leads to the following greedy strategy to choose H_k sequentially. At each step, the maximum reduction in the distance from the current model to the target distribution is sought. Specifically, let p_{k-1} be the current model. $H_k = \arg\max \mathrm{KL}(p_k \| p_{k-1})$ is chosen, which can be implemented by the maximum likelihood of the following exponential family model:

$$p_k(\mathbf{I}) = \frac{1}{Z_k(\lambda_k)} \exp\{\lambda_k H_k(\mathbf{I})\} \, p_{k-1}(\mathbf{I}), \qquad (9.24)$$

in which p_{k-1} plays the role of the current reference distribution, and both H_k and λ_k are obtained by maximizing the likelihood function of (9.24) as a function of H_k and λ_k. In the end, a model of the form (9.5) is obtained.

In the above discussion, it is assumed that there is a large dictionary of features $\{H_i, i = 1, \ldots, N\}$, and a small number of them from this large dictionary may be selected. A related strategy for feature selection is via ℓ_1 regularization, as in basis pursuit [26] or Lasso [230]. Specifically, the following full model is assumed, instead of the final selected model:

$$p(\mathbf{I}; \lambda) = \frac{1}{Z(\lambda)} \exp \left\{ \sum_{i=1}^{N} \lambda_i H_i(\mathbf{I}) \right\} q(\mathbf{I}), \qquad (9.25)$$

in which $\lambda = (\lambda_i, i = 1, \ldots, N)$ is a long vector. The vector λ is assumed to be sparse, i.e., only a small number of its components are different from zero. Let $L(\lambda)$ be the log-likelihood of the above full model; model selection can be performed by maximizing the ℓ_1-regularized log-likelihood, $L(\lambda) + \rho|\lambda|$, in which $|\lambda| = \sum_{i=1}^{N} |\lambda_i|$ is the ℓ_1 norm of λ, and ρ is a tuning constant. The maximization of the penalized log-likelihood $L(\lambda) + \rho|\lambda|$ can be accomplished by an epsilon-boosting algorithm [67, 206], in which at each step, the component of $L'(\lambda)$ is chosen that has the maximum magnitude, and then this component is updated by a small amount ϵ.

It is also possible that the dictionary of the features is parameterized by some continuous parameters γ, so the model is

$$p(\mathbf{I}; \theta) = \frac{1}{Z(\theta)} \exp \left\{ \sum_{i=1}^{N} \lambda_i H_i(\mathbf{I}; \gamma) \right\} q(\mathbf{I}), \qquad (9.26)$$

in which $\theta = (\lambda, \gamma)$. Both λ and γ may be learned by maximum likelihood.

2D Toy Example

So far, not much detail has been given about the features. In this section, the idea of information projection using concrete examples of learning two-dimensional distributions, in which the feature statistics are linear projections or filter responses, shall be elaborated.

Figure 9.5 illustrates two examples of information projection. The training examples $\{\mathbf{I}_m, m = 1, \ldots, M\}$, are two-dimensional, i.e., they are images of different two-dimensional points. The scatterplot of the data forms a two-dimensional cloud of points. The features are of the form $H_k(\mathbf{I}) = h(\langle \mathbf{I}, B_k \rangle)$, in which B_k is also a two-dimensional vector, just like $\mathbf{I}$. $\langle \mathbf{I}, B_k \rangle$ is the projection of $\mathbf{I}$ on B_k. One may also call it a filter response, in which B_k plays the role of a filter. $h(r)$ is a one-hot indicator vector. Specifically, the range of $\langle \mathbf{I}, B_k \rangle$ is divided into a finite number of L bins, so that $h(r) = (h_l(r), l = 1, \ldots, L)$. $h_l(r) = 1$ if r falls into the l-th bin, and $h_l(r) = 0$ otherwise. Thus, $\sum_{m=1}^{M} H_k(\mathbf{I}_m)/M = \sum_{m=1}^{M} h(\langle \mathbf{I}_m, B_k \rangle)/M$ is the histogram of the projected points $\{\langle \mathbf{I}_m, B_k \rangle, m = 1, \ldots, M\}$ projected onto B_k. It may be assumed that the squared length $|B_k|^2 = 1$ and that the direction B_k may be discretized in the two-dimensional domain.

Consider starting from the uniform distribution over the two-dimensional domain of $\mathbf{I}$, assumed to be the unit square. Then, apply the model pursuit strategy by

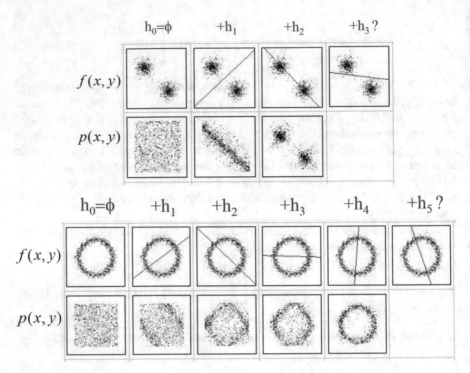

Fig. 9.5 Reprinted with permission from [155]. This is an example of learning two-dimensional distributions by information projection. In each step, the marginal distribution of the data points projected onto a selected vector is matched

selecting $B_k, k = 1, \ldots, K$. Each time, a B_k is selected, and the marginal histogram of the projected points is matched. After a number of steps, a model

$$p(\mathbf{I}; \mathbf{B}, \lambda) = \frac{1}{Z} \exp \left\{ \sum_{k=1}^{K} \lambda_k h(\langle \mathbf{I}, B_k \rangle) \right\} q(\mathbf{I}) \qquad (9.27)$$

is pursued, in which $\lambda = (\lambda_k, k = 1, \ldots, K)$ and $\mathbf{B} = (B_k, k = 1, \ldots, K)$ is the learned dictionary of projections or filters.

Figure 9.5 illustrates the learning process. In each example, the first row displays the target distribution f, as well as the selected direction or filter B_k for each k. The second row displays the learned model p as more directions are added. After adding only a small number of filters, the learned model p is very similar to f. The learning method is related to projection pursuit [65].

In Fig. 9.5, the starting point is the uniform distribution. One can also start from a Gaussian white noise model with a small variance.

In addition to the pursuit strategy, the dictionary $\mathbf{B}$ may also be learned directly by maximum likelihood, by taking derivatives with respect to both λ and $\mathbf{B}$. In order to take the derivative with respect to B_k, h needs to be made continuous and differentiable. A possible choice is the rectified linear unit $h(r) = \max(0, r - b)$, in which b is the threshold, which can also be estimated by maximum likelihood.

Although the two toy examples are simple, they are very illustrative and have deep implications. The model (9.27) can be extended to model large images by making the filters B_k convolutional, i.e., B_k is a localized image patch (e.g., 7×7) and B_k is applied around each pixel. If this model is learned from natural images, Gabor filters and differences of Gaussian filters will be learned. Model (9.27) is the simplest FRAME model.

Learning Shape Patterns

In addition to learning image appearance patterns, shape patterns may also be learned by information projection. Figure 9.6 illustrates an example of learning

Fig. 9.6 Reprinted with permission from [276]. This is an example of learning a sequence of models for shapes by adding shape statistics

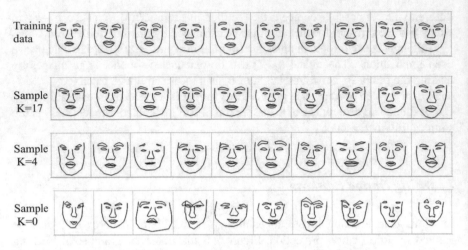

Fig. 9.7 Reprinted with permission from [155]. This is an example of learning a sequence of models for face shapes by adding shape statistics

generic object shapes by adding shape statistics. Figure 9.7 illustrates an example of learning specific face shapes by adding relevant statistics.

Relation to Discriminative Learning

Suppose negative examples from the reference distribution $q(\mathbf{I})$ are observed, and positive examples from the model $p(\mathbf{I}; \lambda)$ in (9.5) are observed. Let α be the prior probability that a positive example is observed. Then the posterior probability that an example $\mathbf{I}$ is a positive example is

$$p(+|\mathbf{I}) = \frac{1}{1 + \exp\left\{-\sum_{k=1}^{K} \lambda_k H_k(\mathbf{I}) - b\right\}}, \tag{9.28}$$

in which $b = \log[\alpha/(1-\alpha)]$. This is a logistic regression model. If examples from multiple categories are observed, a multinomial logistic regression will be obtained.

The learning method in the previous section can be considered a generative version of AdaBoost [59].

Chapter 10
Information Scaling

One important property of natural image data that distinguishes vision from other sensory tasks such as speech recognition is that scale plays an important role in image formation and interpretation. Specifically, visual objects can appear at a wide range of scales in the images due to the change of viewing distance as well as camera resolution. The same objects appearing at different scales produce different image data with different statistical properties. Figure 10.1 shows two examples of information scaling, where the change of scale causes the change of image properties, which may trigger the change of the modeling scheme for image representation.

In this section, we study the change of statistical properties, in particular, some information theoretical properties, of the image data over scale. We show that the entropy rate, defined as entropy per pixel, of the image data changes over scale. Moreover, the inferential uncertainty of the outside scene that generates the image data also changes with scale. We call these changes information scaling.

10.1 Image Scaling

To give the reader some concrete ideas, we first study information scaling empirically by experimenting with the so-called dead leaves model.

Model and Assumptions

The dead leaves model [171] was used by Lee et al. [145] in their investigation in image statistics of natural scenes, which was also previously used to model natural images. For our purpose, we may consider that the model describes an ivy wall covered by a large number of leaves of similar sizes. See Fig. 10.2 for some

© Springer Nature Switzerland AG 2023
S.-C. Zhu, Y. N. Wu, *Computer Vision*, https://doi.org/10.1007/978-3-030-96530-3_10

Fig. 10.1 Images of the same objects can appear very different at different viewing distances or camera resolution, a phenomenon we call information scaling

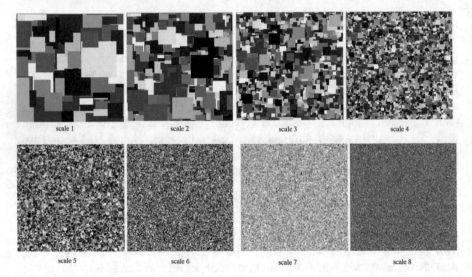

Fig. 10.2 First published in the Quarterly of Applied Mathematics in Volume 66:81–122, 2008, published by Brown University. Reprinted with permission from [257]. Pictures of the simulated ivy wall taken at 8 viewing distances. The viewing distance of the $i + 1$-st image is twice that of the i-th image

examples. We assume that the leaves are of a squared shape and are uniformly colored. Each leaf is represented by:

1. Its length or width r, which follows a distribution $f(r) \propto 1/r^3$ over a finite range $[r_{min}, r_{max}]$.
2. Its color or shade a, which follows a uniform distribution over $[a_{min}, a_{max}]$.
3. Its position (x, y, z), where the wall serves as the (x, y) plane, and $z \in [0, z_{max}]$ is the distance between the leaf and the wall. We assume that z_{max} is very small so that z matters only for deciding the occlusions among the leaves.

For the collection of leaves $\{(r_k, a_k, x_k, y_k, z_k)\}$, we assume that r_k are independent of each other, and so are a_k. (x_k, y_k, z_k) follows a Poisson process in

$R^2 \times [0, z_{max}]$. We assume that the intensity of the Poisson process λ is large enough so that the leaves completely cover the wall. As noted by Lee et al. [146], $\{(r_k, a_k, x_k, y_k, z_k)\}$ is a Poisson process in the joint domain $[r_{min}, r_{max}] \times [a_{min}, a_{max}] \times R^2 \times [0, z_{max}]$ with respect to the measure $f(r)drda\lambda dxdydz$.

Lee et al. [146] show that this Poisson process is scale invariant under the assumption that $[r_{min}, r_{max}] \to [0, \infty]$. Specifically, under the scaling transformation $x' = x/s$ and $y' = y/s$, where s is a scaling parameter, we have $r' = r/s$, and the Poisson process will be distributed in $[r_{min}/s, r_{max}/s] \times [a_{min}, a_{max}] \times R^2 \times [0, z_{max}]$ with respect to the measure $f(sr')sdr'da\lambda sdx'sdy'dz$, which is equal to $f(r')dr'da\lambda dx'dy'dz'$ because $f(r) \propto 1/r^3$. As $[r_{min}, r_{max}] \to [0, \infty]$, $[r_{min}/s, r_{max}/s] \to [0, \infty]$ too, so the Poisson process is invariant under the scaling transformation. The assumption of Lee et al. [145] appears to hold for most of the studies of natural image statistics.

However, in our experiment, $[r_{min}, r_{max}]$ is assumed to be a relatively narrow range. Under the scaling transformation, this range will change to $[r_{min}/s, r_{max}/s]$, which is far from being invariant. From this perspective, we may consider that Lee et al. [146] and the papers cited above are concerned with the marginal statistics by integrating over the whole range of scale. Our work, however, is concerned with conditional statistics given a narrow range of scale, especially how such conditional statistics change under the scaling transformation. While it is important to look at the marginal statistics over the whole range of scales, it is perhaps even more important to study the conditional statistics at different scales in order to model different image patterns. Moreover, the conditional statistics at different scales may have to be accounted for by different regimes of statistical models.

Image Formation and Scaling

Let $O_k \subset R^2$ be the squared area covered by leaf k in the (x, y) domain of the ivy wall. Then the scene of the ivy wall can be represented by a function $W(x, y) = a_{k(x,y)}$, where $k(x, y) = \arg\max_{k:(x,y) \in O_k} z_k$, i.e., the most forefront leaf that covers (x, y). $W(x, y)$ is a piecewise constant function defined on R^2.

Now let us see what happens if we take a picture of $W(x, y)$ from a distance d. Suppose the scope of the domain covered by the camera is $\Omega \subset R^2$, where Ω is a finite rectangular region. As noted by Mumford and Gidas [176], a camera or a human eye only has a finite array of sensors or photoreceptors. Each sensor receives light from a small neighborhood of Ω. As a simple model of the image formation process, we may divide the continuous domain Ω into a rectangular array of squared windows of length or width σd, where σ is decided by the resolution of the camera. Let $\{\Omega_{ij}\}$ be these squared windows, with $(i, j) \in D$, where D is a rectangular lattice.

The image $\mathbf{I}$ is defined on D. Let $s = d\sigma$ be the scale parameter of the image formation process; then

$$\mathbf{I}_s(i, j) = \frac{1}{s^2} \int_{\Omega_{ij}} W(x, y) dx dy, \ (i, j) \in D, \tag{10.1}$$

which is the average of $W(x, y)$ within window Ω_{ij}. Equation (10.1) can also be written as

$$w_s(x, y) = \frac{1}{s^2} \int W(x', y') g((x - x')/s, (y - y')/s) dx' dy' = W * g_s \tag{10.2}$$

$$\mathbf{I}_s(i, j) = w_s(u + is, v + js), \tag{10.3}$$

where g is a uniform density function within the window $[-1/2, 1/2] \times [-1/2, 1/2]$, and $g_s(x, y) = g(x/s, y/s)/s^2$. $(u, v) \in [0, s)^2$ denotes the small shifting of the rectangular lattice. There are two operations involved. Equation (10.2) is smoothing: w_s is a smoothed version of W. Equation (10.3) is subsampling: $\mathbf{I}_s$ is a discrete sampling of w_s. To be more general, g in Eq. (10.2) can be any density function, for instance, Gaussian density function.

The scale parameter s can be changed by either changing the viewing distance d or the camera resolution σ. If we increase s by increasing the viewing distance or zooming out the camera, then both the size of the scope Ω and the size of the windows Ω_{ij} will increase proportionally. So the resulting image $\mathbf{I}_s$ will change. For example, if we double s to $2s$, then $\mathbf{I}_{2s}$ will cover a scope 4 times as large as the scope of $\mathbf{I}_s$. Because each squared window of size $2s$ contains 4 squared windows of size s, if we look within the portion of $\mathbf{I}_{2s}$ that corresponds to $\mathbf{I}_s$, then the intensity of a pixel in $\mathbf{I}_{2s}$ is the block average of the intensities of the corresponding 2×2 pixels in $\mathbf{I}_s$.

If g is a Gaussian kernel, then the set of $\{w_s(x, y), s > 0\}$ forms a scale space. The scale space theory can account for the change of image intensities due to scaling. But it does not explain the change of statistical properties of the image data under the scaling transformation.

Empirical Observations on Information Scaling

Figure 10.2 shows a sequence of 8 images of W taken at 8 viewing distances. The images are generated according to Eq. (10.1). The viewing distance of the $i + 1$-st image is twice that of the i-th image. So the viewing distance of the last image is 128 times that of the first image. Within this wide range of viewing distance, the images display markedly different statistical properties even though they are generated by the same W. The reason is that the square leaves appear at different scales in different images:

(1) For an image taken at a near distance, such as image (1), the window size of a pixel is much less than the average size of the leaves, i.e., $s \ll r$. The image can be represented deterministically by a relatively small number of

occluding squares, or by local geometric structures such as edges, corners, etc. The constituent elements of the image are squares or local geometrical structures, instead of pixels.

(2) For an image at an intermediate distance, the window size of a pixel becomes comparable to the average size of leaves, i.e. $s \approx r$. The image becomes more complex. For images (4) and (5), they cannot be represented by a small number of geometrical structures anymore. The basic elements have to be pixels themselves. If a simple interpretation of the image is sought after, this interpretation has to be some sort of simple summary that cannot code the image intensities deterministically. The summary can be in the form of some spatial statistics of image intensities.

(3) For an image at a far distance, the window size of a pixel can be much larger than the average size of the squares, i.e., $s \gg r$. Each pixel covers a large number of leaves, and its intensity value is the average of many leaves. The image is approaching the white noise.

Computer vision algorithms always start from the analysis of local image patches, often at multiple resolutions. We take some local 7×7 image patches from the images at different scales shown in Fig. 10.3. These local image patches exhibit very different characteristics. Patches from near-distance images are highly structured, corresponding to simple regular structures such as edges and corners, etc. As the distance increases, the patches become more irregular and random. So the local analysis in a computer vision system should be prepared to deal with such local image patches with different regularities and randomness.

Change of Compression Rate

We perform some empirical studies on the change of statistical properties of the image data over scale. What we care about most is the complexity or randomness of the image, and we measure the complexity rate or randomness empirically by JPEG 2000 compression rate. Generally speaking, for a simple and regular image, there are a lot of redundancies in the image intensities. So only a small number of bits are needed to store the image without any loss of information up to the discretization precision. For a complex and random image, there is not much regularity or redundancy in the data. Therefore, a large number of bits are required to store the image. The reason we use JPEG 2000 to measure the complexity rate is two folded. First, JPEG 2000 is the state-of-the-art image compression standard and currently gives the best approximation to image complexity. Second, given the popularity of JPEG 2000, our results should also be interesting to the image compression community.

When the image is compressed by JPEG 2000, the size of the compressed image file is recorded in terms of the number of bits. This number is then divided by the number of pixels to give the compression rate in terms of bits per pixel. Figure 10.4a

Scale

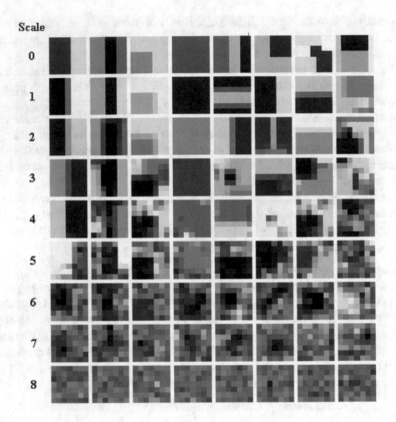

Fig. 10.3 First published in the Quarterly of Applied Mathematics in Volume 66:81–122, 2008, published by Brown University. Reprinted with permission from [257]. The 7×7 local patches taken from the images at different scales

plots this measurement in the order of viewing distance for images in Fig. 10.2. At a close distance, the randomness is small, meaning that the image is quite regular. Then the randomness starts to increase over distance because more and more leaves are covered by the scope of the camera. At a far distance, however, the randomness begins to decrease because the local averaging operation reduces the marginal variance and eventually smoothes the image into a constant image because of the law of large number. In this plot, there are three curves. They correspond to three different r_{min} in our simulation study, while r_{max} is always fixed at the same value. For smaller r_{min}, the corresponding curve shifts to the left because the average size of the leaves is smaller.

We also use a simple measure of smoothness as an indicator of randomness or complexity rate. We compute pairwise differences between intensities of adjacent pixels $\nabla_x \mathbf{I}(i, j) = \mathbf{I}(i, j) - \mathbf{I}(i-1, j)$ and $\nabla_y \mathbf{I}(i, j) = \mathbf{I}(i, j) - \mathbf{I}(i, j-1)$. $\nabla \mathbf{I}(i, j) = (\nabla_x \mathbf{I}(i, j), \nabla_y \mathbf{I}(i, j))$ is the gradient of $\mathbf{I}$ at (i, j). The gradient is a very useful local feature that can be used for edge detection [257]. It is also extensively used

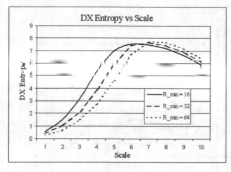
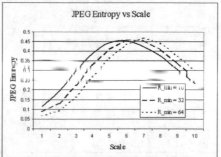

Fig. 10.4 First published in the Quarterly of Applied Mathematics in Volume 66:81–122, 2008, published by Brown University. Reprinted with permission from [257]. The change of statistical properties over scale. (**a**) JPEG compression rate. (**b**) Entropy of marginal histogram of $\nabla_x \mathbf{I}$

in image processing. We make a marginal histogram of $\{\nabla_x \mathbf{I}(i, j), (i, j) \in D\}$ and compute the entropy of the histogram. Figure 10.4b plots this entropy over the order of distance for images in Fig. 10.2. The plot behaves similarly to the plot of the JPEG 2000 compression rate.

Variance Normalization

The local averaging operation in Eq. (10.1) reduces the marginal variance of the image intensities. A more appropriate measurement of randomness should be the compression rate of variance-normalized image, which is invariant of linear trans-formations of image intensities. Specifically, for an image $\mathbf{I}$, let σ^2 be the marginal variance of $\mathbf{I}$. Let $\mathbf{I}'(i, j) = \mathbf{I}(i, j)/\sigma$. Then $\mathbf{I}'$ is the variance-normalized version of $\mathbf{I}$, and the marginal variance of $\mathbf{I}'$ is 1. We compute the JPEG compression rates of variance-normalized versions of the images in Fig. 10.2. Figure 10.5a displays the variance-normalized JPEG compression rate over the order of distance for the three runs of the simulation study. The compression rate increases monotonically toward an upper bound represented by the horizontal line. This suggests that the scaling process increases the randomness and transforms a regular image to a random image. The upper bound is the JPEG compression rate of the Gaussian white noise process with variance 1.

The convergence of the compression rate of the variance-normalized image to that of the Gaussian white noise image is due to the effect of the central limit theorem. As another illustration, we compute the kurtosis of the marginal distribution of $\{\nabla_x \mathbf{I}(x), x \in D\}$. The kurtosis decreases monotonically toward 0, meaning that the image feature becomes closer to Gaussian distribution.

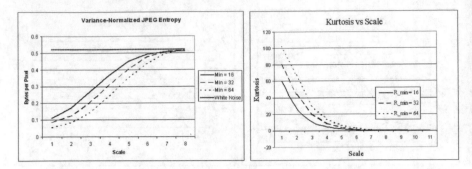

Fig. 10.5 First published in the Quarterly of Applied Mathematics in Volume 66:81–122, 2008, published by Brown University. Reprinted with permission from [257]. (**a**) The change of JPEG compression rate of the variance-normalized versions of the images in Fig. 10.2. (**b**) The change of kurtosis

Basic Information Theoretical Concepts

Let $\mathbf{I}(x, y)$ be an image with $(x, y) \in D$, where D is the discrete lattice of pixels (in what follows, we use (x, y) instead of (i, j) to denote discrete pixels). Let $p(\mathbf{I})$ be the distribution of $\mathbf{I}$. We are interested in the following statistical properties [29]:

(1) *Entropy and entropy rate:* The entropy of p is defined as

$$\mathcal{H}(p) = \mathrm{E}_p[-\log p(\mathbf{I})] = -\int p(\mathbf{I}) \log p(\mathbf{I}) d\mathbf{I}, \qquad (10.4)$$

and the entropy rate of p is defined as $\bar{\mathcal{H}}(p) = \mathcal{H}(p)/|D|$, where $|D|$ is the number of pixels in lattice D.

(2) *Relative entropy and relative entropy rate:* For two distributions p and q, the relative entropy or the Kullback–Leibler divergence between p and q is defined as

$$\mathrm{KL}(p\|q) = \mathrm{E}_p \left[\log \frac{p(\mathbf{I})}{q(\mathbf{I})} \right] = -\mathcal{H}(p) - \mathrm{E}_p[\log q(\mathbf{I})] \geq 0. \qquad (10.5)$$

The relative entropy rate is $\hat{k}(p\|q) = \mathrm{KL}(p\|q)/|D|$.

(3) *Relative entropy with respect to Gaussian white noise:* For an image distribution p, let

$$\frac{1}{|D|} \sum_{(x,y) \in D} \mathrm{E}[\mathbf{I}(x, y)^2] = \sigma^2 \qquad (10.6)$$

be the marginal variance. Let q be the Gaussian white noise distribution with mean 0 and variance σ^2, i.e., $\mathbf{I}(x, y) \sim \mathrm{N}(0, \sigma^2)$ independently. Then

$$KL(p\|q) = -\mathcal{H}(p) - E_p[\log q(\mathbf{I})] = \mathcal{H}(q) - \mathcal{H}(p) \geq 0. \quad (10.7)$$

The second equation in (10.7) follows from $E_p[\log q(\mathbf{I})] = E_q[\log q(\mathbf{I})]$ because $\log q(\mathbf{I})$ is linear in $\sum_{x,y} \mathbf{I}(x, y)^2$, which has the same expectations under both p and q. Because $\mathcal{H}(q) \geq \mathcal{H}(p)$ according to (10.7), the Gaussian white noise distribution has the maximum entropy among all the image distributions with the same marginal variance.

(4) *Entropy rate of variance-normalized image:* Continue from (10.7) and calculate the entropy rate of Gaussian white noise explicitly, we obtain the relative entropy rate

$$\hat{k}(p\|q) = \log \sqrt{2\pi e} - [\bar{\mathcal{H}}(p(\mathbf{I})) - \log \sigma] = \log \sqrt{2\pi e} - \bar{\mathcal{H}}(p(\mathbf{I}')), (10.8)$$

where $\mathbf{I}' = \mathbf{I}/\sigma$ is the variance-normalized version of image $\mathbf{I}$, and $p(\mathbf{I}')$ denotes the distribution of $\mathbf{I}'$. So the entropy rate of the variance-normalized image $\bar{\mathcal{H}}(p(\mathbf{I}'))$ determines the relative entropy rate $\hat{k}(p\|q)$ of $p(\mathbf{I})$ with respect to the Gaussian white noise $q(\mathbf{I})$. In other words, $\bar{\mathcal{H}}(p(\mathbf{I}'))$ measures the departure of p from the Gaussian white noise hypothesis.

Change of Entropy Rate

For simplicity, let us study what happens if we double the viewing distance or zoom out the image by a factor of 2. Suppose the current image is $\mathbf{I}(x, y)$, $(x, y) \in D$. If we double the viewing distance, the window covered by a pixel will double its size. So the original $\mathbf{I}$ will be reduced to a smaller image $\mathbf{I}_-$ defined on a reduced lattice D_-, and each pixel of $\mathbf{I}_-$ will be the block average of four pixels of $\mathbf{I}$. More specifically, the process can be accounted for by two steps, similar to Eqs. (10.2) and (10.3):

(1) *Local smoothing:* Let the smoothed image be $\mathbf{J}$; then $\mathbf{J}(x, y) = \sum_{u,v} \mathbf{I}(x + u, y+v)/4$, where $(u, v) \in \{(0, 0), (0, 1), (1, 0), (1, 1)\}$. We can write $\mathbf{J} = \mathbf{I} * g$, where g is the uniform distribution over $\{(0, 0), (0, -1), (-1, 0), (-1, -1)\}$. In general, g can be any kernel with appropriate bandwidth, such as a Gaussian distribution function.

(2) *Subsampling:* $\mathbf{I}_-^{(u,v)}(x, y) = \mathbf{J}(2x + u, 2y + v)$, where, again, $(u, v) \in \{(0, 0), (0, 1), (1, 0), (1, 1)\}$. Any of the four $\mathbf{I}_-^{(u,v)}$ can be regarded as a subsampled version of $\mathbf{J}$.

Theorem 2 *Smoothing effect: Let D be an $M \times N$ lattice, and $\mathbf{I}$ is defined on D. Let $\mathbf{J} = \mathbf{I} * g$, where g is a local averaging kernel or a probability distribution. As* $\min(M, N) \to \infty$,

$$\bar{\mathcal{H}}(p(\mathbf{J})) - \bar{\mathcal{H}}(p(\mathbf{I})) \to \frac{1}{4\pi^2} \int_0^{2\pi} \int_0^{2\pi} \log|\hat{g}(\omega)| d\omega \le 0, \qquad (10.9)$$

where $\omega = (\omega_x, \omega_y)$ is the spatial frequency, and $\hat{g}(\omega) = \sum_{x,y} g(x, y)$ $\exp\{-i(\omega_x x + \omega_y y)\}$ is the Fourier transform of the kernel g, where the sum is over the support of g.

Proof Let $\mathbf{I}$ be the image defined on the integer lattice $[0, M - 1] \times [0, N - 1]$. The discrete Fourier transform of $\mathbf{I}$ is

$$\hat{\mathbf{I}}(\omega) = \sum_{x=0}^{M-1} \sum_{y=0}^{N-1} \mathbf{I}(x, y) \exp\{-i(\omega_x x + \omega_y y)\}, \qquad (10.10)$$

where $\omega_x \in \{2\pi m/M, m = 0, ..., M-1\}$ and $\omega_y \in \{2\pi n/N, n = 0, ..., N-1\}$. The Fourier transforms of $\mathbf{J}$ and g can be similarly defined. Because $\hat{\mathbf{I}}$ and $\hat{\mathbf{J}}$ are obtained from $\mathbf{I}$ and $\mathbf{J}$, respectively, by the same linear transformation, $\mathcal{H}(p(\hat{\mathbf{J}})) - \mathcal{H}(p(\hat{\mathbf{I}})) = \mathcal{H}(p(\mathbf{J})) - \mathcal{H}(p(\mathbf{I}))$.

For convolution with periodic boundary condition, $\hat{\mathbf{J}}(\omega) = \hat{\mathbf{I}}(\omega)\hat{g}(\omega)$. So

$$\bar{\mathcal{H}}(p(\mathbf{J})) - \bar{\mathcal{H}}(p(\mathbf{I})) = \frac{1}{|D|} \left[\mathcal{H}(p(\hat{\mathbf{J}})) - \mathcal{H}(p(\hat{\mathbf{I}})) \right] \qquad (10.11)$$

$$= \frac{1}{MN} \sum_{\omega} \log|\hat{g}(\omega)| = \frac{1}{4\pi^2} \sum_{\omega} \log|\hat{g}(\omega)| \Delta\omega \qquad (10.12)$$

$$\to \frac{1}{4\pi^2} \int_0^{2\pi} \int_0^{2\pi} \log|\hat{g}(\omega)| d\omega, \qquad (10.13)$$

as $\min(M, N) \to \infty$, where $\Delta\omega = (2\pi/M) \times (2\pi/N)$.

A smoothing kernel g is a probability distribution function, $\hat{g}$ is the characteristic function of g, and

$$\hat{g}(\omega) = \sum_{x,y} g(x, y) \exp\{-i(\omega_x x + \omega_y y)\} \qquad (10.14)$$

$$= E_g \left[\exp\{-i(\omega_x X + \omega_y Y)\} \right], \qquad (10.15)$$

where $(X, Y) \sim g(x, y)$. Then,

$$|\hat{g}(\omega)|^2 = \left| E_g \left[\exp\{-i(\omega_x X + \omega_y Y)\} \right] \right|^2 \qquad (10.16)$$

$$\le E_g \left[|\exp\{-i(\omega_x X + \omega_y Y)\}|^2 \right] = 1. \qquad (10.17)$$

Thus, $\int \log|\hat{g}(\omega)| d\omega \le 0$ QED.

The above theorem tells us that there is always loss of information under the smoothing operation. This is consistent with the intuition in scale space theory, where the increase in scale results in the loss of fine details in the image. The change of entropy rate under linear filtering was first derived in the classical paper of Shannon [215, 216].

Next, let us study the effect of subsampling. There are four subsampled versions $\mathbf{I}_-^{(u,v)}(x, y) = \mathbf{J}(2x + u, 2y + v)$, where $(u, v) \in \{(0, 0), (0, 1), (1, 0), (1, 1)\}$. Each $\mathbf{I}_-^{(u,v)}$ is defined on a subsampled lattice D_-, with $|D_-| = |D|/4$. □

Theorem 3 *Subsampling effect: The average entropy rate of* $\mathbf{I}_-^{(u,v)}$ *is no less than the entropy rate of* $\mathbf{J}$,

$$\frac{1}{4} \sum_{u,v} \bar{\mathcal{H}}(p(\mathbf{I}_-^{(u,v)})) - \bar{\mathcal{H}}(p(\mathbf{J})) = \bar{\mathcal{M}}(\mathbf{I}_-^{(u,v)}, \forall(u, v)) \geq 0, \qquad (10.18)$$

where $\mathcal{M}(\mathbf{I}_-^{(u,v)}, \forall(u, v)) = \mathrm{KL}(p(\mathbf{J}) \| \prod_{u,v} p(\mathbf{I}_-^{(u,v)}))$ *is defined as the mutual information among the four subsampled versions, and* $\bar{\mathcal{M}} = \mathcal{M}/|D|$.

Proof

$$\sum_{u,v} \mathcal{H}(p(\mathbf{I}_-^{(u,v)})) - \mathcal{H}(p(\mathbf{J})) = \mathrm{E}\left[\log \frac{p(\mathbf{J})}{\prod_{u,v} p(\mathbf{I}_-^{(u,v)})}\right] \qquad (10.19)$$

$$= \mathrm{KL}(p(\mathbf{J}) \| \prod_{u,v} p(\mathbf{I}_-^{(u,v)})) \qquad (10.20)$$

$$= \mathcal{M}(\mathbf{I}_-^{(u,v)}, \forall(u, v)) \geq 0, \qquad (10.21)$$

where the expectation is with respect to the distribution of $\mathbf{J}$, which is also the joint distribution of $\mathbf{I}^{(u,v)}$ QED. □

The scaling of the entropy rate is a combination of Eqs. (10.9) and (10.18):

$$\left\{\frac{1}{4} \sum_{u,v} \bar{\mathcal{H}}(p(\mathbf{I}_-^{(u,v)})) - \bar{\mathcal{H}}(p(\mathbf{I}))\right\} - \left\{\bar{\mathcal{M}}(\mathbf{I}_-^{(u,v)}) + \frac{1}{4\pi^2} \int \log |\hat{g}(\omega)| d\omega\right\} \to 0.$$

$$(10.22)$$

For regular image patterns, the mutual information per pixel can be much greater than $-\int \log |\hat{g}(\omega)| d\omega/4\pi^2$, so the entropy rate increases with distance, or in other words, the image becomes more random. For very random patterns, the reverse is true. When the mutual information rate equals $-\int \log |\hat{g}(\omega)| d\omega/4\pi^2$, we have scale invariance. More careful analysis is needed to determine when this is true.

Next, we study the change of entropy rate of variance-normalized image $\bar{\mathcal{H}}(p(\mathbf{I}'))$. For simplicity, let us assume that $p(\mathbf{I})$ comes from a stationary process, and $\mathbf{I}_-$ can be any subsampled version of $\mathbf{J} = \mathbf{I} * g$, which is also stationary. Let $\sigma^2 = \text{Var}[\mathbf{I}(x, y)]$ and $\sigma_-^2 = \text{Var}[\mathbf{I}_-(x, y)]$ be the marginal variances of $\mathbf{I}$ and $\mathbf{I}_-$, respectively. Let $\mathbf{I}' = \mathbf{I}/\sigma$ and $\mathbf{I}'_- = \mathbf{I}_-/\sigma_-$ be the variance-normalized versions of $\mathbf{I}$ and $\mathbf{I}_-$, respectively. It is easy to show that

$$\rho^2 = \frac{\sigma_-^2}{\sigma^2} = \frac{1}{4} \sum_{u,v} \text{corr}(\mathbf{I}(x, y), \mathbf{I}(x + u, y + v)) \le 1, \ (u, v) \in \{(0, 0), (0, 1), (1, 0), (1, 1)\}, (10.23)$$

so the smoothing operation reduces the marginal variance. Therefore, we can modify (10.22) into

$$\bar{\mathcal{H}}(p(\mathbf{I}'_-)) - \bar{\mathcal{H}}(p(\mathbf{I}')) \approx \bar{\mathcal{M}}(\mathbf{I}_-^{(u,v)}) - \log \rho + \frac{1}{4\pi^2} \int \log |\hat{g}(\omega)| d\omega, (10.24)$$

where the difference between the left-hand side and right-hand side converges to 0 as $|D| \to \infty$. In (10.24), the term $-\log \rho$ is positive, and it compensates for the loss of entropy rate caused by smoothing, i.e., $\int \log |\hat{g}(\omega)| d\omega / 4\pi^2$, which is negative. As a matter of fact, the first two terms, i.e., the mutual information term and the $-\log \rho$ term on the right- hand side of (10.24), balance each other, in the sense that if one is small, then the other tends to be large. However, we have not been able to identify conditions under which the right-hand side of (10.24) is always positive, which would have established the monotone increase of the entropy rate of variance-normalized image or monotone decrease of the departure from Gaussian white noise.

10.2   Perceptual Entropy

The above analysis of entropy rate is only about the observed image $\mathbf{I}$ alone. The goal of computer vision is to interpret the observed image in order to recognize the objects in the outside world. In this subsection, we shall go beyond the statistical properties of the observed image itself and study the interaction between the observed image and the outside scene that produces the image.

Again, we would like to use the dead leaves model to convey the basic idea. Suppose our attention is restricted to a finite scope $\Omega \subset \mathbf{R}^2$, and let $W = ((x_i, y_i, r_i, a_i), i = 1, ..., N)$ be the leaves in Ω that are not completely occluded by other leaves. Then we have $W \sim p(W)$ and $\mathbf{I} = \gamma(W)$, where $p(W)$ comes from the Poisson process that generates the dead leaves, and γ represents the transformation defined by Eq. (10.1) for a scale parameter s.

For convenience, we assume that both W and $\mathbf{I}$ are properly discretized. For any joint distribution $p(W, \mathbf{I})$, the conditional entropy $\mathcal{H}(p(W \mid \mathbf{I}))$ is defined as

$$\mathcal{H}(p(W \mid \mathbf{I})) = -\sum_{W,\mathbf{I}} p(W, \mathbf{I}) \log p(W \mid \mathbf{I}). \tag{10.25}$$

$\mathcal{H}(p(W \mid \mathbf{I}))$ measures the inferential uncertainty or the imperceptibility of W from the image $\mathbf{I}$.

Proposition 1 *If $W \sim p(W)$ and $\mathbf{I} = \gamma(W)$, then $\mathcal{H}(p(W|\mathbf{I})) = \mathcal{H}(p(W)) - \mathcal{H}(p(\mathbf{I}))$. That is, imperceptibility = scene entropy - image entropy.*

This proposition is easy to prove. The marginal distribution of $\mathbf{I}$ is $p(\mathbf{I}) = \sum_{W:\gamma(W)=\mathbf{I}} p(W)$. The posterior distribution of W given $\mathbf{I}$ is $p(W|\mathbf{I}) = p(W, \mathbf{I})/p(\mathbf{I}) = p(W)/p(\mathbf{I})$. Here, $p(W, \mathbf{I}) = p(W)$ because $\mathbf{I}$ is determined by W. Following the definition in (10.25), $\mathcal{H}(p(W \mid \mathbf{I})) = -\sum_W p(W)(\log p(W) - \log p(\mathbf{I})) = \mathcal{H}(p(W)) - \mathcal{H}(p(\mathbf{I}))$. Here $\mathrm{E}_W[\log p(\mathbf{I})] = \mathrm{E}_{\mathbf{I}}[\log p(\mathbf{I})]$ since $\mathbf{I}$ is determined by W.

If we increase the viewing distance or equivalently zoom out the camera while fixing the scope $\Omega \subset \mathbb{R}^2$, i.e., fixing W, then we obtain a zoomed-out version $\mathbf{I}_- = R(\mathbf{I})$, where R represents the zooming-out operation of smoothing and subsampling and is a many-to-one transformation. During the process of zooming out, the total entropy of the image will decrease, i.e., $\mathcal{H}(p(\mathbf{I}_-)) \leq \mathcal{H}(p(\mathbf{I}))$, even though the entropy per pixel can increase as we have shown in the previous subsection. Therefore, we have the following result.

Proposition 2 *If $W \sim p(W)$, $\mathbf{I} = \gamma(W)$, and $\mathbf{I}_- = R(\mathbf{I})$, where R is a many-to-one mapping, then $\mathcal{H}(p(W|\mathbf{I}_-)) \geq \mathcal{H}(p(W|\mathbf{I}))$, i.e., the imperceptibility increases as the image is reduced.*

What does this result tell us in terms of interpreting image $\mathbf{I}$ or $\mathbf{I}_-$? Although the model $W \sim p(W)$ and $\mathbf{I} = \gamma(W)$ is the right physical model for all the scale s, this model is meaningful in interpreting $\mathbf{I}$ only within a limited range, say $s \leq s_{\text{bound}}$, so that the imperceptibility $\mathcal{H}(p(W \mid \mathbf{I}))$ is below a small threshold. In this regime, the representation $\mathbf{I} = \gamma(W)$ is good for both recognition and coding. For recognition, $\mathcal{H}(p(W \mid \mathbf{I}))$ is small, so W can be accurately determined from $\mathbf{I}$. For coding, we can first code W according to $p(W)$, with a coding cost $\mathcal{H}(p(W))$. Then we code $\mathbf{I}$ using $\mathbf{I} = \gamma(W)$ without any coding cost. The total coding cost would be just $\mathcal{H}(p(W))$. If the imperceptibility $\mathcal{H}(p(W \mid \mathbf{I}))$ is small, $\mathcal{H}(p(W)) \approx \mathcal{H}(p(\mathbf{I}))$, so coding W will not incur coding overhead.

But if s is very large, the imperceptibility $\mathcal{H}(p(W \mid \mathbf{I}))$ can be large according to Proposition 2. In this case, the representation $\mathbf{I} = \gamma(W)$ is not good for either recognition or coding. For recognition, W cannot be estimated with much certainty. For coding, if we still code W first and code $\mathbf{I}$ by $\mathbf{I} = \gamma(W)$, this will not be an efficient coding, since $\mathcal{H}(p(W))$ can be much larger than $\mathcal{H}(p(\mathbf{I}))$, and the difference is imperceptibility $\mathcal{H}(p(W \mid \mathbf{I}))$.

Then what should we do? The regime of $s > s_{\text{bound}}$ is quite puzzling for vision modeling. Our knowledge of geometry, optics, and mechanics enables us to model every phenomenon in our physical environment. Such models may be sufficient for computer graphics as far as generating physically realistic images is concerned. For instance, a garden scene can be constructed by simulating billions of leaves and grass strands, and the image can be produced by projecting these billions of objects onto the image with perspective geometry. A river scene, a fire scene, or a smoke scene can be obtained using computational fluid dynamics. A piece of cloth can be generated using a dense set of particles that follow the law of mechanics. Realistic lighting can be simulated by ray tracing and optics. But such models are hardly meaningful for vision because the imperceptibilities of the underlying elements or variables are intolerable. When we look at a garden scene, we never really perceive every leaf or every strand of grass. When we look at a river scene, we do not perceive the constituent elements used in fluid dynamics. When we look at a scene with sophisticated lighting and reflection, we do not trace back the light rays. In those situations where physical variables are not perceptible due to scaling or other aspects of the image formation process, it is quite a challenge to come up with good models for the observed images. Such models do not have to be physically realistic, but they should generate visually realistic images so that such models can be employed to interpret the observed image at a level of sophistication that is comparable to human vision.

Suppose the image $\mathbf{I}$ is reduced to an image $\mathbf{I}_- = R(\mathbf{I})$, so that W cannot be reliably inferred. Then, instead of pursuing a detailed description W from $\mathbf{I}_-$, we may choose to estimate some aspects of W from $\mathbf{I}_-$. For instance, in the simulated ivy wall example, we may estimate properties of the overall distribution of colors of leaves, as well as the overall distribution of their sizes, etc. Let us call it $W_- = \rho(W)$, with ρ being a many-to-one reduction function. It is possible that we can estimate W_- from $\mathbf{I}_-$ because of the following result.

Proposition 3 *Let $W \sim p(W)$, $\mathbf{I} = \gamma(W)$, and $W_- = \rho(W)$, $\mathbf{I}_- = R(\mathbf{I})$, where both ρ and R are many-to-one mappings; we have*

$$(1) \quad \mathcal{H}(p(W_-|\mathbf{I}_-)) \leq \mathcal{H}(p(W|\mathbf{I}_-)). \tag{10.26}$$

$$(2) \quad p(\mathbf{I}_-|W_-) = \frac{\sum_{W:\rho(W)=W_-;R(\gamma(W))=\mathbf{I}_-} p(W)}{\sum_{W:\rho(W)=W_-} p(W)}. \tag{10.27}$$

Result (1) tells us that even if W is imperceptible from $\mathbf{I}_-$, W_- may still be perceptible. Result (2) tells us that although W defines $\mathbf{I}$ deterministically via $\mathbf{I} = \gamma(W)$, W_- may only define $\mathbf{I}_-$ statistically via a probability distribution $p(\mathbf{I}_-|W_-)$. While W represents deterministic structures, W_- may only represent some texture properties. Thus, we have a transition from a deterministic representation of the image intensities $\mathbf{I} = \gamma(W)$ to a statistical characterization $\mathbf{I}_- \sim p(\mathbf{I}_-|W_-)$. See Fig. 10.6 for an illustration.

For an image $\mathbf{I}$, we may extract $F(\mathbf{I})$, which can be a dimension reduction or a statistical summary, so that $F(\mathbf{I})$ contains as much information about $\mathbf{I}$ as possible as far as W or W_- is concerned. In the following proposition, we shall not distinguish between $(W, \mathbf{I})$ and $(W_-, \mathbf{I}_-)$ for notational uniformity.

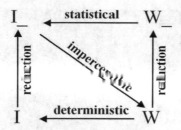

Fig. 10.6 First published in the Quarterly of Applied Mathematics in Volume 66:81–122, 2008, published by Brown University. Reprinted with permission from [257]. The transition from deterministic representation to statistical description

Proposition 4 *Let* $F = F(\mathbf{I})$:

(1) If $W \sim p(W)$, $\mathbf{I} = \gamma(W)$, *then* $\mathrm{KL}(p(W|\mathbf{I})\|p(W|F)) = \mathcal{H}(p(\mathbf{I}|F))$.

(2) If $W \sim p(W)$ *and* $[\mathbf{I}|W] \sim p(\mathbf{I}|W)$, *then* $\mathrm{KL}(p(W|\mathbf{I})\|p(W|F)) = \mathcal{M}(W, \mathbf{I}|F)$, *where* $\mathcal{M}(W, \mathbf{I}|F) = \mathrm{E}_{W,\mathbf{I}}\{\log[p(W, \mathbf{I}|F)/(p(W|F)p(\mathbf{I}|F))]\}$ *is the mutual information between* W *and* $\mathbf{I}$ *given* F.

Result (1) tells us that for $F(\mathbf{I})$ to contain as much information about W as possible, we want to make $\mathcal{H}(p(\mathbf{I}|F))$ to be as small as possible, so that F can be used to reconstruct $\mathbf{I}$ accurately. Result (2) tells us that if we want to estimate W, we want F to be sufficient about $\mathbf{I}$ as far as W is concerned. $\mathcal{M}(W, \mathbf{I}|F)$ can be considered a measurement of sufficiency.

Now let us study this issue from the coding perspective. Suppose the image $\mathbf{I}$ follows a true distribution $f(\mathbf{I})$, and we use a model $w \sim p(w)$, and $[\mathbf{I} \mid w] \sim p(\mathbf{I} \mid w)$ to code $\mathbf{I} \sim f(\mathbf{I})$. Here the variable w is augmented solely for the purpose of coding. It might be some $w = W_- = \rho(W)$, or it may not have any correspondence to the reality W. In the coding scheme, for an image $\mathbf{I}$, we first estimate w by a sample from the posterior distribution $p(w|\mathbf{I})$, and then we code w by $p(w)$ with coding length $-\log p(w)$. After that, we code $\mathbf{I}$ by $p(\mathbf{I}|w)$ with coding length $-\log p(\mathbf{I}|w)$. So the average coding length is $-\mathrm{E}_f\left[\mathrm{E}_{p(w|\mathbf{I})}(\log p(w) + \log p(\mathbf{I}|w))\right]$.

Proposition 5 *The average coding length is* $\mathrm{E}_f[\mathcal{H}(p(w|\mathbf{I}))] + \mathrm{KL}(f(\mathbf{I})\|p(\mathbf{I})) + \mathcal{H}(f)$, *where* $p(\mathbf{I}) = \sum_w p(w)p(\mathbf{I} \mid w)$ *is the marginal distribution of* $\mathbf{I}$ *under the model. So, coding redundancy = imperceptibility + model bias.*

The above proposition provides a selection criterion for models with latent variables. The imperceptibility term comes up because we assume a coding scheme where w must be coded first, and then $\mathbf{I}$ is coded based on w. Given the latent variable structure of the model, it is very natural to assume such a coding scheme

Fig. 10.7 The image contains patterns of different complexities, from very simple patterns such as geometric patterns to very random patterns such as leaves at far distance

A Continuous Spectrum

Image patterns of different entropy regimes are not only connected by image scaling, they co-exist and blend seamlessly in a single image (Fig. 10.7). For instance, imagine we are in the wood of maple trees and taking a picture. The patterns displayed in Fig. 10.1 may appear together in the picture we take because the maple leaves can appear at different distances from the camera when the picture is taken. In addition, even for the same objects in a fixed image, when we analyze this image in multiple resolutions, we may recognize patterns from different regimes. The close connection between different regimes calls for a common theoretical framework for modeling patterns in these regimes. In particular, it calls for the integration of the MRFs and sparse coding models that work well in the high-entropy regime and low-entropy regime, respectively.

10.3   Perceptual Scale Space

When an image is viewed at varying resolutions, it is known to create discrete perceptual jumps or transitions amid the continuous intensity changes. Wang and Zhu [248] studied a perceptual scale space theory that differs from the traditional image scale space theory in two aspects: (i) In representation, the perceptual scale space adopts a full generative model. From a Gaussian pyramid, it computes a sketch pyramid where each layer is a primal sketch representation—an attribute graph whose elements are image primitives for the image structures. Each primal sketch graph generates the image in the Gaussian pyramid, and the changes between the primal sketch graphs in adjacent layers are represented by a set of basic and

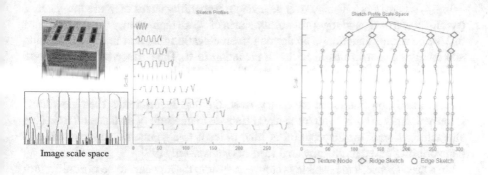

Fig. 10.8 Reprinted with permission from [248]. Scale space of a 1D signal. (**a**) A toaster image from which a line is taken as the 1D signal. (**b**) Trajectories of zero-crossings of the 2nd derivative of the 1D signal. The finest scale is at the bottom. (**c**) The 1D signal at different scales. The black segments on the curves correspond to primal sketch primitives (step edge or bar). (**d**) A symbolic representation of the sketch in scale space with three types of transitions

composite graph operators to account for the perceptual transitions. (ii) In computation, the sketch pyramid and graph operators are inferred, as hidden variables, from the images through Bayesian inference by stochastic algorithm, in contrast to the deterministic transforms or feature extraction, such as computing zero-crossings, extremal points, and inflection points in the image scale space. Studying the perceptual transitions under the Bayesian framework makes it convenient to use the statistical modeling and learning tools for (a) modeling the Gestalt properties of the sketch graph, such as continuity and parallelism, etc.; (b) learning the most frequent graph operators, i.e., perceptual transitions, in image scaling; and (c) learning the prior probabilities of the graph operators conditioning on their local neighboring sketch graph structures (Fig. 10.8).

In experiments, they learn the parameters and decision thresholds through human experiments, and they show that the sketch pyramid is a more parsimonious representation than a multi-resolution Gaussian/Wavelet pyramid. They also demonstrate an application on adaptive image display—showing a large image on a small screen (say PDA) through a selective tour of its image pyramid. In this application, the sketch pyramid provides a means for calculating information gain in zooming-in different areas of an image by counting a number of operators expanding the primal sketches, such that the maximum information is displayed in a given number of frames.

10.4 Energy Landscape

The distribution of images in the image space $\Omega_I \subset R^N$ defined by an energy function $U : \Omega_I \to R$ can be understood as a manifold, or landscape, of dimension R^N in the high-dimensional space R^{N+1}. The energy function is given by $U =$

$-\log f$, where f is a density over Ω_I. High-probability states of f are low-energy (stable) states of U, and low-probability states of f are high-energy (unstable) states of U. The energy function is analogous to an elevation function that maps latitude and longitude coordinates in R^2 to an elevation in R. The surface of the landscape given by the elevation map is a 2D manifold in 3D space, and the same intuition extends to higher dimensions. The energy function U defines a geodesic distance measure and non-Euclidean geometry over Ω_I that incorporates the "elevation" information given by U to alter the usual Euclidean measure of distance on R^N. This characterization of the distribution of images leads to a geometric understanding of perceptibility based on the physical idea of *metastability* [99].

In a broad sense, a metastable system is a system that appears to be in equilibrium when viewed over short time periods but which deviates substantially from the short-scale quasi-equilibrium over long time periods. The concept of metastability provides a framework for understanding the structure of a density f by observing quasi-equilibrium behavior in the physical system associated with the landscape of U.

One can computationally simulate a diffusion process in the physical system of U by obtaining MCMC samples with a steady state f. In virtually all situations, the MCMC sampler is theoretically ergodic with respect to Ω_I, meaning that the sampling process on f will eventually visit every image $\mathbf{I}$ in the image space with probability 1 if the sampling is continued for a sufficient number of steps. On the other hand, it is well known that local MCMC samplers have difficulty mixing between separate modes, which is usually considered a major drawback of MCMC methods. However, the slow mixing and high autocorrelation of local MCMC samples is actually a manifestation of metastable phenomena in the landscape of U that provide a means of understanding f.

A density f that models realistic images will have an astronomical abundance of local modes that represent the diverse variety of possible appearances along the data manifold in the image space. On the other hand, groups of related minima that represent similar images often merge in the landscape of U to form macroscopic basins or funnels that capture consistent clusters or concepts of images found in the training data. The macroscopic basins are contained regions that permit diffusion interiorly but are separated by energy barriers that dramatically decrease the probability of cross-basin diffusion. Therefore, one can identify concepts within an image density f by identifying the metastable regions of the energy landscape. It is important to remember that metastable behavior often exists across a continuous spectrum of time scales, so there is no ground truth for the correct metastable structure and/or conceptual clustering of f. Instead, the metastable description provides a natural way to explore the concepts within a density at a range of degrees of visual similarity, permitting "coarser" or "finer" mappings depending on the context.

We can intuitively understand the structure of the energy landscape of natural images by considering Ω_I as a "universe" of images. The high-energy/low-

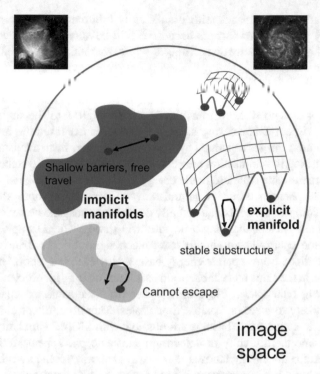

Fig. 10.9 The meaningful structures of image density can be intuitively understood as high-density regions in our universe. The majority of the image universe consists of empty space that represents appearances not observed within the training set. Texton-scale images are analogous to star clusters with a stable substructure that enables the recognition of a distinct appearance. Texture-scale images are analogous to nebulae whose mass covers a wide area but which contain little recognizable substructure. The gravitational pull of these structures is analogous to the metastable behavior of MCMC chains, which enables efficient mapping of macroscopic energy features

probability regions of U are empty space, which accounts for the vast majority of the volume of Ω_I. The low-energy regions and local modes of U represent high-density regions. Low-energy regions that represent texton-scale images are concentrated structures such as stars, while low-energy regions representing texture are more diffused and loose structures such as nebulas. Groups of related stars and nebulae form galaxies that represent general concepts within the image data, as in Fig. 10.9.

The discussion above reveals an important connection between visual percepti-bility of difference among a population of images with density f and metastable structures in the energy landscape U. A realistic image following f will have its own associated local minima region that captures a single stable appearance. The barriers between realistic images should depend on the degree of similarity between images. Images that share a similar appearance will be separated by lower energy barriers because it is possible to smoothly transit between similar images without encountering low-probability/high-energy/unrealistic images along the interpolation path. Images with dramatically different appearances should be separated by much higher barriers because it will be necessary to encounter low-probability/high-energy/unrealistic images while smoothly transforming between differing appearances.

For example, it is possible to smoothly transit between two images of the digit 1 while still maintaining the appearance of the digit 1 throughout the interpolation path, but it is not possible to transition from the digit 1 to the digit 0 (or any other digit) without encountering an image that does not resemble a digit at all. Groups of similar images separated by low-energy barriers form metastable regions of U, establishing the connection between visual perception of differences between images and metastable structure in the energy landscape.

Metastable phenomena are a natural way of representing both structure and variation in complex concepts, but actually detecting metastable behavior in a given energy landscape is a challenging task. Intuitively, two MCMC samples initialized from the same energy basin should meet much more quickly than two MCMC samples initialized from separate energy basins. However, the "short" mixing time of two chains in the same metastable basin is still far too long for efficient simulation in high-dimensional spaces. No chains, either from the same or separate energy basins, are likely to meet in feasible time scales. Therefore, direct observation of membership in a metastable basin is not possible with MCMC simulation.

To overcome the difficulty of detecting metastable phenomena in their natural state, we perturb the energy landscape in a way that will accelerate mixing within a metastable basin while preserving the long mixing times between separate energy basins. If the perturbation is sufficiently small, then it is reasonable to expect metastable phenomena in the original and altered landscape will be very similar. On the other hand, the perturbation must be strong enough to overcome shallow energy barriers that exist within a metastable basin to encourage fast mixing between modes within the same macroscopic basin.

Drawing inspiration from the magnetized Ising model, we modify the original energy landscape with an L_2 penalty toward a known low-energy image. The target of the L_2 penalty acts as a "representative" of the energy basin to which it belongs. Given a candidate mode x_0 and a target mode x^*, initialize an MCMC sample X from x_0 and update the sample using the magnetized energy

$$U_{\mathrm{mag}}(x) = U(x) + \alpha\|x - x^*\|_2, \tag{10.28}$$

where U is the original energy function and α is the strength of magnetization. Sampling is continued until either the MCMC sample X comes with a small Euclidean distance ε of the target state x^* or until an upper limit on the number of steps is reached. Since $\frac{d}{dx}\|x - x^*\|_2 = \frac{x-x^*}{\|x-x^*\|_2}$, the gradient of the magnetization penalty $\alpha\|x - x^*\|_2$ is a vector with magnitude α pointing toward x^* for every point $x \neq x^*$. This shows that the gradient of U_{mag} differs from the gradient of U by a magnitude of at most α throughout the state space, allowing us to uniformly control the degree of landscape perturbation in a single parameter. We call this method attraction–diffusion (AD) (Fig. 10.10).

Fig. 10.10 Illustration of attraction-diffusion

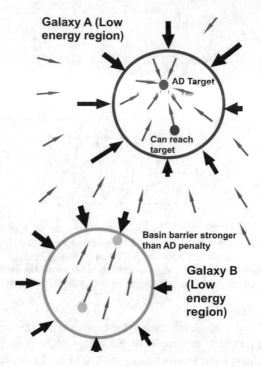

It is clear that in the limiting case $\alpha \to 0$, sampling on U_{mag} is identical to sampling on U, and that X will never come within a close distance of x^* even when x_0 and x^* are in the same metastable basin. On the other hand, in the limiting case, $\alpha \to \infty$, all probability mass is focused on x^* and X will quickly reach the target state. In between these limiting cases, there exists a spectrum of values for α for which the gradients from the original energy U and the magnetization penalty $\alpha\|x - x^*\|_2$ affect sampling at approximately the same magnitude. Within this spectrum, one observes that sometimes the diffusion paths reach their target x^* and sometimes the diffusion paths never reach their target. If we consider successful travel in the magnetized landscape to approximately represent metastable membership in the original landscape, then it becomes possible to reason about metastable structures in the original landscape based on the finite-step behavior of MCMC samples in the magnetized landscape. We note that metastable structures at finer resolutions can be detected when α is relatively small, while larger values of α will only preserve the most prominent barriers in the landscape. Like real-world concepts, metastable concepts only exist within a certain spectrum of identity or difference depending on the context of the situation.

Now that we are equipped with a method for determining whether two minima belong to the same metastable basin, it becomes possible to efficiently identify the different metastable structures in an arbitrary high-dimensional energy landscape using computational methods. This process allows to extend techniques for mapping

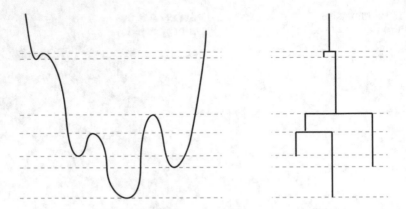

Fig. 10.11 First published in the Quarterly of Applied Mathematics in Volume 77:269–321, 2019, published by Brown University. Reprinted with permission from [99]. Construction of a disconnectivity graph (right) from a 1D energy landscape (left). The DG encodes the depth of local minima and the energy at which basins merge. The same procedure can be used to hierarchically cluster image concepts using an energy function U over the image space

and visualizing energy landscapes from the chemical physics literature to energy functions defined over the image space. The physical system defined by a low-temperature diffusion process has metastable structures that correspond to concepts learned by the energy function.

The analogy between stable states of a physical system (e.g., states with predominantly aligned spins in the Ising model) and recognizable concepts within groups of images allows us to hierarchically cluster the image space by first identifying metastable regions and then grouping regions based on the energy spectrum at which basins merge. The results can be displayed in a tree diagram known as a disconnectivity graph (DG). A disconnectivity graph displays: (1) the minimum energy within each basin (leaves of the tree), (2) the energy level at which basins merge in the energy landscape, also called the energy barrier (branches of the tree), as shown in Fig. 10.11. The key concept underlying this procedure is the idea that perceptibility can be grounded in metastable phenomena in an energy landscape that represents perceptual memory (Fig. 10.12).

We demonstrate the principles of the above discussion by mapping a learned energy landscape U that has been trained to model MNIST. It is important to note that the true density of an image concept is never known and that mapping must be done on a *learned* density. U is a computational representation of a large set of observed images, much like our memory creates perceptual models of structured images that we regularly observe. The macroscopic energy structures of U can effectively distinguish between different image groups (digits) that are recognizable to humans. Each digit has at least one stable basin that represents the digit's appearance, and relations between image basins follow visually intuitive relations between images.

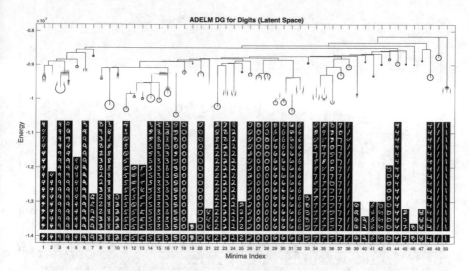

Fig. 10.12 First published in the Quarterly of Applied Mathematics in Volume 77:269–321, 2019, published by Brown University. Reprinted with permission from [99]. Map of the energy landscape of an image density trained with image patches from an ivy texture at four scales. The closest two scales develop recognizable and separate basins in the landscape that represent distinguishable patterns. The furthest two scales do not contain perceptible subgroups and form flat nebula-like basins that encode texture appearance

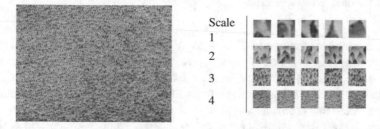

Fig. 10.13 First published in the Quarterly of Applied Mathematics in Volume 77:269–321, 2019, published by Brown University. Reprinted with permission from [99]. Ivy texture image and image patches from four scales. The first two scales contain images that can be clustered by a human. The images cross the perceptibility threshold from texton representation to texture representation between Scale 2 and Scale 3

The link between perceptibility and metastability can also be clearly observed when mapping the density of image patches from a single texture at a variety of different scales. The density f models 32×32 pixel image patches of the same ivy texture image from four different scales (see Fig. 10.13). At the closest scale, the images are composed of simple bar and stripe features. The second-closest scale features the composition of about 2 or 3 leaves in different arrangements. At the furthest two scales, it is difficult to distinguish distinct image groups, and images from these scales are perceived as textures by a human.

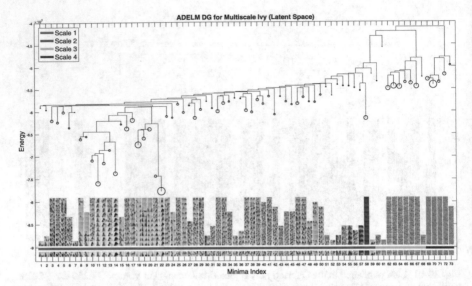

Fig. 10.14 First published in the Quarterly of Applied Mathematics in Volume 77:269–321, 2019, published by Brown University. Reprinted with permission from [99]. Map of the energy landscape of an image density trained with image patches from an ivy texture at four scales. The closest two scales develop recognizable and separate basins in the landscape that represent distinguishable patterns. The furthest two scales do not contain perceptible subgroups and form flat nebula-like basins that encode texture appearance

The metastable structure of $U = -\log f$ shares many similarities with human perceptibility (see Fig. 10.14). Image groups from the closest scale are easiest to recognize, and the landscape forms a handful of strong basins that represent the different bar and stripe configurations found at close range. A rich variety of separate metastable basins appear at Scale 2 to encode many distinct compositions of a few leaves. The landscape represents images from Scales 3 and 4 with a large macroscopic basin with little substructure. Between Scale 2 and Scale 3, a phase transition occurs where the identifiability of leaf compositions changes from distinguishable (texton scale) to indistinguishable (texture scale). The behaviors of human perceptibility and image landscape metastability are therefore quite similar across multi-scale image data.

Chapter 11
Deep Image Models

In this chapter, we will present the deep FRAME model or deep energy-based model as a recursive multi-layer generalization of the original FRAME model. We shall also present the generator model that can be considered a nonlinear multi-layer generalization of the factor analysis model. Such multi-layer models capture the fact that visual patterns and concepts appear at multiple layers of abstractions.

11.1 Deep FRAME and Deep Energy-Based Model

The original FRAME (Filters, Random fields, And Maximum Entropy) model [259, 279, 281] is a Markov random field model of stationary spatial processes such as stochastic textures. The log probability density function of the model is the sum of translation invariant potential functions that are point-wise one-dimensional nonlinear transformations of linear filter responses.

The deep generalization of the FRAME model is inspired by the successes of deep convolutional neural networks [138, 144], and as the name suggests, it can be considered a deep version of the FRAME model. The filters used in the original FRAME model are linear filters that capture local image features. In [161], the linear filters are replaced by the nonlinear filters at a certain convolutional layer of a pre-trained deep ConvNet. Such filters can capture more complex patterns, and the deep FRAME model built on such filters can be more expressive.

Furthermore, instead of using filters from a pre-trained ConvNet, we can also learn the filters from scratch. The resulting model is also known as a deep convolutional energy-based model [32, 180, 266]. Such a model can be considered a recursive multi-layer generalization of the original FRAME model. The log probability density function of the original FRAME model consists of nonlinear transformations of linear filter responses. If we repeat this structure recursively, we get the deep convolutional energy-based model with multiple layers of linear

© Springer Nature Switzerland AG 2023
S.-C. Zhu, Y. N. Wu, *Computer Vision*, https://doi.org/10.1007/978-3-030-96530-3_11

filtering followed by point-wise nonlinear transformations. Xie et al. [266] show that it is possible to learn such a model from natural images.

The deep FRAME model can be written as an exponential tilting of a reference distribution such as the uniform measure or the Gaussian white noise model. If the reference distribution is the Gaussian white noise model, the local modes of the probability density follow an auto-encoder. We call it the Hopfield auto-encoder because it defines the local energy minima of the model [108]. In the Hopfield auto-encoder, the bottom-up filters detect the patterns corresponding to the filters, and then the detection results are used as the coefficients in the top-down representation where the filters play the role of basis functions.

The learning of the deep FRAME model and deep energy-based model follows an analysis by synthesis scheme [85]. We can use Markov chain Monte Carlo (MCMC) such as the Langevin dynamics to sample from the current model to generate synthetic images, which runs gradient descent on the energy function of the model while adding Gaussian white noises for Brownian motion or diffusion. For deep FRAME models, the gradient can be efficiently computed by back-propagation. Then we update the model parameters based on the statistical difference between the observed images and the synthetic images so that the model shifts its probability density function from the synthetic images to the observed images. In the zero-temperature limit, this learning and sampling algorithm admits an adversarial interpretation, where the learning step and the sampling step play a minimax game based on a value function.

ConvNet Filters

The convolutional neural network (CNN or ConvNet) [144] is a specialized neural network devised for analyzing data such as images, where the linear transformations take place around each pixel, i.e., they are filters or convolutions. See Fig. 11.1 for an illustration.

A ConvNet consists of multiple layers of linear filtering and point-wise nonlinear transformation, as expressed by the following recursive formula:

$$\left[F_j^{(l)} * \mathbf{I}\right](y) = h\left(\sum_{k=1}^{N_{l-1}} \sum_{x \in \mathcal{S}_l} w_{k,x}^{(l,j)} \left[F_k^{(l-1)} * \mathbf{I}\right](y+x) + b_{l,j}\right), \quad (11.1)$$

or

$$\mathbf{I}_j^{(l)}(y) = h\left(\sum_{k=1}^{N_{l-1}} \sum_{x \in \mathcal{S}_l} w_{k,x}^{(l,j)} \mathbf{I}_k^{(l-1)}(y+x) + b_{l,j}\right), \quad (11.2)$$

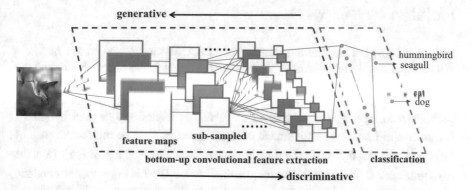

Fig. 11.1 Reprinted with permission from [160]. Convolutional neural networks consist of multiple layers of filtering and subsampling operations for bottom-up feature extraction, resulting in multiple layers of feature maps and their subsampled versions. The top layer features are used for classification via multinomial logistic regression. The discriminative direction is from image to category, whereas the generative direction is from category to image

where $l = 1, \ldots, L$ indexes the layer, and $\mathbf{I}_j^{(l)} = F_j^{(l)} * \mathbf{I}$ are filtered images or feature maps at layer l. In Fig. 11.1, the feature maps are illustrated by the square shapes. Each $[F_j^{(l)} * \mathbf{I}](x)$ is called a filter response or a feature extracted at layer l.

$\{F_j^{(l)}, j = 1, \ldots, N_l\}$ are the filters at layer l, and $\{F_k^{(l-1)}, k = 1, \ldots, N_{l-1}\}$ are the filters at layer $l - 1$. j and k are used to index the filters at layers l and $l - 1$, respectively, and N_l and N_{l-1} are the numbers of filters at layers l and $l - 1$, respectively. The filters are locally supported, so the range of $\mathbf{I}$ in $\sum_x$ is within local support $\mathcal{S}_l$ (such as a 7×7 image patch). We let $\mathbf{I}^{(0)} = \mathbf{I}$. The filter responses at layer l are computed from the filter responses at layer $l - 1$, by linear filtering defined by the weights $w_{k,x}^{(l,j)}$ as well as the bias term $b_{l,j}$, followed by the nonlinear transformation $h(\cdot)$. The most commonly used nonlinear transformation in the modern ConvNets is the rectified linear unit (ReLU), $h(r) = \max(0, r)$ [138]. $\{F_j^{(l)}\}$ are nonlinear filters because we incorporate $h(\cdot)$ in the computation of the filter responses. We call $\mathbf{I}_j^{(l)} = F_j^{(l)} * \mathbf{I}$ the filtered image or the feature map of filter j at layer l. We denote $\mathbf{I}^{(l)} = (\mathbf{I}_j^{(l)}, j = 1, \ldots, N_l)$, which consists of a total of N_l feature maps at layer l, and $j = 1, \ldots, N_l$. Sometimes, people call $\mathbf{I}^{(l)}$ as a whole feature map or filtered image with N_l channels, where each $\mathbf{I}_j^{(l)}$ corresponds to one channel. For a colored image, $\mathbf{I}^{(0)} = \mathbf{I}$ has 3 channels for RGB.

The filtering operations are often followed by subsampling and local max pooling (e.g., $\mathbf{I}(x_1, x_2) \leftarrow \max_{(s_1, s_2) \in \{0,1\}^2} \mathbf{I}(2x_1 + s_1, 2x_2 + s_2)$). See Fig. 11.1 for an illustration of subsampling. After a number of layers with subsampling, the filtered images or feature maps are reduced to 1×1 at the top layer. These features are then used for classification (e.g., does the image contain a hummingbird or a seagull or a dog?) via multinomial logistic regression.

FRAME with ConvNet Filters

Instead of using linear filters as in the original FRAME model, we can use the filters at a certain convolutional layer of a pre-trained ConvNet. We call such a model the deep FRAME model.

Suppose there exists a bank of filters $\{F_k^{(l)}, k = 1, \ldots, K\}$ at a certain convolutional layer l of a pre-trained ConvNet, as recursively defined by (11.1). For an image $\mathbf{I}$ defined on the image domain $\mathcal{D}$, let $F_k^{(l)} * \mathbf{I}$ be the feature map of filter $F_k^{(l)}$, and let $[F_k^{(l)} * \mathbf{I}](x)$ be the filter response of $\mathbf{I}$ to $F_k^{(l)}$ at position x (x is the two-dimensional coordinate). We assume that $[F_k^{(l)} * \mathbf{I}](x)$ is the response obtained after applying the nonlinear transformation or rectification function $h(\cdot)$. Then the non-stationary deep FRAME model becomes

$$p(\mathbf{I}; \theta) = \frac{1}{Z(\theta)} \exp\left\{ \sum_{k=1}^{K} \sum_{x \in \mathcal{D}} w_{k,x} [F_k^{(l)} * \mathbf{I}](x) \right\} q(\mathbf{I}), \qquad (11.3)$$

where $q(\mathbf{I})$ is the Gaussian white noise model (4.43), and $\theta = (w_{k,x}, \forall k, x)$ are the unknown parameters to be learned from the training data. Model (11.3) shares the same form as model (4.42) with linear filters, except that the rectification function $h(r) = \max(0, r)$ in model (4.42) is already absorbed in the ConvNet filters $\{F_k^{(l)}\}$ in model (11.3). We can also make model (11.3) stationary by letting $w_{k,x} = w_k$ for all x.

Learning and Sampling

The basic learning algorithm estimates the unknown parameters θ from a set of aligned training images $\{\mathbf{I}_i, i = 1, \ldots, n\}$ that come from the same object category. Again the weight parameters can be estimated by maximizing the log-likelihood function and can be computed by the stochastic gradient ascent algorithm [274]:

$$w_{k,x}^{(t+1)} = w_{k,x}^{(t)} + \gamma_t \left[\frac{1}{n} \sum_{i=1}^{n} \left[F_k^{(l)} * \mathbf{I}_i \right](x) - \frac{1}{\tilde{n}} \sum_{i=1}^{\tilde{n}} \left[F_k^{(l)} * \tilde{\mathbf{I}}_i \right](x) \right],$$

$$(11.4)$$

for every $k \in \{1, \ldots, K\}$ and $x \in \mathcal{D}$, where γ_t is the learning rate, and $\{\tilde{\mathbf{I}}_i, i = 1, \ldots, \tilde{n}\}$ are the synthetic images sampled from $p(\mathbf{I}; \theta^{(t)})$ using MCMC. This is an analysis by synthesis scheme that seeks to match the average filter responses of the synthetic images to those of the observed images.

In order to sample from $p(\mathbf{I}; \theta)$, we adopt the Langevin dynamics [80, 157]. Writing the energy function

Fig. 11.2 Reprinted with permission from [161]. Generating object patterns. In each row, the left half displays 4 of the training images (224 × 224), and the right half displays 4 of the synthetic images. In the last row, the learned model generates hybrid patterns of lion and tiger

$$U(\mathbf{I}, \theta) = -\sum_{k=1}^{K} \sum_{x \in \mathcal{D}} w_{k,x} \left[F_k^{(l)} * \mathbf{I} \right](x) + \frac{1}{2\sigma^2} \|\mathbf{I}\|^2, \tag{11.5}$$

the Langevin dynamics iterates

$$\mathbf{I}_{\tau+1} = \mathbf{I}_\tau - sU'(\mathbf{I}_\tau, \theta) + \sqrt{2s}\,e_\tau, \tag{11.6}$$

where $U'(\mathbf{I}, \theta) = \partial U(\mathbf{I}, \theta)/\partial \mathbf{I}$. This gradient can be computed by back-propagation. In (11.6), s is a small step size, and $e_\tau \sim N(0, I_D)$, independently across τ, where I_D is the identity matrix of dimension $D = |\mathcal{D}|$, i.e., the dimensionality of $\mathbf{I}$. e_τ is a Gaussian white noise image whose pixel values follow $N(0, 1)$ independently. Here we use τ to denote the time steps of the Langevin sampling process because t is used for the time steps of the learning process. The Langevin sampling process (11.6) is an inner loop within the learning process (11.4). Between every two consecutive updates of θ in the learning process, we run a finite number of steps of the Langevin dynamics starting from the images generated by the previous iteration of the learning algorithm.

The Langevin dynamics was first applied to the FRAME model by Zhu and Mumford [279], where the gradient descent component is interpreted as the Gibbs Reaction And Diffusion Equation (GRADE), and the patterns are formed via the reactions and diffusions controlled by different types of filters.

We first learn a non-stationary FRAME model (11.3) from images of aligned objects. The images were collected from the Internet. For each category, the number of training images was around 10. We used $\tilde{n} = 16$ parallel chains for Langevin sampling with 100 Langevin steps between every two consecutive updates of the parameters. Figure 11.2 shows some experiments using filters from the 3rd convolutional layer of the VGG ConvNet [219], a commonly used pre-trained

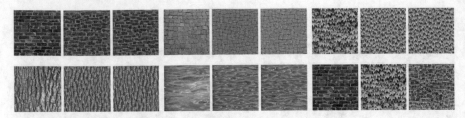

Fig. 11.3 Reprinted with permission from [161]. Generating texture patterns. For each category, the first image (224 × 224) is the training image, and the next 2 images are generated images, except for the last 3 images, where the first 2 are the training images, and the last one is the generated image that mixes brick wall and ivy

ConvNet trained on ImageNet ILSVRC2012 dataset [39]. For each experiment on each row, the left half displays 4 of the training images, and the right half displays 4 of the synthetic images generated by the Langevin dynamics. The last experiment is about learning the hybrid pattern of lion and tiger. The model re-mixes local image patterns seamlessly.

Figure 11.3 shows results from experiments on the stationary model for texture images. The model does not require image alignment. It re-shuffles the local patterns seamlessly. Each experiment is illustrated by 3 images, where the first image is the training image, and the other 2 images are generated by the learning algorithm. In the last 3 images, the first 2 images are training images, and the last image is generated by the learned model that mixes the patterns of brick wall and ivy.

Learning a New Layer of Filters

On top of the existing pre-trained convolutional layer of filters $\{F_k^{(l)}, k = 1, \ldots, K\}$, we can build another layer of filters $\{F_j^{(l+1)}, j = 1, \ldots, J\}$, according to the recursive formula (11.1), so that

$$\left[F_j^{(l+1)} * \mathbf{I}\right](y) = h\left(\sum_{k,x} w_{k,x}^{(j)}\left[F_k^{(l)} * \mathbf{I}\right](y + x) + b_j\right), \qquad (11.7)$$

where $h(r) = \max(0, r)$. The set $\{F_j^{(l+1)}\}$ is like a dictionary of "words" to describe different types of objects or patterns in the training images.

Due to the recursive nature of ConvNet, the deep FRAME model (11.3) based on filters $\{F_k^{(l)}\}$ corresponds to a single filter in $\{F_j^{(l+1)}\}$ at a particular position y (e.g., the origin $y = 0$) where we assume that the object appears. In [32], we show that the rectification function $h(r) = \max(0, r)$ can be justified by a mixture model where the object can either appear at a position or not. The bias term is related to $-\log Z(\theta)$.

Model (11.3) is used to model images where the objects are aligned and are of the same category. For images of non-aligned objects from multiple categories, we can extend the model (11.3) to a convolutional version with a whole new layer of multiple filters

$$p(\mathbf{I}; \theta) = \frac{1}{Z(\theta)} \exp\left\{ \sum_{j=1}^{J} \sum_{x \in \mathcal{D}} \left[F_j^{(l+1)} * \mathbf{I} \right](x) \right\} q(\mathbf{I}), \tag{11.8}$$

where $\{F_j^{(l+1)}\}$ are defined by (11.7), and $\theta = (w_{k,x}^{(j)}, \forall j, k, x)$. This model is a product of experts model [100, 207], where each $[F_j^{(l+1)} * \mathbf{I}](x)$ is an expert about a mixture of an activation or inactivation of a pattern of type j at position x. The stationary model for textures is a special case of this model.

Suppose we observe images of non-aligned objects from multiple categories $\{\mathbf{I}_i, i = 1, \ldots, n\}$, and we want to learn a new layer of filters $\{F_j^{(l+1)}, j = 1, \ldots, J\}$ by fitting the model (11.8) with (11.7) to the observed images, where $\{F_j^{(l+1)}\}$ model different types of objects in these images. This is an unsupervised learning problem because we do not know where the objects are. The model can still be learned by the analysis by synthesis scheme as before.

Let $L(\theta) = \frac{1}{n} \sum_{i=1}^{n} \log p(\mathbf{I}_i; \theta)$ be the log-likelihood where $p(\mathbf{I}; \theta)$ is defined by (11.8) and (11.7). Then the gradient ascent learning algorithm is based on

$$\frac{\partial L(\theta)}{\partial w_{k,x}^{(j)}} = \frac{1}{n} \sum_{i=1}^{n} \sum_{y \in \mathcal{D}} s_{j,y}(\mathbf{I}_i) \left[F_k^{(l)} * \mathbf{I}_i \right](y + x) - \mathrm{E}_\theta \left[\sum_{y \in \mathcal{D}} s_{j,y}(\mathbf{I}) \left[F_k^{(l)} * \mathbf{I} \right](y + x) \right],$$

$$\tag{11.9}$$

where

$$s_{j,y}(\mathbf{I}) = h' \left(\sum_{k,x} w_{k,x}^{(j)} \left[F_k^{(l)} * \mathbf{I} \right](y + x) + b_j \right) \tag{11.10}$$

is a binary on/off detector of object j at position y on image $\mathbf{I}$, because for $h(r) = \max(0, r)$, $h'(r) = 0$ if $r \leq 0$, and $h'(r) = 1$ if $r > 0$. The gradient (11.9) admits an EM [38] interpretation that is typical in unsupervised learning algorithms that involve hidden variables. Specifically, $s_{j,y}()$ detects the object of type j that is modeled by $F_j^{(l+1)}$ at location y. This step can be considered a hard-decision E-step. With the objects detected, the parameters of $F_j^{(l+1)}$ are then refined in a similar way as in (11.4), which can be considered the M-step. That is, we learn $F_j^{(l+1)}$ only from image patches where objects of type j are detected.

Figure 11.4 displays two experiments. In each experiment, the first image (224×224) is the training image, and the rest 2 images are generated by the learned

Fig. 11.4 Reprinted with permission from [161]. Learning a new layer of filters without requiring object bounding boxes or image alignment. For each experiment, the first image (224 × 224) is the training image, and the next 2 images are generated by the learned model

model. In the first scenery experiment, we learn 10 filters at the 4th convolutional layer, based on the pre-trained VGG filters at the 3rd layer. The size of each Conv4 filter to be learned is $11 \times 11 \times 256$. In the second sunflower experiment, we learn 20 filters of size $7 \times 7 \times 256$. Clearly, these learned filters capture the local objects or patterns and re-shuffle them seamlessly.

Deep Convolutional Energy-Based Model

Instead of relying on the pre-trained filters from an existing ConvNet, we can also learn the filters $\{F_k^{(l)}, k = 1, \ldots, K\}$ from scratch. The resulting model is a deep convolutional energy-based model (EBM) [32, 180, 266],

$$p(\mathbf{I}; \theta) = \frac{1}{Z(\theta)} \exp\{f(\mathbf{I}; \theta)\}q(\mathbf{I}), \tag{11.11}$$

where $f(\mathbf{I}; \theta)$ is defined by a ConvNet, and θ collects all the weight and bias parameters of the ConvNet. In model (11.8) with (11.7), we have

$$f(\mathbf{I}; \theta) = \sum_{j=1}^{J} \sum_{x \in \mathcal{D}} \left[F_j^{(l+1)} * \mathbf{I} \right](x). \tag{11.12}$$

Using more compact notation, we can define $f(\mathbf{I}; \theta)$ recursively by

$$\mathbf{I}^{(l)} = h(w_l \mathbf{I}^{(l-1)} + b_l), \tag{11.13}$$

for $l = 1, \ldots, L$, where $h(\cdot)$ is applied element-wise. $\mathbf{I}^{(0)} = \mathbf{I}$, and $f(\mathbf{I}; \theta) = \mathbf{I}^{(L)}$. $\mathbf{I}^{(l)}$ consists of all the filtered images or feature maps at layer l, and the rows of w_l consist of all the filters as well as all the locations where the filters operate on $\mathbf{I}^{(l-1)}$ to extract the features in $\mathbf{I}^{(l)}$. We assume that at the final layer L, $\mathbf{I}^{(L)}$ is reduced to a number (i.e., a 1×1 feature map). $\theta = (w_l, b_l, l = 1, \ldots, L)$. We can compare the compact Eq. (11.13) with the more detailed Eq. (11.2).

For piecewise linear $h(\cdot)$, such as $h(r) = \max(0, r)$, the function $f(\mathbf{I}; \theta)$ is piecewise linear [174, 192]. Specifically, $h(r) = \max(0, r) = 1(r > 0)r$, where $1(r > 0)$ is the indicator function that returns 1 if $r > 0$ and 0 otherwise. Then

$$\mathbf{I}^{(l)} = s_l(\mathbf{I}; \theta)(w_l \mathbf{I}^{(l-1)} + b_l), \tag{11.14}$$

where

$$s_l(\mathbf{I}; \theta) = \text{diag}(1(w_l \mathbf{I}^{(l-1)} + b_l > 0)), \tag{11.15}$$

i.e., a diagonal matrix of binary indicators (the indicator function is applied element-wise) [192]. Let $s = (s_l, l = 1, \ldots, L)$ consist of indicators at all the layers; then

$$f(\mathbf{I}; \theta) = \mathbf{B}_{s(\mathbf{I}; \theta)} \mathbf{I} + a_{s(\mathbf{I}; \theta)} \tag{11.16}$$

is piecewise linear, where

$$\mathbf{B}_s = \prod_{l=L}^{1} s_l w_l, \tag{11.17}$$

and a_s can be similarly calculated. $s(\mathbf{I}; \theta)$ partitions the image space of $\mathbf{I}$ into exponentially many pieces [192] according to the value of $s(\mathbf{I}; \theta)$. The partition is recursive because $s_l(\mathbf{I}; \theta)$ depends on $s_{l-1}(\mathbf{I}; \theta)$. The boundaries between the pieces are all linear. On each piece with $s(\mathbf{I}; \theta) = s$, where s on the right-hand side denotes a particular value of $s(\mathbf{I}; \theta)$, $f(\mathbf{I}; \theta)$ is a linear function $f(\mathbf{I}; \theta) = \mathbf{B}_s \mathbf{I} + a_s$. The binary switches in $s(\mathbf{I}; \theta)$ reconfigure the linear transformation according to (11.17).

$f(\mathbf{I}; \theta)$ generalizes three familiar structures in statistics:

(1) Generalized linear model (GLM). A GLM structure is a composition of a linear combination of the input variables and a nonlinear link function. A ConvNet

can be viewed as a recursion of this structure, where each component of $\mathbf{I}^{(l)}$ is a GLM transformation of $\mathbf{I}^{(l-1)}$, with h being the link function.

(2) Linear spline. A one-dimensional linear spline is of the form $y = \beta_0 + \sum_{k=1}^{d} \beta_k \max(0, x - a_k)$, where a_k are the knots. The ConvNet $f(\mathbf{I}; \theta)$ can be viewed as a multi-dimensional linear spline. The number of linear pieces is exponential in the number of layers [192]. Such a structure can approximate any continuous nonlinear function by a large number of linear pieces.

(3) CART [21] and MARS [66]. In the classification and regression tree (CART) and the multivariate adaptive regression splines (MARS), the input domain is recursively partitioned. The linear pieces mentioned above are also recursively partitioned according to the values of $s_l(\mathbf{I}; \theta)$ for $l = 1, \ldots, L$. Moreover, MARS also makes use of the hinge function $\max(0, r)$.

For Gaussian reference $q(\mathbf{I})$, the energy function is

$$U(\mathbf{I}; \theta) = -f(\mathbf{I}; \theta) + \frac{1}{2\sigma^2} \|\mathbf{I}\|^2. \tag{11.18}$$

We can continue to use Langevin dynamics (12.96) to sample from $p(\mathbf{I}; \theta)$.

The parameter θ can be learned by the stochastic gradient ascent algorithm [274]

$$\theta^{(t+1)} = \theta^{(t)} + \gamma_t \left[\frac{1}{n} \sum_{i=1}^{n} \frac{\partial}{\partial \theta} f(\mathbf{I}_i; \theta^{(t)}) - \frac{1}{\tilde{n}} \sum_{i=1}^{\tilde{n}} \frac{\partial}{\partial \theta} f(\tilde{\mathbf{I}}_i; \theta^{(t)}) \right], \tag{11.19}$$

where again γ_t is the learning rate, and $\{\tilde{\mathbf{I}}_i, i = 1, \ldots, \tilde{n}\}$ are the synthetic images sampled from $p(\mathbf{I}; \theta^{(t)})$. This is again an analysis by synthesis scheme. This step shifts the probability density function $p(\mathbf{I}; \theta)$, or more specifically, the high-probability regions or the low-energy regions, from the synthetic images $\{\tilde{\mathbf{I}}_i\}$ to the observed images $\{\mathbf{I}_i\}$.

In the sampling step, we need to compute $\partial f(\mathbf{I}; \theta)/\partial \mathbf{I}$. In the learning step, we need to compute $\partial f(\mathbf{I}; \theta)/\partial \theta$. Both derivatives can be calculated by the chain rule back-propagation, and they share the computations of $\partial \mathbf{I}^{(l)}/\partial \mathbf{I}^{(l-1)}$.

Our experiments show that the model is quite expressive. For example, we learn a 3-layer model. The first layer has 100 15×15 filters with a subsampling size of 3 pixels. The second layer has 64 5×5 filters with a subsampling size of 1. The third layer has 30 3×3 filters with a subsampling size of 1. We learn a model (11.11) for each texture category from a single training image. Figure 11.5 displays some results. For each category, the first image is the training image, and the rest are 2 of the images generated by the learning algorithm. We use $\tilde{n} = 16$ parallel chains for Langevin sampling. The number of Langevin iterations between every two consecutive updates of parameters is 10. The training images are of the size 224×224, whose intensities are within $[0, 255]$. We fix $\sigma^2 = 1$ in the reference distribution q.

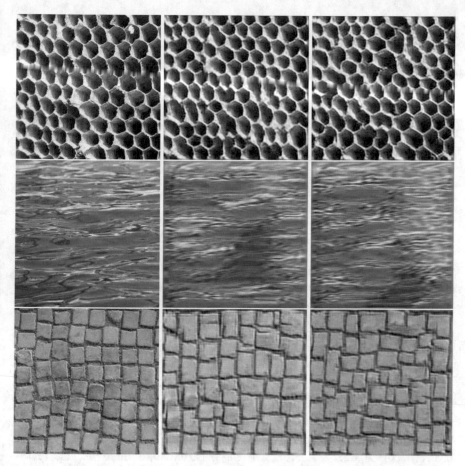

Fig. 11.5 Reprinted with permission from [266]. Generating texture patterns. For each category, the first image (224×224) is the training image, and the rest are 2 of the images generated by the learning algorithm

While the sparse FRAME model is interpretable in terms of the symbolic sketch of the images, the deep FRAME model is not interpretable with its multiple layers of dense connections in linear filtering.

Hopfield Auto-Encoder

Consider the sparse FRAME model. Let us assume that the reference distribution $q(\mathbf{I})$ is white noise with mean 0 and variance $\sigma^2 = 1$. The energy function is

$$U(\mathbf{I}) = \frac{1}{2}\|\mathbf{I}\|^2 - \sum_{j=1}^{m} w_j h(\langle \mathbf{I}, B_{k_j, x_j}\rangle). \tag{11.20}$$

This energy function can be multi-modal, and each local minimum $\hat{\mathbf{I}}$ satisfies $U'(\hat{\mathbf{I}}) = 0$. Thus,

$$\hat{\mathbf{I}} = \sum_{j=1}^{m} w_i h'(\langle \hat{\mathbf{I}}, B_{k_j,x_j} \rangle) B_{k_j,x_j}. \tag{11.21}$$

This reveals an auto-encoder [16, 242] hidden in the local modes:

$$\text{Encoding}: c_j = w_j h'(\langle \hat{\mathbf{I}}, B_{k_j,x_j} \rangle), \tag{11.22}$$

$$\text{Decoding}: \hat{\mathbf{I}} = \sum_{i=1}^{n} c_j B_{k_j,x_j}, \tag{11.23}$$

where (11.22) encodes $\hat{\mathbf{I}}$ by (c_j), and (11.23) reconstructs $\hat{\mathbf{I}}$ from (c_j). B_{k_j,x_j} serves as both bottom-up filter in (11.22) and top-down basis function in (11.23). We call this auto-encoder the Hopfield auto-encoder because $\hat{\mathbf{I}}$ is a local minimum of the energy function (11.20). Hopfield [108] proposes that the local energy minima may be used for content-addressable memory.

The Hopfield auto-encoder also presents itself in the deep convolutional energy-based model (11.11) [266]. The energy function of the model is $\|\mathbf{I}\|^2/2 - f(\mathbf{I}; \theta)$. The local minima satisfy the Hopfield auto-encoder $\hat{\mathbf{I}} = f'(\hat{\mathbf{I}}; \theta)$, or more specifically,

$$\text{Encoding}: s = s(\hat{\mathbf{I}}; \theta), \tag{11.24}$$

$$\text{Decoding}: \hat{\mathbf{I}} = \mathbf{B}_s. \tag{11.25}$$

The encoding process is a bottom-up computation of the indicators at different layers $s_l = s_l(\mathbf{I}; \theta)$, for $l = 1, \ldots, L$, where w_l plays the role of filters, see Eq. (11.15). The decoding process is a top-down computation for reconstruction, where s_l plays the role of coefficients, and w_l plays the role of basis functions. See Eq. (11.17). The encoding process detects the patterns corresponding to the filters, and then the decoding process reconstructs the image using the detected filters as the basis functions.

Multi-grid Sampling and Modeling

In the high-dimensional space, e.g., image space, the model can be highly multi-modal. The MCMC in general and the Langevin dynamics in particular may have difficulty traversing different modes, and it may be very time-consuming to converge. A simple and popular modification of the maximum likelihood learning is the contrastive divergence (CD) learning [100], where we obtain the synthesized example by initializing a finite-step MCMC from the observed example. The CD learning is related to score matching estimator [114, 115] and auto-encoder

[6, 227, 241]. Such a method has the ability to handle large training datasets via mini-batch training. However, bias may be introduced in the learned model parameters in that the synthesized images can be far from the fair examples of the current model. A further modification of CD is persistent CD [231], where at the initial learning epoch the MCMC is still initialized from the observed examples, while in each subsequent learning epoch, the finite-step MCMC is initialized from the synthesized example of the previous epoch. The resulting synthesized examples can be less biased by the observed examples. However, the persistent chains may still have difficulty traversing different modes of the learned model.

A multi-grid sampling and learning method is developed in [69] to address the above challenges under the constraint of finite budget MCMC. Specifically, each training image is repeatedly downscaled to get its multi-grid versions. A separate energy-based model is learned at each grid. Within each iteration of the learning algorithm, for each observed training image, the corresponding synthesized images are generated at multiple grids. We can initialize the finite-step MCMC sampling from the minimal 1×1 version of the training image, and the synthesized image at each grid serves to initialize the finite-step MCMC that samples from the model of the subsequent finer grid. See Fig. 11.6 for an illustration, where the images are sampled sequentially at 3 grids, with 30 steps of Langevin dynamics at each grid. After obtaining the synthesized images at the multiple grids, the models at the multiple grids are updated separately and simultaneously based on the differences between the synthesized images and the observed training images at different grids.

Unlike the original CD or persistent CD, the learned models are capable of generating new synthesized images from scratch with a fixed budget MCMC because we only need to initialize the MCMC by sampling from the one-dimensional histogram of the 1×1 versions of the training images. See Fig. 11.7 for generated examples by models learned from large image datasets.

The learned energy-based model is a bottom-up ConvNet that consists of multiple layers of features. These features can be used for subsequent tasks such as classification. The learned models can also be used as a prior distribution for inpainting, as illustrated by Fig. 11.8. See [69] for experiment details and numerical evaluations.

Adversarial Interpretation

The deep convolutional energy-based model (11.11) can be written as

$$p(\mathbf{I}; \theta) = \frac{1}{Z(\theta)} \exp[-U(\mathbf{I}; \theta)], \tag{11.26}$$

where the energy function $U(\mathbf{I}; \theta) = -f(\mathbf{I}; \theta) + \frac{1}{2\sigma^2} \|\mathbf{I}\|^2$. The update of θ is based on $L'(\theta)$ that can be approximated by

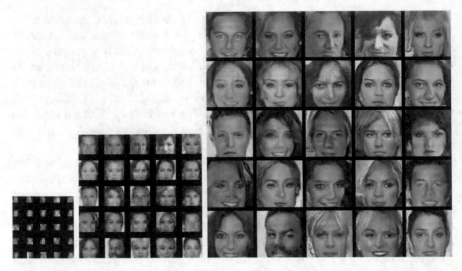

Fig. 11.6 Reprinted with permission from [69]. Synthesized images at multi-grids. From left to right: 4×4 grid, 16×16 grid, and 64×64 grid. A synthesized image at each grid is obtained by 30-step Langevin sampling initialized from the synthesized image at the previous coarser grid, beginning with the 1×1 grid

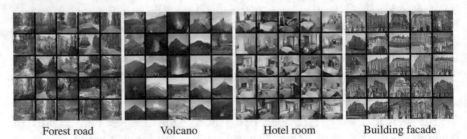

Forest road   Volcano   Hotel room   Building facade

Fig. 11.7 Reprinted with permission from [69]. Synthesized images from models learned by multi-grid method from 4 categories of MIT places205 datasets

$$\frac{1}{\tilde{n}} \sum_{i=1}^{\tilde{n}} \frac{\partial}{\partial \theta} U(\tilde{\mathbf{I}}_i; \theta) - \frac{1}{n} \sum_{i=1}^{n} \frac{\partial}{\partial \theta} U(\mathbf{I}_i; \theta), \tag{11.27}$$

where $\{\tilde{\mathbf{I}}_i, i = 1, \ldots, \tilde{n}\}$ are the synthetic images that are generated by the Langevin dynamics. At the zero-temperature limit, the Langevin dynamics becomes gradient descent:

$$\tilde{\mathbf{I}}_{\tau+1} = \tilde{\mathbf{I}}_\tau - \delta \frac{\partial}{\partial \tilde{\mathbf{I}}} U(\tilde{\mathbf{I}}_\tau; \theta). \tag{11.28}$$

Consider the value function

Fig. 11.8 Reprinted with permission from [69]. Inpainting examples on CelebA dataset. In each block from left to right: the original image, masked input, inpainted image by the multi-grid method

$$V(\tilde{\mathbf{I}}_i, i = 1, \dots, \tilde{n}; \theta) = \frac{1}{\tilde{n}} \sum_{i=1}^{\tilde{n}} U(\tilde{\mathbf{I}}_i; \theta) - \frac{1}{n} \sum_{i=1}^{n} U(\mathbf{I}_i; \theta). \tag{11.29}$$

The updating of θ is to increase V by shifting the low-energy regions from the synthetic images $\{\tilde{\mathbf{I}}_i\}$ to the observed images $\{\mathbf{I}_i\}$, whereas the updating of $\{\tilde{\mathbf{I}}_i, i = 1, \dots, \tilde{n}\}$ is to decrease V by moving the synthetic images toward the low-energy regions. This is an adversarial interpretation of the learning and sampling algorithm. It can also be considered a generalization of the herding method [249] from exponential family models to general energy-based models.

11.2 Generator Network

This section studies the problem of learning and inference in the generator network [81]. The generator network is based on a top-down ConvNet. It is a nonlinear generalization of the factor analysis model. We develop the alternating back-propagation algorithm for maximum likelihood learning of the generator network.

Factor Analysis

Let $\mathbf{I}$ be a D-dimensional observed example, such as an image. Let z be the d-dimensional vector of continuous latent factors, $z = (z_k, k = 1, \dots, d)$. The traditional factor analysis model is $\mathbf{I} = Wz + \epsilon$, where W is $D \times d$ matrix, and ϵ is a D-dimensional error vector or the observational noise. We assume that $z \sim \mathrm{N}(0, I_d)$, where I_d stands for the d-dimensional identity matrix. We also assume that $\epsilon \sim \mathrm{N}(0, \sigma^2 I_D)$, i.e., the observational errors are Gaussian white noises, and ϵ is independent of z. There are three perspectives to view W:

(1) *Basis vectors.* Write $W = (W_1, \dots, W_d)$, where each W_k is a D-dimensional column vector. Then $\mathbf{I} = \sum_{k=1}^{d} z_k W_k + \epsilon$, i.e., W_k are the basis vectors and z_k are the coefficients.

(2) *Loading matrix.* Write $W = (w_1, \ldots, w_D)^\top$, where $w_j^\top$ is the j-th row of W. Then $\mathbf{I}_j = \langle w_j, z \rangle + \epsilon_j$, where $\mathbf{I}_j$ and ϵ_j are the j-th components of $\mathbf{I}$ and ϵ, respectively. Each $\mathbf{I}_j$ is a loading of the d factors, where w_j is a vector of loading weights, indicating which factors are important for determining $\mathbf{I}_j$. W is called the loading matrix.

(3) *Matrix factorization.* Suppose we observe $\mathbf{I} = (\mathbf{I}_1, \ldots, \mathbf{I}_n)$, whose factors are $Z = (z_1, \ldots, z_n)$; then $\mathbf{I} \approx WZ$.

The factor analysis model can be learned by the Rubin–Thayer EM algorithm, which involves alternating regressions of z on $\mathbf{I}$ in the E-step and of $\mathbf{I}$ on z in the M-step [154, 209].

The factor analysis model is the prototype of many subsequent models that generalize the prior model of z:

(1) *Independent component analysis.* [116] $d = D$, $\epsilon = 0$, and z_k are assumed to follow independent heavy-tailed distributions.

(2) *Sparse coding.* [190] $d > D$, and z is assumed to be a redundant but sparse vector, i.e., only a small number of z_k are non-zero or significantly different from zero.

(3) *Nonnegative matrix factorization.* [148] It is assumed that $z_k \geq 0$.

Nonlinear Factor Analysis

The generator network is a nonlinear generalization of factor analysis. It generalizes the linear mapping in factor analysis to a nonlinear mapping that is defined by a convolutional neural network (ConvNet or CNN) [47, 138, 144]. It has been shown that the generator network is capable of generating realistic images [40, 199].

The generator network has the following properties:

(1) *Analysis*: The model disentangles the variations in the observed examples into independent variations of latent factors.

(2) *Synthesis*: The model can synthesize new examples by sampling the factors from the known prior distribution and transforming the factors into the synthesized examples.

(3) *Embedding*: The model embeds the high-dimensional non-Euclidean manifold formed by the observed examples into the low-dimensional Euclidean space of the latent factors so that linear interpolation in the low-dimensional factor space results in nonlinear interpolation in the data space.

Specifically, the generator network model [81] retains the assumptions that $d < D$, $z \sim N(0, I_d)$, and $\epsilon \sim N(0, \sigma^2 I_D)$ as in traditional factor analysis but generalizes the linear mapping Wz to a nonlinear mapping $g(z; \theta)$, where g is a ConvNet, and θ collects all the connection weights and bias terms of the ConvNet. Then the model becomes

$$\mathbf{I} = g(z; \theta) + \epsilon,$$

$$z \sim N(0, I_d), \; \epsilon \sim N(0, \sigma^2 I_D), \; d < D. \tag{11.30}$$

The reconstruction error is $\|\mathbf{I} - g(z; \theta)\|^2$.

Although $g(z; \theta)$ can be any nonlinear mapping, the ConvNet parameterization of $g(z; \theta)$ makes it particularly close to the original factor analysis. Specifically, we can write the top-down ConvNet as follows:

$$z^{(l-1)} = g_l(W_l z^{(l)} + b_l), \tag{11.31}$$

where g_l is element-wise nonlinearity at layer l, W_l is the weight matrix, b_l is the vector of bias terms at layer l, and $\theta = (W_l, b_l, l = 1, \ldots, L)$. $z^{(0)} = g(z; \theta)$, and $z^{(L)} = z$. The top-down ConvNet (11.31) can be considered a recursion of the original factor analysis model, where the factors at the layer $l - 1$ are obtained by the linear superposition of the basis vectors or basis functions that are column vectors of W_l, with the factors at the layer l serving as the coefficients of the linear superposition. In the case of ConvNet, the basis functions are shift-invariant versions of one another, like wavelets.

Learning by Alternating Back-Propagation

The factor analysis model can be learned by the Rubin–Thayer EM algorithm [38, 209], where both the E-step and the M-step are based on multivariate linear regression. Inspired by this algorithm, we propose an alternating back-propagation algorithm for learning the generator network that iterates the following two steps:

(1) *Inferential back-propagation*: For each training example, infer the continuous latent factors by Langevin dynamics.
(2) *Learning back-propagation*: Update the parameters given the inferred latent factors by gradient descent.

The Langevin dynamics [179] is a stochastic sampling counterpart of gradient descent. The gradient computations in both steps are based on back-propagation. Because of the ConvNet structure, the gradient computation in step (1) is actually a by-product of the gradient computation in step (2).

The alternating back-propagation algorithm follows the tradition of alternating operations in unsupervised learning, such as alternating linear regression in the EM algorithm for factor analysis, alternating least squares algorithm for matrix factorization [127, 136], and alternating gradient descent algorithm for sparse coding [190]. All these unsupervised learning algorithms alternate an inference step and a learning step, as is the case with alternating back-propagation.

Specifically, the joint density is $p(z, \mathbf{I}; \theta) = p(z)p(\mathbf{I}|z; \theta)$, and

$$\log p(z, \mathbf{I}; \theta) = -\frac{1}{2\sigma^2}\|\mathbf{I} - g(z; \theta)\|^2 - \frac{1}{2}\|z\|^2 + \text{constant}, \qquad (11.32)$$

where the constant term is independent of z and θ.

The marginal density is obtained by integrating out the latent factors z, i.e., $p(\mathbf{I}; \theta) = \int p(z, \mathbf{I}; \theta)dz$. The inference of z given $\mathbf{I}$ is based on the posterior density $p(z|\mathbf{I}; \theta) = p(z, \mathbf{I}; \theta)/p(\mathbf{I}; \theta) \propto p(z, \mathbf{I}; \theta)$ as a function of z.

For the training data $\{\mathbf{I}_i, i = 1, \ldots, n\}$, the generator model can be trained by maximizing the log-likelihood

$$L(\theta) = \frac{1}{n}\sum_{i=1}^{n}\log p(\mathbf{I}_i; \theta). \qquad (11.33)$$

The gradient of $L(\theta)$ is obtained according to the following identity:

$$\frac{\partial}{\partial\theta}\log p(\mathbf{I}; \theta) = \frac{1}{p(\mathbf{I}; \theta)}\frac{\partial}{\partial\theta}\int p(\mathbf{I}, z; \theta)dz \qquad (11.34)$$

$$= \frac{1}{p(\mathbf{I}; \theta)}\int\left[\frac{\partial}{\partial\theta}\log p(\mathbf{I}, z; \theta)\right]p(\mathbf{I}, z; \theta)dz \qquad (11.35)$$

$$= \int\left[\frac{\partial}{\partial\theta}\log p(\mathbf{I}, z; \theta)\right]\frac{p(\mathbf{I}, z; \theta)}{p(\mathbf{I}; \theta)}dz \qquad (11.36)$$

$$= \mathrm{E}_{p(z|\mathbf{I};\theta)}\left[\frac{\partial}{\partial\theta}\log p(z, \mathbf{I}; \theta)\right]. \qquad (11.37)$$

The above identity underlies the EM algorithm, where $\mathrm{E}_{p(z|\mathbf{I};\theta)}$ is the expectation with respect to the posterior distribution of the latent factors $p(z|\mathbf{I}; \theta)$, and is computed in the E-step. The usefulness of identity (11.37) lies in the fact that the derivative of the complete-data log-likelihood $\log p(z, \mathbf{I}; \theta)$ on the right-hand side can be obtained in closed form. In the EM algorithm, the M-step maximizes the expectation of $\log p(z, \mathbf{I}; \theta)$ with respect to the current posterior distribution of the latent factors. In general, the expectation in (11.37) is analytically intractable and has to be approximated by MCMC that samples from the posterior $p(z|\mathbf{I}; \theta)$, such as the Langevin inference dynamics, which iterates

$$z_{\tau+1} = z_\tau + s\frac{\partial}{\partial z}\log p(z_\tau, \mathbf{I}; \theta) + \sqrt{2s}e_\tau, \qquad (11.38)$$

where τ indexes the time step, s is the step size, and e_τ denotes the noise term, $e_\tau \sim \mathrm{N}(0, I_d)$.

We take the derivative of $\log p(z, \mathbf{I}; \theta)$ in (11.38) because this derivative is the same as the derivative of the log-posterior $\log p(z|\mathbf{I}; \theta)$ since $p(z|\mathbf{I}; \theta)$ is proportional to $p(z, \mathbf{I}; \theta)$ as a function of z.

The Langevin inference solves a ℓ_2 penalized nonlinear least squares problem so that z_i can reconstruct $\mathbf{I}_i$ given the current θ. The Langevin inference process performs explaining-away reasoning, where the latent factors in z compete with each other to explain $\mathbf{I}$.

The stochastic gradient algorithm of [274] can be used for learning, where in each iteration, for each z_i, only a single copy of z_i is sampled from $p(z_i|\mathbf{I}_i, \theta)$ by running a finite number of steps of Langevin dynamics starting from the current value of z_i, i.e., the warm start. With z_i sampled from $p(z_i | \mathbf{I}_i, \theta)$ for each training image $\mathbf{I}_i$ by the Langevin inference process, the Monte Carlo approximation to $L'(\theta)$ is

$$L'(\theta) \approx \frac{1}{n} \sum_{i=1}^{n} \frac{\partial}{\partial \theta} \log p(z_i, \mathbf{I}_i; \theta)$$

$$= \frac{1}{n} \sum_{i=1}^{n} \frac{1}{\sigma^2} (\mathbf{I}_i - g(z_i; \theta)) \frac{\partial}{\partial \theta} g(z_i; \theta). \tag{11.39}$$

The updating of θ solves a nonlinear regression problem so that the learned θ enables better reconstruction of $\mathbf{I}_i$ by the inferred z_i. Given the inferred z_i, the learning of θ is a supervised learning problem [47].

Han et al. [94] describe the training algorithm that iterates the following two steps: (1) Inference back-propagation: update z_i by running a finite number of steps of Langevin dynamics. (2) Learning back-propagation: update θ by one step of gradient descent.

Both the inferential back-propagation and the learning back-propagation are guided by the residual $\mathbf{I}_i - g(z_i; \theta)$. The inferential back-propagation is based on $\partial g(z; \theta)/\partial z$, whereas the learning back-propagation is based on $\partial g(z; \theta)/\partial \theta$.

The Langevin dynamics can be extended to Hamiltonian Monte Carlo [179] or more sophisticated versions [80].

Figure 11.9 illustrates the results of modeling textures where we learn a separate model from each texture image. The factors z at the top layer form a $\sqrt{d} \times \sqrt{d}$ image, with each pixel following $N(0, 1)$ independently. The $\sqrt{d} \times \sqrt{d}$ image z is then transformed into $\mathbf{I}$ by the top-down ConvNet. In order to obtain the synthesized image, we randomly sample a 7×7 image z from $N(0, I)$ and then expand the learned network to generate the 448×448 synthesized image.

Figure 11.10 illustrates the learned D-dimensional manifold. We learn a model where $z = (z_1, \ldots, z_d)$ has $d = 100$ components from 1000 face images randomly selected from the CelebA dataset [156]. The left panel of Fig. 11.10 displays the images generated by the learned model. The right panel displays the interpolation results. The images at the four corners are generated by the z vectors of four images randomly selected from the training set. The images in the middle are obtained by first interpolating the z's of the four corner images using the sphere interpolation [44] and then generating the images by the learned ConvNet.

Figure 11.11 depicts the reconstructions of face images by linear PCA and nonlinear generator model. For PCA, we learn the d eigenvectors from the training images and then project the testing images on the learned eigenvectors for reconstruction. The generator is learned by alternating back-propagation. We infer

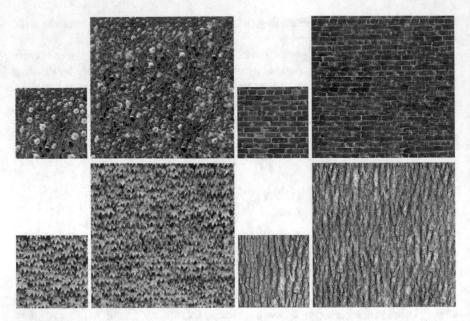

Fig. 11.9 Reprinted with permission from [94]. Modeling texture patterns. For each example, *Left:* the 224 × 224 observed image. *Right:* the 448 × 448 generated image

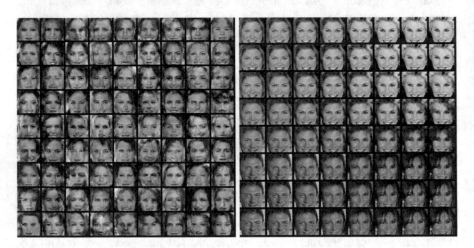

Fig. 11.10 Reprinted with permission from [94]. Modeling object patterns. *Left:* each image is obtained by first sampling $z \sim N(0, I_{100})$ and then generating the image by $g(z; \theta)$ with the learned θ. *Right:* interpolation. The images at the four corners are reconstructed from the inferred z vectors of four images randomly selected from the training set. Each image in the middle is obtained by first interpolating the z vectors of the four corner images and then generating the image by $g(z; \theta)$

Fig. 11.11 Reprinted with permission from [94]. Comparison between generator as nonlinear and PCA as linear factor analysis. *Row 1:* original testing images. *Row 2:* reconstructions by PCA eigenvectors learned from training images. *Row 3:* reconstructions by the generator learned from training images

the d-dimensional latent factors z using inferential back-propagation and then reconstruct the testing image by $g(z; \theta)$ using the inferred z and the learned θ.

Nonlinear Generalization of AAM Model

Active appearance models (AAMs) [28, 137] use a linear model to jointly capture the appearance and geometric variations in an image. For the face images, the appearance information mainly includes colors, illuminations, and identities, while the geometric information mainly includes the shapes and viewing angles. Given a set of landmark points, the AAM model can learn the eigenvectors from these landmarks to extract the geometric information. With the known landmark points, the faces can be aligned into a canonical shape and view, by warping the faces with the mean landmarks. Under this canonical geometric state, the appearance information can be extracted by the principal component analysis.

Can we extract the appearance and geometric knowledge from the images without any landmarks or any supervised information? Moreover, can we disentangle the appearance and geometric information by an unsupervised method? The deformable generator model can solve this problem.

The deformable generator model [271, 272] disentangles the appearance and geometric information of an image into two independent latent vectors with two generator networks: one appearance generator and one geometric generator. The appearance generator produces the appearance details, including color, illumination, identity, or category, of an image. The geometric generator produces a displacement of the coordinates of each pixel and performs geometric warping, such as stretching and rotation, on the image generated by the appearance generator to obtain the final generated image, as shown in Fig. 11.12. The geometric operation only modifies the positions of pixels in an image without changing the color and illumination. Therefore, the color and illumination information and the geometric information

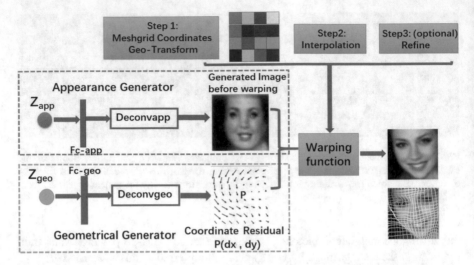

Fig. 11.12 Reprinted with permission from [271]. An illustration of the proposed model. The model contains two generator networks: one appearance generator and one geometric generator. The two generators are connected with a warping function to produce the final image. The warping function includes a geometric transformation operation for image coordinates and a differentiable interpolation operation. The refining operation is optional for improving the warping function

are disentangled by the geometric generator and the appearance generator in the model.

The model can be expressed as

$$\mathbf{I} = G(z^a, z^g; \theta) + \epsilon$$
$$= F_w(g_a(z^a; \theta_a), g_g(z^g; \theta_g)) + \epsilon, \tag{11.40}$$

where $z^a \sim \mathrm{N}(0, I_{d_a})$, $z^g \sim \mathrm{N}(0, I_{d_g})$, and $\epsilon \sim \mathrm{N}(0, \sigma^2 I_D)$ are independent. F_w is the warping function, which uses the deformation field generated by the geometric generator $g_g(z^g; \theta_g)$ to warp the image generated by the appearance generator $g_a(z^a; \theta_a)$ to synthesize the final output image $\mathbf{I}$. This deformable generator model can be learned by extending the alternating back-propagation algorithm.

The abstracted geometric knowledge can be transferred. For the unseen images as in Fig. 11.13, we can first infer their appearance and geometric latent vectors, and then the geometric knowledge can be transferred by recombining the inferred geometric latent vector from the target image with the inferred appearance latent vector from the source image.

For high-resolution images, we train the deformable generator with 40K faces from FFHQ [126], which are cropped to 256×256 pixels. Similar results corresponding to Fig. 11.13 are demonstrated in Fig. 11.14.

Fig. 11.13 Reprinted with permission from [271]. Transferring and recombining geometric and appearance vectors. The first row shows seven faces from the CelebA dataset. The second row shows the faces generated by transferring and recombining the second through seventh faces' geometric vectors z_g with the first face's appearance vector z_a in the first row. The third row shows the faces generated by transferring and recombining the second through seventh faces' appearance vectors z_a with the first face's geometric vector z_g in the first row. The deformable generator model is trained on the 10,000 face images from the CelebA dataset, which are cropped to 64×64 pixels, and the faces in the training data have a diverse variety of colors, illuminations, identities, viewing angles, shapes, and expressions

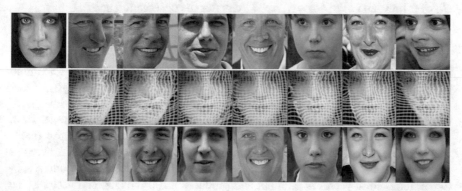

Fig. 11.14 Reprinted with permission from [271]. Transferring and recombining the abstracted geometric and appearance knowledge. The first row shows 8 unseen faces from FFHQ. The second row shows the generated faces by transferring and recombining the 2nd–8th faces' geometric vectors with the first face's appearance vector. The third row shows the generated faces by recombining the 2nd–8th faces' appearance vectors with the first face's geometric vector in the first row. The deformable generator model is trained on the 40,000 face images randomly selected from the FFHQ dataset [126]. The training images are cropped to 256×256 pixels, and the faces have different colors, illuminations, identities, viewing angles, shapes, and expressions

Dynamic Generator Model

Let $\mathbf{I} = (\mathbf{I}_t, t = 1, \ldots, T)$ be the observed video sequence, where $\mathbf{I}_t$ is a frame at time t. The dynamic generator model consists of the following two components:

$$z_t = F_\alpha(z_{t-1}, u_t), \tag{11.41}$$

$$\mathbf{I}_t = G_\beta(z_t) + \epsilon_t, \tag{11.42}$$

where $t = 1, \ldots, T$ (11.41) is the transition model, and (11.42) is the emission model. z_t is the d-dimensional hidden state vector. $u_t \sim N(0, I)$ is the noise vector of a certain dimensionality. The Gaussian noise vectors $(u_t, t = 1, \ldots, T)$ are independent of each other. The sequence of $(z_t, t = 1, \ldots, T)$ follows a nonlinear auto-regressive model, where the noise vector u_t encodes the randomness in the transition from z_{t-1} to z_t in the d-dimensional state space. F_α is a feedforward neural network or multi-layer perceptron, where α denotes the weight and bias parameters of the network. We can adopt a residual form [97] for F_α to model the change of the state vector. $\mathbf{I}_t$ is the D-dimensional image, which is generated by the d-dimensional hidden state vector z_t. G_β is a top-down convolutional network (sometimes also called deconvolution network), where β denotes the weight and bias parameters of this top-down network. $\epsilon_t \sim N(0, \sigma^2 I_D)$ is the residual error. We let $\theta = (\alpha, \beta)$ denote all the model parameters.

Let $u = (u_t, t = 1, \ldots, T)$. u consists of the latent random vectors that need to be inferred from $\mathbf{I}$. Although $\mathbf{I}_t$ is generated by the state vector z_t, $z = (z_t, t = 1, \ldots, T)$ are generated by u. In fact, we can write $\mathbf{I} = H_\theta(u) + \epsilon$, where H_θ composes F_α and G_β over time, and $\epsilon = (\epsilon_t, t = 1, \ldots, T)$ denotes the observation noises.

Let $p(u)$ be the prior distribution of u. Let $p_\theta(\mathbf{I}|u) \sim N(H_\theta(u), \sigma^2 I)$ be the conditional distribution of $\mathbf{I}$ given u, where I is the identity matrix whose dimension matches that of $\mathbf{I}$. The marginal distribution is $p_\theta(\mathbf{I}) = \int p(u)p_\theta(\mathbf{I}|u)du$ with the latent variable u integrated out. We estimate the model parameter θ by the maximum likelihood method that maximizes the observed data log-likelihood $\log p_\theta(\mathbf{I})$.

We learn the model for dynamic textures, which are sequences of images of moving scenes that exhibit stationarity in time. We learn a separate model from each example. Each observed video clip is prepared to be of the size 64 pixels × 64 pixels × 60 frames. The transition model is a feedforward neural network with three layers. The network takes a 100-dimensional state vector s_{t-1} and a 100-dimensional noise vector u_t as input and produces a 100-dimensional vector r_t, so that $z_t = \tanh(z_{t-1} + r_t)$. The numbers of nodes in the three layers of the feedforward neural network are $\{20, 20, 100\}$. The emission model is a top-down deconvolution neural network or generator model that maps the 100-dimensional state vector (i.e., $1 \times 1 \times 100$) to the image frame of size $64 \times 64 \times 3$ by 6 layers of deconvolutions with a kernel size of 4 and up-sampling factor of 2 from top to bottom. The numbers of channels at different layers of the generator are $\{512, 512, 256, 128, 64, 3\}$. Batch normalization [118] and ReLU layers are added between deconvolution layers, and tanh activation function is used at the bottom layer to make the output image intensities fall within $[-1, 1]$. Once the model is learned, we can synthesize dynamic textures from the learned model.

To speed up the training process and relieve the burden of computer memory, we can use truncated back-propagation through time in training the model. That is, we divide the whole training sequence into different non-overlapped chunks and run

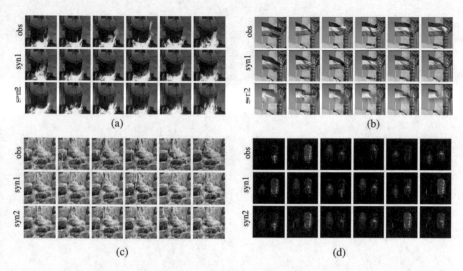

Fig. 11.15 Reprinted with permission from [260]. Generating dynamic textures. For each category, the first row displays 6 frames of the observed sequence, and the second and third rows show the corresponding frames of two synthesized sequences generated by the learned model, (**a**) burning fire heating a pot, (**b**) flapping flag, (**c**) waterfall, (**d**) flashing lights

forward and backward passes through chunks of the sequence instead of the whole sequence. We carry hidden states z_t forward in time forever, but only back-propagate for the length (the number of image frames) of the chunk. In this experiment, the length of the chunk is set to be 30 image frames.

An "infinite length" dynamic texture can be synthesized from a typically "short" input sequence by just drawing "infinite" i.i.d. samples from Gaussian distribution. Figure 11.15 shows five results. For each example, the first row displays 6 frames of the observed 60-frame sequence, while the second and third rows display 6 frames of two synthesized sequences of 120 frames in length, which are generated by the learned model.

Chapter 12
A Tale of Three Families: Discriminative, Descriptive, and Generative Models

12.1 Introduction

Three Families of Probabilistic Models

This chapter gives a general introduction to three families of probabilistic models and their connections. Most of the models studied in the previous chapters, as well as most of the models in the current machine learning and deep learning literature, belong to these three families of models.

The first class consists of discriminative models or classifiers that are commonly used in supervised learning. The second class consists of descriptive models—also known as energy-based models—that define unnormalized probability density functions in the data space. These models are generalizations of the FRAME model introduced in the previous chapters. The third class consists of generative models that are directed top-down models that involve latent variables. The generative models are generalizations of factor analysis and its variants. They are also called directed graphical models.

About the names of the models, we use the term "generative models" in a much narrower sense than in the current literature. They refer to top-down models that consist of latent variables that follow simple prior distributions so that the examples can be directly generated. As to the "descriptive models," they refer to the energy-based models or deep FRAME model introduced in the previous chapter. They only describe the examples in terms of their probability densities, but they cannot directly generate the examples. The generative task is left to iterative MCMC sampling algorithms. Therefore, these models are not literally generative as they do not explicitly define a generative process, and that is why we call them descriptive.

© Springer Nature Switzerland AG 2023
S.-C. Zhu, Y. N. Wu, *Computer Vision*, https://doi.org/10.1007/978-3-030-96530-3_12

$$
\begin{array}{ccc}
\textit{Top-down mapping} & \textit{Bottom-up mapping} & \textit{Bottom-up mapping} \\
\text{latent variables } z & \text{log density } f_\theta(x) & \text{logit score (density ratio)} \\
\Downarrow & \Uparrow & \Uparrow \qquad (12.1) \\
\text{example } x \approx g_\theta(z) & \text{example } x & \text{example } x \\
\text{(a) \textbf{Generative} (sampler)} & \text{(b) \textbf{Descriptive} (density)} & \text{(b) \textbf{Discriminative} (classifier)}
\end{array}
$$

Density vs. Sampler A descriptive model specifies the probability density function explicitly, up to a normalizing constant. A discriminative model specifies the ratios between two or more densities via the Bayes rule. A generative model, on the other hand, does not specify a data density explicitly. It specifies a sampler or a sampling process that transforms latent variables with a known distribution, e.g., Gaussian white noise variables, to the observed example. By analogy to reinforcement learning, a density is like a value network or a critic, and a sampler is like a policy network or an actor.

The above diagram illustrates the three families of probabilistic models. A generative model is based on a top-down mapping from the latent variables z to the example x. A descriptive model is based on a bottom-up mapping from the example x to the log of the unnormalized density. A discriminative model is based on a bottom-up mapping from the example x to the logit score that is also the ratio between the densities of positive and negative classes in the binary classification situation (which can be easily generalized to the multi-class situation). All the mappings can be parameterized by deep neural networks.

In the previous chapter, we introduced the descriptive models and generative models for image and video data and the associated maximum likelihood learning algorithms. This chapter will give a more general treatment. We shall still emphasize the maximum likelihood learning algorithm. Meanwhile, we shall also present various joint training schemes, such as variational learning and adversarial learning. We shall make this chapter self-contained so that readers who are interested in the development in the deep learning era can read this chapter in isolation.

Notation We shall adopt the notation commonly used in the current literature. We use x to denote the training example, e.g., an image or a sentence. We use z to denote the latent variables in the generative model. We use y to denote the output in the discriminative model, e.g., image category. We use θ to denote the model parameters. We use the notations ∇_x and ∇_θ to denote $\frac{\partial}{\partial x}$ and $\frac{\partial}{\partial \theta}$, respectively. For a function $h(\theta)$, its derivative at a fixed value, say, θ_t, is denoted $\nabla_\theta h(\theta_t)$. We use D_{KL} to denote the Kullback–Leibler divergence.

Supervised, Unsupervised, and Self-supervised Learning

Supervised learning refers to the situation where both the input x and the output y are given, and we want to learn to predict y based on x. More formally, we learn a discriminative or predictive model $p(y|x)$ by maximum likelihood, i.e., we maximize the average of $\log p(y|x)$ over the model parameters where the average is over the training set $\{x, y\}$. The limitation of supervised learning is that it can be expensive and time-consuming to obtain y in the form of label or annotation.

Unsupervised learning refers to the situation where only the input x is given, but the output y is unavailable. In that case, we can learn a descriptive model or a generative model, again by maximum likelihood, but we maximize the average of $\log p(x)$ over the model parameters, where the average is over the training set $\{x\}$, instead of the average of $\log p(y|x)$, as y is not available. The descriptive model specifies $p(x)$ up to an unknown normalizing constant, and it is closely related to the discriminative model through the Bayes rule. For the generative model, $p(x)$ is implicit because it involves integrating out the latent variables z. The latent z can be inferred from the input x.

Semi-supervised learning refers to the situation between supervised and unsupervised learning, where there are a small number of labeled examples where both x and y are given, and there are a large number of unlabeled examples where only x is given. In that case, we can again learn the model by maximum likelihood, where we maximize the sum of $\log p(y|x)$ over the labeled examples and $\log p(x)$ over the unlabeled examples. Thus probabilistic modeling provides a unified likelihood-based framework for supervised, unsupervised, and semi-supervised learning.

There is also self-supervised learning, which is to translate unsupervised learning into supervised learning. Specifically, even if we are only given x without y, we can nonetheless create a task where we artificially introduce y for a modification of x that depends on y, and we then learn $p(y|x)$ instead of $p(x)$. This type of learning can be more formally treated as learning descriptive model by various conditional likelihoods.

MCMC for Synthesis and Inference

Although likelihood-based learning with probabilistic models is a principled framework for supervised, unsupervised, and semi-supervised learning, the bottleneck for likelihood-based learning for unsupervised learning is that the derivative of the log-likelihood function $\log p(x)$ usually involves intractable integrals or expectations, which require expensive MCMC sampling. A lot of effort has been spent on getting around this obstacle.

We may use short-run MCMC, i.e., running MCMC such as Langevin dynamics or Hamiltonian Monte Carlo (HMC) [179] from a fixed initial noise distribution for a fixed number of steps, for inference and synthesis computations. This is affordable on modern computing platforms. It can also be justified as a modification or perturbation of the maximum likelihood learning.

Short-run MCMC is convenient for learning models with multiple layers of latent variables organized in complex architectures because top-down feedback and lateral inhibition between the latent variables at different layers can automatically emerge in short-run MCMC. The short-run Langevin dynamics can also be compared with attractor dynamics that is a commonly assumed framework for modeling neural computations [7, 108, 198]. One can also run persistent Markov chains, i.e., in each learning iteration, we initialize finite-step MCMC from the samples generated in the previous learning iteration.

Deep Networks as Function Approximators

All three classes of models can be parameterized by deep neural networks [138, 144], which are compositions of multiple layers of linear transformations and coordinate-wise nonlinear transformations.

Specifically, consider a nonlinear transformation $f(x)$ that can be decomposed recursively as $s_l = W_l h_{l-1} + b_l$, and $h_l = r_l(s_l)$, for $l = 1, \ldots, L$, with $f(x) = h_L$ and $h_0 = x$. W_l is the weight matrix at layer l, and b_l is the bias vector at layer l. Both s_l and h_l are vectors of the same dimensionality, and r_l is a one-dimensional nonlinear function, or rectification function, that is applied coordinate-wise or element-wise.

The nonlinear rectification is crucial for $f(x)$ to approximate nonlinear mapping. In the past, the nonlinear rectification $r_l()$ is usually sigmoid transformation, which is approximately a two-piece constant function. This makes $f(x)$ approximately piecewise constant function. Modern deep networks usually use $r_l(s) = \max(0, s)$, the rectified linear unit or ReLU, which makes $f(x)$ piecewise linear.

There are two special classes of neural networks. One consists of convolutional neural networks [138, 144], which are commonly applied to images, where the same linear transformations are applied around each pixel locally. The other class consists of recurrent neural networks [103], which are commonly applied to sequence data such as speech and natural language. Recently, the transformer model [239] has become the most prominent architecture.

Deep neural networks are powerful function approximators that can approximate highly nonlinear high-dimensional continuous functions by interpolating training examples. Modern deep networks are highly overparameterized, meaning that the number of parameters greatly exceeds the number of training examples. Thus they have enough capacity to fit the training data, yet they tend not to overfit the training data because the networks are learned by stochastic gradient descent algorithm where the gradient is computed via back-propagation. The stochastic gradient descent algorithm provides implicit regularization [11, 221].

Learned Computation

Because of the strong approximation capacity, the boundary between representation and computation is rather blurred because a deep network can approximate an iterative algorithm. Sometimes this is called learned computation.

In fact, the residual network [97] can be considered a finite-step iterative algorithm. It is of the form $x_{l+1} = x_l + f_l(x_l)$, where l indexes the layer. Meanwhile, l may also be interpreted as time step of an iterative algorithm, i.e., we can also write $x_{t+1} = x_t + f_t(x_t)$, which is to model iterative updating or refinement. In general, it can be interpreted as a mixture of both, i.e., there is actually a small number of layers, and each layer is computed by a finite-step iterative algorithm.

The transformer model [239] can also be considered a finite-step iterative algorithm that iteratively updates the vector representations of the words of an input sentence through the self-attention mechanism where the words pay attention to and gather information from each other. The graph convolutional network [134] can learn the iterative message passing mechanism where the nodes of the graph send messages to each other.

In the above iterative updating mechanisms, there is no need to know the objective functions of these iterative mechanisms. They can be embedded into a classifier and be trained by the classification loss via back-propagation through time.

Amortized Computation for Synthesis and Inference Sampling

Even if there is an objective function, we can still learn a deep network that directly maps the input to an approximate solution. Sometimes this is called amortized computation, which seeks to approximate an iterative algorithm of multiple time steps.

In the case of the generative model, recall that we can use short-run MCMC as an approximate sampler for synthesis and inference. We can also learn a network that produces the samples directly. In the case of posterior sampling, this is referred to as variational inference model [133]. In fact, the short-run MCMC can be considered a noise-injected residual network.

When there are multiple layers of latent variables, designing a network for approximate inference sampling can be a non-trivial task, whereas short-run MCMC remains automatic.

Distributed Representation and Embedding

Deep neural networks are based on continuous vectors and weight matrices. They are highly interpolative and amendable to gradient-based computations. On the other hand, high-level reasoning can also be highly symbolic, with symbols, logic, and grammar. For a dictionary of symbols, each symbol can be represented by a one-hot vector, and a small subset of symbols selected from the dictionary can be represented by a sparse vector. This is in contrast to the vectors in deep networks, which are continuous and dense. Such vectors are called distributed representations. They are also commonly referred to as embeddings. For instance, the word2vec model [172, 193] represents each word by a dense vector, and this means we embed the words in a continuous Euclidean space. A modern deep network such as transformer [239] or graph neural network [134] can be viewed as a team of vectors, which are operated on by learned matrices so that they can pass on messages to each other. For discrete or symbolic inputs or outputs such as words or tokens, they can be encoded into vectors or decoded from the vectors.

It is still unclear how to unify symbolic and dense representations. Sometimes this is referred to as the contrast between symbolism and connectionism. It is likely that there is a duality or complementarity between sparse vectors and dense vectors, and each is more convenient and efficient than the other depending on the scenario.

Perturbations of Kullback–Leibler Divergence

A unifying theoretical device for studying various learning methods is to perturb the Kullback–Leibler divergence for maximum likelihood by other Kullback–Leibler divergences. This scheme consists of three Kullback–Leibler divergences: (1) KL-divergence underlying maximum likelihood learning. This is the target of the perturbations. (2) KL-divergence underlying synthesis sampling. (3) KL-divergence underlying inference sampling. (2) and (3) are perturbations that are applied to (1). The sign in front of (2) is negative, and the sign in front of (3) is positive.

The above theoretical framework explains the following learning algorithms: (1) The maximum likelihood learning algorithm. (2) The learning algorithm based on short-run MCMC for synthesis and inference. (3) The learning methods based on learned networks for synthesis and inference, including adversarial learning [81] and variational learning [133].

Kullback–Leibler Divergence in Two Directions

To be more specific, recall that for two probability densities $p(x)$ and $q(x)$, we define

$$D_{\mathrm{KL}}(p\|q) = \mathrm{E}_p\left[\log \frac{p(x)}{q(x)}\right] = \int p(x) \log \frac{p(x)}{q(x)} dx. \tag{12.2}$$

The KL-divergence appears in two scenarios:

(1) **Maximum likelihood learning**. Suppose the training examples $x_i \sim p_{\mathrm{data}}(x)$ are independent for $i = 1, \ldots, n$. Suppose we want to learn a model $p_\theta(x)$. The log-likelihood function is

$$L(\theta) = \frac{1}{n} \sum_{i=1}^{n} \log p_\theta(x_i) \to E_{p_{\text{data}}}[\log p_\theta(x)]. \tag{12.3}$$

Thus for big n, maximizing $L(\theta)$ is equivalent to minimizing

$$D_{\text{KL}}(p_{\text{data}} \| p_\theta) = -\text{entropy}(p_{\text{data}}) - E_{p_{\text{data}}}[\log p_\theta(x)] = -\text{entropy}(p_{\text{data}}) - L(\theta), \tag{12.4}$$

where $E_{p_{\text{data}}}$ can be approximated by averaging over $\{x_i\}$. We can think of it as projecting p_{data} onto the model space $\{p_\theta, \forall \theta\}$.

For the rest of this chapter, for notational simplicity, we will not distinguish between $E_{p_{\text{data}}}$ and sample average over $\{x_i\}$, and we will treat $D_{\text{KL}}(p_{\text{data}} \| p_\theta)$ as the loss function for maximum likelihood learning.

(2) **Variational approximation**. Suppose we have a target distribution p_{target}, and we know p_{target} up to a normalizing constant, e.g., $p_{\text{target}}(x) = \exp(f(x))/Z$, where we know $f(x)$ but the normalizing constant $Z = \int \exp(f(x))dx$ is analytically intractable. Suppose we want to approximate it by a distribution q_ϕ. We can find ϕ by minimizing

$$D_{\text{KL}}(q_\phi \| p_{\text{target}}) = E_{q_\phi}[\log q_\phi(x)] - E_{q_\phi}[f(x)] + \log Z. \tag{12.5}$$

This time, we place q_ϕ on the left-hand side and p_{target} on the right-hand side of the KL-divergence, because p_{target} is accessible only through $f(x)$. The above minimization does not require knowledge of $\log Z$.

The behaviors of $\min_\theta D_{\text{KL}}(p_{\text{data}} \| p_\theta)$ in (1) and $\min_\phi D_{\text{KL}}(q_\phi \| p_{\text{target}})$ in (2) are different. In (1), p_θ tends to cover all the modes of p_{data} because $D_{\text{KL}}(p_{\text{data}} \| p_\theta)$ is the expectation with respect to p_{data}. In (2), q_ϕ tends to focus on some major modes of p_{target}, while ignoring the minor modes, because $D_{\text{KL}}(q_\phi \| p_{\text{target}})$ is the expectation with respect to q_ϕ.

In the perturbation scheme mentioned in the previous subsection, the KL-divergence for maximum likelihood is (12.4). The perturbations are of the form in (12.5).

12.2 Descriptive Energy-Based Model

Model and Origin

Let x be a training example, e.g., an image or a sentence. A descriptive model specifies an unnormalized probability density function

$$p_\theta(x) = \frac{1}{Z(\theta)} \exp(f_\theta(x)), \tag{12.6}$$

where $f_\theta(x)$ is parameterized by a deep network, with θ collecting all the weight and bias parameters. $Z(\theta) = \int \exp(f_\theta(x))dx$ is the normalizing constant.

Such a model originated from statistical mechanics and is called the Gibbs distribution, where x is the state or configuration of a physical system, and $-f_\theta(x)$ is the energy function of the state so that the lower energy states are more likely to be observed. For that reason, the above model is also called energy-based model (EBM) in the literature [32, 70, 99, 161, 180, 182, 184, 266, 269, 270].

In classical mechanics, the configuration $x(t)$ evolves deterministically over time t according to a partial differential equation. Then where does the probability distribution come from? We may consider the ensemble or population $(x(t), t \in [t_0, t_1])$, for a long enough burn-in time t_0 and long enough duration $t_1 - t_0$. For a random time $t \sim \text{Uniform}[t_0, t_1]$, $x(t)$ follows a probability distribution $p(x)$, and it can be modeled by a Gibbs distribution.

The quantity $Z(\theta)$ is called the partition function in statistical mechanics. An important identity is

$$\nabla_\theta \log Z(\theta) = \mathbb{E}_{p_\theta}[\nabla_\theta f_\theta(x)]. \tag{12.7}$$

The non-differentiability of $\log Z(\theta)$ underlies the phase transition phenomena in statistical physics.

The descriptive model has strong expressive power because it only needs to specify a scalar-valued function $f_\theta(x)$. $f_\theta(x)$ is like an objective function (or value function, or constraints, or rules). The descriptive model is only responsible for specifying the objective function and is not responsible for optimizing the objective function or providing near-optimal solutions. The latter task is left to MCMC sampling. As a result, a simple descriptive model $p_\theta(x)$ or the objective function $f_\theta(x)$ can explain rich patterns and complex behaviors.

The descriptive model has been used for inverse reinforcement learning, where $-f_\theta(x)$ serves as the cost function [283]. It has also been used for Markov logic network [204], where $f_\theta(x)$ combines logical rules.

Gradient-Based Sampling

For high-dimensional x, such as image, sampling from $p_\theta(x)$ requires MCMC, such as Langevin dynamics or Hamiltonian Monte Carlo. The Langevin dynamics iterates

$$x_{t+1} = x_t + s\nabla_x f_\theta(x_t) + \sqrt{2s}e_t, \tag{12.8}$$

where s is the step size and $e_t \sim \mathrm{N}(0, I)$ is the Gaussian white noise term. The Langevin dynamics has a gradient ascent term $\nabla_x f_\theta(x_t)$, and e_t is the diffusion term for randomness. As $s \to 0$ and $t \to \infty$, the distribution of x_t converges to $p_\theta(x)$.

We can write the Langevin dynamics in continuous time as

$$x_{t+\Delta t} = x_t + \frac{1}{2}\nabla_x f_\theta(x_t)\Delta t + \sqrt{\Delta t}\, e_t, \tag{12.9}$$

or more formally,

$$dx_t = \frac{1}{2}\nabla_x f_\theta(x_t)dt + dB_t, \tag{12.10}$$

where dB_t plays the role of $\sqrt{\Delta t}\, e_t$.

Let p_t be the distribution of x_t. Then according to the Fokker–Planck equation, we have

$$\nabla_t p_t(x) = \frac{1}{2}[\nabla_x(f_\theta(x)p_t(x)) + \nabla_x^2 p_t(x)]. \tag{12.11}$$

$p_\theta(x)$ is the solution to $\nabla_t p_t(x) = 0$, i.e., the stationary distribution. In terms of variational approximation,

$$D_{\mathrm{KL}}(p_t \| p_\theta) = -\mathrm{entropy}(p_t) - \mathrm{E}_{p_t}[f_\theta(x)] + \log Z(\theta) \to 0 \tag{12.12}$$

monotonically as $t \to \infty$ under fairly general conditions. The gradient term in the Langevin dynamics increases $f_\theta(x)$ or decreases energy, while the noise term e_t increases the entropy of p_t.

Intuitively, imagine a population of x's that are distributed according to $p_\theta(x)$. The deterministic gradient ascent term in the Langevin dynamics pushes the points toward the local modes of the log density, making the distribution of the points more focused on the local modes of the density. Meanwhile, the random diffusion term in the Langevin dynamics adds random noises to the points, making the distribution of the points more diffused from the local modes of the density. The two terms balance each other so that the overall distribution of the points after each Langevin iteration remains unchanged.

Hamiltonian Monte Carlo (HMC) [105, 179] is a more powerful gradient-based MCMC sampling method. Similar to gradient descent with momentum, it can navigate the high curvature regions of the energy landscape more smoothly and efficiently. The step size in HMC can be adaptively selected based on the acceptance rate calculated from the energy function [105].

In order to traverse local modes and facilitate fast mixing of the Markov chain, one can add a temperature parameter to interpolate the multi-modal target density and a simple unimodal reference density such as Gaussian white noise distribution. One can then use simulated annealing [135] or more principled and effective MCMC schemes such as simulated tempering [168], parallel tempering [48, 77], or replica exchange [226] to sample from multi-modal density.

Maximum Likelihood Estimation (MLE)

The descriptive model $p_\theta(x)$ can be learned by maximum likelihood estimation (MLE). The log-likelihood is the average of

$$\log p_\theta(x) = f_\theta(x) - \log Z(\theta), \tag{12.13}$$

where the average is over the training set $\{x\}$. The gradient of $\log p_\theta(x)$ with respect to θ is

$$\delta_\theta(x) = \nabla_\theta \log p_\theta(x) = \nabla_\theta f_\theta(x) - E_{p_\theta(x)}[\nabla_\theta f_\theta(x)], \tag{12.14}$$

where

$$\nabla_\theta \log Z(\theta) = E_{p_\theta(x)}[\nabla_\theta f_\theta(x)]. \tag{12.15}$$

Suppose we observe training examples $\{x_i, i = 1, \ldots, n\} \sim p_{\text{data}}$, where p_{data} is the data distribution. For large n, the sample average over $\{x_i\}$ approximates the expectation with respect to p_{data}. For notational convenience, we treat the sample average and the expectation as the same.

The log-likelihood is

$$L(\theta) = \frac{1}{n} \sum_{i=1}^{n} \log p_\theta(x_i) \doteq E_{p_{\text{data}}}[\log p_\theta(x)]. \tag{12.16}$$

The derivative of the log-likelihood is

$$L'(\theta) = E_{p_{\text{data}}}[\delta_\theta(x)] = E_{p_{\text{data}}}[\nabla_\theta f_\theta(x)] - E_{p_\theta}[\nabla_\theta f_\theta(x)] \tag{12.17}$$

$$\doteq \frac{1}{n} \sum_{i=1}^{n} \nabla_\theta f_\theta(x_i) - \frac{1}{n} \sum_{i=1}^{n} \nabla_\theta f_\theta(x_i^-), \tag{12.18}$$

where $x_i^- \sim p_\theta(x)$ for $i = 1, \ldots, n$ are the generated examples from the current model $p_\theta(x)$.

The above equation leads to the "analysis by synthesis" learning algorithm. At iteration t, let θ_t be the current model parameters. We generate synthesized examples $x_i^- \sim p_{\theta_t}(x)$ for $i = 1, \ldots, n$. Then we update $\theta_{t+1} = \theta_t + \eta_t L'(\theta_t)$, where η_t is the learning rate, and $L'(\theta_t)$ is the statistical difference between the synthesized examples and observed examples (Fig. 12.1).

Fig. 12.1 Reprinted with permission from [266]. Learning the descriptive model by maximum likelihood: (**a**) goose, (**b**) tiger. For each category, the first row displays four of the training images, and the second row displays four of the images generated by the learning algorithm. $f_\alpha(x)$ is parameterized by a four-layer bottom-up deep network, where the first layer has 100 7×7 filters with subsampling size 2, the second layer has 64 5×5 filters with subsampling size 1, the third layer has 20 3×3 filters with subsampling size 1, and the fourth layer is a fully connected layer with a single filter that covers the whole image. The number of parallel chains for Langevin sampling is 16, and the number of Langevin iterations between every two consecutive updates of parameters is 10. The training images are 224×224 pixels

Objective Function and Estimating Equation of MLE

The maximum likelihood learning minimizes the Kullback–Leibler divergence $D_{\mathrm{KL}}(p_{\mathrm{data}} \| p_\theta)$ over θ. Geometrically, it is to project p_{data} onto the manifold formed by $\{p_\theta, \forall \theta\}$.

The maximum likelihood learning algorithm converges to the solution to the following estimating equation:

$$\mathrm{E}_{p_\theta}\left[\nabla_\theta f_\theta(x)\right] = \mathrm{E}_{p_{\mathrm{data}}}\left[\nabla_\theta f_\theta(x)\right], \tag{12.19}$$

where the model matches the data in terms of the expectation of $\nabla_\theta f_\theta(x)$.

For the FRAME model or in general the exponential family model, $f_\theta(x) = \langle \theta, h(x) \rangle$ for feature vector $h(x)$; hence, $\nabla_\theta f_\theta(x) = h(x)$ and $L'(\theta) = \mathrm{E}_{p_{\mathrm{data}}}[h(x)] - \mathrm{E}_{p_\theta}[h(x)]$. The maximum likelihood estimating equation is $\mathrm{E}_{p_\theta}[h(x)] = \mathrm{E}_{p_{\mathrm{data}}}[h(x)]$, i.e., matching feature statistics. For general $f_\theta(x)$, we may still consider $\nabla_\theta f_\theta(x)$ as a feature vector.

Perturbation of KL-divergence

Define $D(\theta) = D_{\mathrm{KL}}(p_{\mathrm{data}} \| p_\theta)$. It is the loss function of MLE. To understand the MLE learning algorithm, let θ_t be the estimate at iteration t. Let us consider the following perturbation of $D(\theta)$:

$$S(\theta) = D(\theta) - D_{\mathrm{KL}}(p_{\theta_t} \| p_\theta) = D_{\mathrm{KL}}(p_{\mathrm{data}} \| p_\theta) - D_{\mathrm{KL}}(p_{\theta_t} \| p_\theta). \tag{12.20}$$

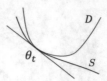

Fig. 12.2 Reprinted with permission from [95]. The surrogate S minorizes (lower bounds) D, and they touch each other at θ_t with the same tangent

$S(\theta)$ is the surrogate objective function for $D(\theta)$ at iteration t. It is simpler than $D(\theta)$, because the $\log Z(\theta)$ term gets canceled, and the gradient can be more easily computed (Fig. 12.2).

The perturbation term $D_{\mathrm{KL}}(p_{\theta_t}\|p_\theta)$, as a function of θ, with θ_t fixed, has the following properties: (1) It achieves minimum zero at $\theta = \theta_t$. (2) Its derivative is zero at $\theta = \theta_t$. As a result, $S(\theta_t) = D(\theta_t)$, and $S'(\theta_t) = D'(\theta_t)$. Geometrically, $S(\theta)$ and $D(\theta)$ touch each other at θ_t, and they are co-tangent at θ_t. Since

$$S(\theta) = -\mathrm{E}_{p_{\mathrm{data}}}[f_\theta(x)] + \mathrm{E}_{p_{\theta_t}}[f_\theta(x)] - \mathrm{entropy}(p_{\mathrm{data}}) + \mathrm{entropy}(p_{\theta_t}), \quad (12.21)$$

where $\log Z(\theta)$ gets canceled, we have

$$-S'(\theta) = \mathrm{E}_{p_{\mathrm{data}}}[\nabla_\theta f_\theta(x)] - \mathrm{E}_{p_{\theta_t}}[\nabla_\theta f_\theta(x)]. \quad (12.22)$$

Thus

$$-D'(\theta_t) = -S'(\theta_t) = \mathrm{E}_{p_{\mathrm{data}}}[\delta_{\theta_t}(x)] = \mathrm{E}_{p_{\mathrm{data}}}[\nabla_\theta f_{\theta_t}(x)] - \mathrm{E}_{p_{\theta_t}}[\nabla_\theta f_{\theta_t}(x)]. \quad (12.23)$$

This justifies the MLE learning algorithm.

We shall use this perturbation scheme repeatedly, where we perturb the MLE loss function $D(\theta) = D_{\mathrm{KL}}(p_{\mathrm{data}}\|p_\theta)$ to a simpler surrogate objective function $S(\theta)$ by subtracting or adding other KL-divergence terms. This enables us to theoretically understand other learning methods that are modifications of MLE learning.

Self-adversarial Interpretation

$S(\theta) = D_{\mathrm{KL}}(p_{\mathrm{data}}\|p_\theta) - D_{\mathrm{KL}}(p_{\theta_t}\|p_\theta)$ leads to an adversarial interpretation. When we update θ by following the gradient of $S(\theta)$ at $\theta = \theta_t$, we want p_θ to move away from p_{θ_t} and move toward p_{data}. That is, the model p_θ criticizes its current version p_{θ_t} by comparing p_{θ_t} to p_{data}. The model serves as both generator and discriminator if we compare it to GAN (generative adversarial networks). In contrast to GAN [8, 81, 199], the learning algorithm is MLE, which in general does not suffer from

issues such as mode collapsing and instability, as it does not involve the competition
between two separate networks.

Short-Run MCMC for Synthesis

We now consider the learning algorithm based on short-run MCMC [184].
   The short-run MCMC is

$$x_0 \sim p_0(x), \ x_{k+1} = x_k + s\nabla_x f_\theta(x_k) + \sqrt{2s}e_k, \ k = 1, \ldots, K, \quad (12.24)$$

where we initialize the Langevin dynamics from a fixed diffused noise distribution
$p_0(x)$, and we run a fixed number of K steps. Let $\tilde{p}_\theta(x)$ be the distribution of x_K. We
use x_K as the synthesized example for approximate maximum likelihood learning
(Figs. 12.3 and 12.4).
   For each x, we define

$$\tilde{\delta}_\theta(x) = \nabla_\theta f_\theta(x) - E_{\tilde{p}_\theta(x)}[\nabla_\theta f_\theta(x)] \quad (12.25)$$

and modify the learning algorithm to

$$\theta_{t+1} = \theta_t + \eta_t E_{p_{data}}[\tilde{\delta}_{\theta_t}(x)] = \theta_t + \eta_t \left(E_{p_{data}}[\nabla_\theta f_\theta(x)] - E_{\tilde{p}_\theta}[\nabla_\theta f_\theta(x)]\right). \quad (12.26)$$

Fig. 12.3 Reprinted with permission from [184]. Synthesis by short-run MCMC: Generating
synthesized examples by running 100 steps of Langevin dynamics initialized from uniform noise
for CelebA (64×64)

Fig. 12.4 Reprinted with permission from [184]. Synthesis by short-run MCMC: Generating
synthesized examples by running 100 steps of Langevin dynamics initialized from uniform noise
for CelebA (128×128)

Objective Function and Estimating Equation with Short-Run MCMC

The following are justifications for the learning algorithm based on short-run MCMC synthesis:

(1) **Objective function**. Again we use perturbation of KL-divergence. At iteration t, with θ_t fixed, the learning algorithm follows the gradient of the following perturbation of $D(\theta)$ at $\theta = \theta_t$:

$$S(\theta) = D(\theta) - D_{KL}(\tilde{p}_{\theta_t} \| p_\theta) = D_{KL}(p_{data} \| p_\theta) - D_{KL}(\tilde{p}_{\theta_t} \| p_\theta), \quad (12.27)$$

so that $\theta_{t+1} = \theta_t + \eta_t S'(\theta_t)$, where η_t is the step size, and

$$- S'(\theta) = \mathrm{E}_{p_{data}}[\nabla_\theta f_\theta(x)] - \mathrm{E}_{\tilde{p}_{\theta_t}}[\nabla_\theta f_\theta(x)]. \quad (12.28)$$

$$- S'(\theta_t) = \mathrm{E}_{p_{data}}[\tilde{\delta}_{\theta_t}(x)] = \mathrm{E}_{p_{data}}[\nabla_\theta f_{\theta_t}(x)] - \mathrm{E}_{\tilde{p}_{\theta_t}}[\nabla_\theta f_{\theta_t}(x)]. \quad (12.29)$$

Compared to the perturbation of KL-divergence in MLE learning, we use $\tilde{p}_{\theta_t}$ instead of p_{θ_t}. While sampling p_{θ_t} can be impractical if it is multi-modal, sampling $\tilde{p}_{\theta_t}$ is practical and exact because it is a short-run MCMC.

Note that $S'(\theta_t) \neq D'(\theta_t)$, because $\tilde{p}_{\theta_t} \neq p_{\theta_t}$. Thus the learning gradient based on short-run MCMC is biased from that of MLE. As a result, the learned p_θ based on short-run MCMC may be biased from MLE.

$S(\theta)$ indicates that we need to minimize $D_{KL}(\tilde{p}_\theta \| p_\theta)$ in order to minimize the bias relative to the maximum likelihood learning. We can do that by increasing K because $D_{KL}(\tilde{p}_\theta \| p_\theta)$ decreases monotonically to zero as K increases. For fixed K, we can also employ more efficient MCMC, especially those that can traverse local modes, such as parallel tempering [48, 77] or replica exchange [226].

(2) **Estimating equation**. The learning algorithm converges to the solution to the following estimating equation:

$$\mathrm{E}_{\tilde{p}_\theta}[\nabla_\theta f_\theta(x)] = \mathrm{E}_{p_{data}}[\nabla_\theta f_\theta(x)], \quad (12.30)$$

which is a perturbation of the maximum likelihood estimating equation where we replace p_θ by $\tilde{p}_\theta$ (Fig. 12.5).

Thus even if the learned p_θ may be biased from MLE, the resulting short-run MCMC $\tilde{p}_\theta$ can nonetheless be considered a valid model, in that it matches p_{data} in terms of expectations of $\nabla_\theta f_\theta(x)$. Recall in the case of FRAME model where $f_\theta(x) = \langle \theta, h(x) \rangle$, $\nabla_\theta f_\theta(x) = h(x)$, i.e., the learned short-run MCMC $\tilde{p}_\theta$ matches p_{data} in terms of expectations of $h(x)$. In general, $\nabla_\theta f_\theta(x)$ may be considered a

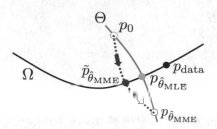

Fig. 12.5 Reprinted with permission from [184]. The blue curve illustrates the model distributions corresponding to different values of parameter θ. The black curve illustrates all the distributions that match p_{data} (black dot) in terms of $E[h(x)]$. The MLE $p_{\hat{\theta}_{\text{MLE}}}$ (green dot) is the intersection between Θ (blue curve) and Ω (black curve). The MCMC (red dotted line) starts from p_0 (hollow blue dot) and runs toward $p_{\hat{\theta}_{\text{MME}}}$ (hollow red dot), but the MCMC stops after K step, reaching $\tilde{p}_{\hat{\theta}_{\text{MME}}}$ (red dot), which is the learned short-run MCMC

Fig. 12.6 Reprinted with permission from [184]. Interpolation by short-run MCMC resembling a generator or flow model: The transition depicts the sequence $F(z_\rho)$ with interpolated noise $z_\rho = \rho z_1 + \sqrt{1 - \rho^2} z_2$, where $\rho \in [0, 1]$ on CelebA (64×64). Left: $F(z_1)$. Right: $F(z_2)$

Fig. 12.7 Reprinted with permission from [184]. Reconstruction by short-run MCMC resembling a generator or flow model: The transition depicts $F(z_t)$ over time t from random initialization $t = 0$ to reconstruction $t = 200$ on CelebA (64×64). Left: Random initialization. Right: Observed examples

generalized version of feature vector $h(x)$. Thus we may justify the learned short-run MCMC $\tilde{p}_\theta$ as a generalized moment matching estimator $\hat{\theta}_{\text{MME}}$. The generalized moment matching explains the synthesis ability of the descriptive model and various learning schemes in general.

The short-run Langevin dynamics can be considered a noise-injected RNN or noise-injected residual network. Specifically, we can write $x_K = F(x_0, e)$, where $e = (e_k, k = 1, \ldots, K)$. We can use it to reconstruct the observed image x by minimizing $\|x - F(x_0, e)\|$ over x_0 and e. As a simple approximation, we can set $e_k = 0$ and write $x_K = F(x_0)$ (Figs. 12.6 and 12.7).

Flow-Based Model

A flow-based model is of the form

$$x = g_\alpha(z); \ z \sim q_0(z), \tag{12.31}$$

where q_0 is a known noise distribution. g_α is a composition of a sequence of invertible transformations where the log determinants of the Jacobians of the transformations can be explicitly obtained. α denotes the parameters. Let $q_\alpha(x)$ be the probability density of the model at a data point x with parameter α. Then under the change of variables, $q_\alpha(x)$ can be expressed as

$$q_\alpha(x) = q_0(g_\alpha^{-1}(x))|\det(\partial g_\alpha^{-1}(x)/\partial x)|. \tag{12.32}$$

More specifically, suppose g_α is composed of a sequence of transformations $g_\alpha = g_{\alpha_1} \circ \cdots \circ g_{\alpha_m}$. The relation between z and x can be written as $z \leftrightarrow h_1 \leftrightarrow \cdots \leftrightarrow h_{m-1} \leftrightarrow x$. And thus we have

$$q_\alpha(x) = q_0(g_\alpha^{-1}(x))\Pi_{t=1}^m|\det(\partial h_{t-1}/\partial h_t)|, \tag{12.33}$$

where we define $z := h_0$ and $x := h_m$ for conciseness. With carefully designed transformations, as explored in flow-based methods, the determinant of the Jacobian matrix $(\partial h_{t-1}/\partial h_t)$ can be computed exactly. The key idea is to choose transformations whose Jacobian is a triangle matrix so that the determinant becomes

$$|\det(\partial h_{t-1}/\partial h_t)| = \Pi|\mathrm{diag}(\partial h_{t-1}/\partial h_t)|. \tag{12.34}$$

The following are the two scenarios for estimating q_α:

(1) Generative modeling by MLE [13, 43, 44, 82, 131, 139, 233], by $\min_\alpha D_{\mathrm{KL}}$ $(p_{\mathrm{data}}\|q_\alpha)$, where $\mathrm{E}_{p_{\mathrm{data}}}$ can be approximated by average over observed examples.
(2) Variational approximation to an unnormalized target density p [130, 132, 202], based on $\min_\alpha D_{\mathrm{KL}}(q_\alpha\|p)$, where

$$D_{\mathrm{KL}}(q_\alpha\|p) = \mathrm{E}_{q_\alpha}[\log q_\alpha(x)] - \mathrm{E}_{q_\alpha}[\log p(x)]$$

$$= \mathrm{E}_z[\log q_0(z) - \log|\det(g_\alpha'(z))|] - \mathrm{E}_{q_\alpha}[\log p(x)]. \tag{12.35}$$

$D_{\mathrm{KL}}(q_\alpha\|p)$ is the difference between energy and entropy, i.e., we want q_α to have low energy but high entropy. $D_{\mathrm{KL}}(q_\alpha\|p)$ can be calculated without inversion of g_α.

When q_α appears on the right of KL-divergence, as in (1), it is forced to cover most of the modes of p_{data}. When q_α appears on the left of KL-divergence, as in (2), it tends to chase the major modes of p while ignoring the minor modes.

The flow-based model has explicit normalized density and can be sampled directly. It is both a density and a sampler.

Flow-Based Reference and Latent Space Sampling

[181] propose to use a flow-based model as the reference distribution for the descriptive model or the energy-based model (EBM) and perform MCMC sampling in latent space.

Instead of using uniform or Gaussian white noise distribution for the reference distribution $q(x)$ in the descriptive model, we can use a flow-based model q_α as the reference model. q_α can be pre-trained by MLE and serves as the backbone of the model so that the model is of the following form:

$$p_\theta(x) = \frac{1}{Z(\theta)} \exp(f_\theta(x)) q_\alpha(x). \qquad (12.36)$$

The resulting model $p_\theta(x)$ is a correction or refinement of q_α or an exponential tilting of $q_\alpha(x)$, and $f_\theta(x)$ is a free-form ConvNet to parameterize the correction. The overall negative energy is $f_\theta(x) + \log q_\alpha(x)$.

In the latent space of z, let $p(z)$ be the distribution of z under $p_\theta(x)$; then

$$p(z)dz = p_\theta(x)dx = \frac{1}{Z(\theta)} \exp(f_\theta(x)) q_\alpha(x)dx. \qquad (12.37)$$

Because $q_\alpha(x)dx = q_0(z)dz$, we have

$$p(z) = \frac{1}{Z(\theta)} \exp(f_\theta(g_\alpha(z))) q_0(z). \qquad (12.38)$$

$p(z)$ is an exponential tilting of the prior noise distribution $q_0(z)$. It is a very simple form that does not involve the Jacobian or inversion of $g_\alpha(z)$.

Instead of sampling $p_\theta(x)$, we can sample $p(z)$ in Eq. (12.38). While $q_\alpha(x)$ is multi-modal, $q_0(z)$ is unimodal. Since $p_\theta(x)$ is a correction of q_α, $p(z)$ is a correction of $p_0(z)$ and can be much less multi-modal than $p_\theta(x)$ that is in the data space. After sampling z from $p(z)$, we can generate $x = g_\alpha(z)$.

The above MCMC sampling scheme is a special case of neutral transport MCMC proposed by Hoffman et al. [104] for sampling from an EBM or the posterior distribution of a generative model. The basic idea is to train a flow-based model as a variational approximation to the target EBM and sample the EBM in the latent space of the flow-based model. In our case, since p_θ is a correction of q_α, we can simply use q_α directly as the approximate flow-based model in the neural transport sampler. The extra benefit is that the distribution $p(z)$ is of an even simpler form than $p_\theta(x)$ because $p(z)$ does not involve the inversion and Jacobian of g_α. As a result, we may use a flow-based backbone model of a more free form such as one based on residual network [13]. We use HMC [179] to sample from $p(z)$ and push the samples forward to the data space through g_α. We can then learn θ by MLE.

Diffusion Recovery Likelihood

Inspired by recent work on diffusion-based models [102, 222, 223], [72] propose a diffusion recovery likelihood method to tackle the challenge of training the descriptive models or energy-based models (EBMs) directly on a dataset by instead learning a sequence of EBMs for the marginal distributions of the diffusion process. Specifically, assume a sequence of noisy observations $\mathbf{x}_0, \mathbf{x}_1, \ldots, \mathbf{x}_T$ such that

$$\mathbf{x}_0 \sim p_{\text{data}}(\mathbf{x}); \quad \mathbf{x}_{t+1} = \sqrt{1 - \sigma_{t+1}^2}\, \mathbf{x}_t + \sigma_{t+1}\epsilon_{t+1}, \quad t = 0, 1, \ldots T - 1. \quad (12.39)$$

The scaling factor $\sqrt{1 - \sigma_{t+1}^2}$ ensures that the sequence is a spherical interpolation between the observed sample and Gaussian white noise. Let $\mathbf{y}_t = \sqrt{1 - \sigma_{t+1}^2}\,\mathbf{x}_t$, and we assume a sequence of marginal EBMs on the perturbed data

$$p_\theta(\mathbf{y}_t) = \frac{1}{Z_{\theta,t}} \exp\left(f_\theta(\mathbf{y}_t, t)\right), \quad (12.40)$$

where $f_\theta(\mathbf{y}_t, t)$ is defined by a neural network conditioned on t. The sequence of marginal EBMs can be learned with recovery likelihoods that are defined as the conditional distributions that invert the diffusion process, which can be derived by Eqs. (12.39) and (12.40):

$$p_\theta(\mathbf{y}_t | \mathbf{x}_{t+1}) = \frac{1}{\tilde{Z}_{\theta,t}(\mathbf{x}_{t+1})} \exp\left(f_\theta(\mathbf{y}_t, t) - \frac{1}{2\sigma_{t+1}^2}\|\mathbf{x}_{t+1} - \mathbf{y}_t\|^2\right), \quad t = 0, 1, \ldots, T - 1. \quad (12.41)$$

Compared to the standard maximum likelihood estimation (MLE) of EBMs, learning marginal EBMs by diffusion recovery likelihood only requires sampling from the conditional distributions in Eq. (12.41), which is much easier than sampling from the marginal distributions due to the additional quadratic term, which makes the conditional EBMs close to unimodal. After learning the marginal EBMs, we can generate synthesized images by a sequence of conditional samples initialized from the Gaussian white noise distribution using MCMC techniques such as Langevin sampling:

$$\mathbf{y}_t^{\tau+1} = \mathbf{y}_t^\tau + \frac{b^2\sigma_t^2}{2}(\nabla_{\mathbf{y}} f_\theta(\mathbf{y}_t^\tau, t) + \frac{1}{\sigma_t^2}(\mathbf{x}_{t+1} - \mathbf{y}_t^\tau)) + b\sigma_t\epsilon^\tau. \quad (12.42)$$

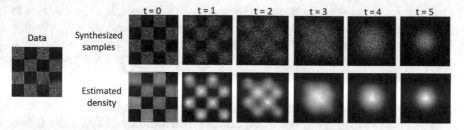

Fig. 12.8 Reprinted with permission from [72]. Illustration of diffusion recovery likelihood on 2D checkerboard example. *Top*: progressively generated samples. *Bottom*: estimated marginal densities

Fig. 12.9 Reprinted with permission from [72]. Generated samples on LSUN 128^2 church_outdoor (*left*), LSUN 128^2 bedroom (*center*), and CelebA 64^2 (*right*)

The framework of recovery likelihood was originally proposed in [17]. Gao et al. [72] adapt it to learning the sequence of marginal EBMs from the diffusion data. Figure 12.8 shows an illustration on a 2D toy example. Figure 12.9 displays uncurated samples generated from learned models on large image datasets.

Diffusion-Based Model

Diffusion-based models [102, 222, 223] prove to be exceedingly powerful in generating photorealistic images. It learns a sampling process instead of an explicit density. Thus it is on the side of the sampler (like a policy network), instead of density (like a value network). The sampling process is similar to the short-run Langevin dynamics for sampling from an energy-based model.

The key idea of the diffusion-based model of [222] is to continuously add noises of infinitesimal variance to the clean image until the resulting image becomes a

Gaussian white noise image. This is a forward diffusion process. Then we learn to reverse this forward process by going from the Gaussian white noise distribution back to the multi-modal data distribution of the clean images. This reverse diffusion process is as if showing the movie of the forward diffusion process in reverse time, and it was inspired by the non-equilibrium thermodynamics [222]. The reverse diffusion process is a denoising process. Adding noises amounts to reducing the precision of the pixel intensities.

There are two slightly different perspectives on the diffusion-based model. One is based on the observation that the conditional distribution $p_\theta(\mathbf{y}_t|\mathbf{x}_{t+1})$ in Eq. (12.41) is approximately Gaussian if σ_{t+1}^2 is infinitesimally small. The conditional Gaussian distribution can be derived by the first-order Taylor expansion of the log density of $\mathbf{y}_t$. Thus the reverse process can be decomposed into a Markov sequence of conditional Gaussian models with infinitesimal variances, and they can be learned within the maximum likelihood or variational inference framework. A single conditional Gaussian model can be learned for the whole reverse process, with time embedding being input to the model. In the learning of the conditional Gaussian model, we can condition on the original clean image for the purpose of variance reduction. More specifically, at each time step of the diffusion process, we can predict the noise image that has been added to the original clean image, and then we can move toward the clean image by removing a small amount of the predicted noise image.

A closely related perspective is to estimate the derivative of the log density of the noisy image at each time step of the diffusion process by score matching [114, 115] via denoising auto-encoder [6, 227, 241]. The derivative of the log density or score is related to the first-order Taylor expansion mentioned above. The derivative or the score enables us to reverse the forward diffusion process via a stochastic differential equation [223].

Intuitively, for a population of points that follow a certain density, if we add small random noise to each point, the resulting population of perturbed points will have a density that is more diffused than the original density. We can achieve the same effect by perturbing each point deterministically via a gradient descent movement on the log density so that the resulting population of the deterministically perturbed points will have the same diffused density resulting from adding random noises. Thus we can reverse the effect of the noise diffusion by deterministic gradient ascent on the log density. This underlies the reversion of the forward diffusion process mentioned above. It also underlies the Langevin dynamics where the gradient ascent and the diffusion term balance each other.

The diffusion-based model is effective for modeling multi-modal data density by the reverse diffusion process starting from a unimodal Gaussian white noise density. The idea is related to simulated annealing [135], simulated tempering [168], parallel tempering [48, 77], or replica exchange [226] for sampling from multi-modal densities.

12.3 Equivalence Between Discriminative and Descriptive Models

Discriminative Model

Let x be an input example, e.g., an image or a text, and let y be a label or annotation of x, e.g., the category that x belongs to in the case of classification. Let us focus on the classification problem, and suppose there are C categories. The commonly used soft-max classifier assumes that

$$p_\theta(y = c|x) = \frac{\exp(f_{c,\theta}(x))}{\sum_{c'=1}^{C} \exp(f_{c',\theta}(x))}, \tag{12.43}$$

where $f_{c,\theta}$ is a deep network, and θ denotes all the weight and bias parameters. For different c, the networks $f_{c,\theta}$ may share a common body and only differ in the head layer.

We can write the above model as

$$p_\theta(y = c|x) = \frac{1}{Z(\theta)(x)} \exp(f_{c,\theta}(x)), \tag{12.44}$$

where

$$Z(\theta)(x) = \sum_{c=1}^{C} \exp(f_{c,\theta}(x)). \tag{12.45}$$

The discriminative model $p_\theta(y|x)$ can be learned by maximum likelihood. The log-likelihood is the average of

$$\log p_\theta(y|x) = f_{y,\theta}(x) - \log Z(\theta)(x), \tag{12.46}$$

where the average is over the training set $\{x, y\}$. The gradient of $\log p_\theta(y|x)$ with respect to θ is

$$\nabla_\theta \log p_\theta(y|x) = \nabla_\theta f_{y,\theta}(x) - \mathrm{E}_{p_\theta(y|x)}[\nabla_\theta f_{y,\theta}(x)], \tag{12.47}$$

where

$$\nabla_\theta \log Z(\theta)(x) = \mathrm{E}_{p_\theta(y|x)}[\nabla_\theta f_{y,\theta}(x)]. \tag{12.48}$$

Let $p_{\mathrm{data}}(x, y)$ be the data distribution of (x, y). The MLE minimizes $D_{\mathrm{KL}}(p_{\mathrm{data}}(y|x) \| p_\theta(y|x))$, where for two conditional distributions $p(y|x)$ and $q(y|x)$, their KL-divergence is defined as

$$D_{KL}(p(y|x)\|q(y|x)) = E_{p(x,y)}\left[\log\frac{p(y|x)}{q(y|x)}\right], \tag{12.49}$$

where the expectation is with respect to $p(x, y) = p(x)p(y|x)$, i.e., we also average over $p(x)$ in addition to $p(y|x)$.

The above calculations are analogous to the calculations for the descriptive model. The difference is that for the discriminative model, the normalizing constant Z and the expectation are summations over y, where y belongs to a finite set of categories, whereas for the descriptive model, the normalizing constant Z and the expectation are integral over x, where x belongs to a high-dimensional space. As a result, the expectation in the descriptive model cannot be calculated in closed form and has to be approximated by MCMC sampling such as Langevin dynamics.

A special case is binary classification, where $y \in \{0, 1\}$. It is usually assumed that

$$f_{0,\theta}(x) = 0, \ \ f_{1,\theta}(x) = f_\theta(x), \tag{12.50}$$

so that

$$p_\theta(y = 1|x) = \frac{1}{1 + \exp(-f_\theta(x))} = \text{sigmoid}(f_\theta(x)), \tag{12.51}$$

and y follows a nonlinear logistic regression on x.

Descriptive Model as Exponential Tilting of a Reference Distribution

A more general version of the descriptive model is of the following form of exponential tilting of a reference distribution [32, 253]:

$$p_\theta(x) = \frac{1}{Z(\theta)} \exp(f_\theta(x))q(x), \tag{12.52}$$

where $q(x)$ is a given reference measure, such as uniform measure or Gaussian white noise distribution. The original form of the descriptive model corresponds to $q(x)$ being a uniform measure. If $q(x)$ is a Gaussian white noise distribution, then the energy function is $-f_\theta(x) + \|x\|^2/2$.

Although $q(x)$ is usually taken to be a simple known distribution, $q(x)$ can also be a model in its own right. We may call it a base model or a backbone model, and $p_\theta(x)$ can be considered a correction of $q(x)$, where $f_\theta(x)$ is the correction term. We may also call $p_\theta(x)$ the energy-based correction of the base model $q(x)$.

Discriminative Model via Bayes Rule

The above exponential tilting leads to the following discriminative model. We can treat p_θ as the positive distribution, and $q(x)$ the negative distribution. Let $y \in \{0, 1\}$, and the prior probability $p(y = 1) = \rho$, so that $p(y = 0) = 1 - \rho$. Let $p(x|y = 1) = p_\theta(x)$, $p(x|y = 0) = q(x)$. Then according to the Bayes rule [32, 121, 143, 149, 234, 253],

$$p(y = 1|x) = \frac{\exp(f_\theta(x) + b)}{1 + \exp(f_\theta(x) + b)}, \tag{12.53}$$

where $b = \log(\rho/(1 - \rho)) - \log Z(\theta)$. This leads to nonlinear logistic regression. Sometimes, people call $f_\theta(x) + b$ the logit or logit score because

$$\log \frac{p(y = 1|x)}{p(y = 0|x)} = \text{logit}(p(y = 1|x)) = f_\theta(x) + b. \tag{12.54}$$

More generally, suppose we have C categories, and

$$p_{c,\theta}(x) = \frac{1}{Z_{c,\theta}} \exp(f_{c,\theta}(x))q(x), \quad c = 1, \ldots, C, \tag{12.55}$$

where $(f_{c,\theta}(x), c = 1, \ldots, C)$ are C networks that may share the same body but with different heads. Suppose the prior probability for category c is ρ_c, then

$$p(y = c|x) = \frac{\exp(f_{c,\theta}(x) + b_c)}{\sum_{c=1}^{C} \exp(f_{c,\theta}(x) + b_c)}, \tag{12.56}$$

where $b_c = \log \rho_c - \log Z_{c,\theta}$. The above is a conventional soft-max classifier. Conversely, if $p(y = c|x)$ is of the above form of soft-max classifier, then $p_{c,\theta}(x)$ is of the form of exponential tilting based on the logit score $f_{c,\theta}(x) + b_c$. Thus the discriminative model and the descriptive model are equivalent to each other.

Noise Contrastive Estimation

The above equivalence suggests that we can learn the descriptive model by fitting a logistic regression. Specifically, suppose we want to learn a descriptive model $p_\theta(x) = \frac{1}{Z(\theta)} \exp(f_\theta(x))q(x)$, where $q(x)$ is a noise distribution, such as Gaussian white noise distribution. We can treat the observed examples as the positive examples, so that for each positive x, $y = 1$, and we generate negative examples from the noise distribution $q(x)$, so that for each negative example $x \sim q(x)$, $y = 0$. Then we learn a discriminator in the form of logistic regression to distinguish between the positive and negative examples, and then $\text{logit}(p(y = 1|x)) = f_\theta(x) + b$, where $b = \log(\rho/(1 - \rho)) - \log Z(\theta)$, where ρ is the proportion of the positive examples. We can learn both θ and b by fitting a logistic regression, where b is treated as an independent bias or intercept term, even though $\log Z(\theta)$ depends on θ. This enables us to learn f_θ and estimate $\log Z(\theta)$. This is called noise contrastive estimation (NCE) [92].

The problem with the above scheme is that the noise distribution and the data distribution usually do not have much overlap, especially if x is of high dimensionality. As a result, $f_\theta(x)$ cannot be well learned.

The introspective learning method [121, 234] tries to remedy the above problem with sampling. After learning $f_\theta(x)$ by noise contrastive estimation, we want to inspect whether $f_\theta(x)$ is well learned. We then treat the current $p_\theta(x)$ as our new $q(x)$, and we draw negative samples from it. If it is well learned, then the negative samples will be close to the positive examples. To check that, we fit a logistic regression again on the positive examples and negative examples from the new $q(x)$. Then we learn a new $\Delta f_\theta(x)$ by the new logistic regression. This $\Delta f_\theta(x)$ can then be added to the previous learned $f_\theta(x)$ to obtain the new $f_\theta(x)$. We can keep repeating this process until $\Delta f_\theta(x)$ is small.

In general, while the descriptive model learns the probability density function, the discriminative model learns the ratios between the probability densities of different classes. If we know the density of a base class, such as the Gaussian white noise, we can learn the densities of other classes by noise contrastive estimation. Noise contrastive estimation is a form of self-supervised learning.

Noise contrastive estimation (NCE) based on diffusion sequence is explored in [203].

Flow Contrastive Estimation

Gao et al. [71] propose an improvement of noise contrastive estimation (NCE) [92] based on the flow-based model. The basic idea is to transform the noise so that the resulting distribution is closer to the data distribution. This is exactly what the flow model achieves. That is, a flow model transforms a known noise distribution $q_0(z)$ by a composition of a sequence of invertible transformations $g_\alpha(\cdot)$. However, in practice, we find that a pre-trained $q_\alpha(x)$, such as learned by MLE, is not strong enough for learning an EBM $p_\theta(x)$ because the synthesized data from the MLE of $q_\alpha(x)$ can still be easily distinguished from the real data by an EBM. Thus, we propose to iteratively train the EBM and flow model, in which case the flow model is adaptively adjusted to become a stronger contrast distribution or a stronger training opponent for EBM. This is achieved by a parameter estimation scheme similar to GAN [81, 199], where $p_\theta(x)$ and $q_\alpha(x)$ play a minimax game with a unified value function: $\min_\alpha \max_\theta V(\theta, \alpha)$,

$$V(\theta, \alpha) = \mathrm{E}_{p_{\text{data}}} \left[\log \frac{p_\theta(x)}{p_\theta(x) + q_\alpha(x)} \right] + \mathrm{E}_z \left[\log \frac{q_\alpha(g_\alpha(z))}{p_\theta(g_\alpha(z)) + q_\alpha(g_\alpha(z))} \right],$$

$$(12.57)$$

where $\mathrm{E}_{p_{\text{data}}}$ is approximated by averaging over observed samples $\{x_i, i = 1, \ldots, n\}$, while E_z is approximated by averaging over negative samples $\{\tilde{x}_i, i = 1, \ldots, n\}$ drawn from $q_\alpha(x)$, with $z_i \sim q_0(z)$ independently for $i = 1, \ldots, n$. In the experiments, we choose Glow [131] as the flow-based model. The algorithm can start from either a randomly initialized Glow model or a pre-trained one by MLE. Here we assume equal prior probabilities for observed samples and negative samples. It can be easily modified to the situation where we assign a higher prior probability to the negative samples, given the fact we have access to an infinite amount of free negative samples.

The objective function can be interpreted from the following perspectives:

(1) Noise contrastive estimation for EBM. The update of θ can be seen as noise contrastive estimation of $p_\theta(x)$, but with a flow-transformed noise distribution $q_\alpha(x)$ that is adaptively updated. The training is essentially a logistic regression. However, unlike regular logistic regression for classification, for each x_i or $\tilde{x}_i$, we must include $\log q_\alpha(x_i)$ or $\log q_\alpha(\tilde{x}_i)$ as an example-dependent bias term. This forces $p_\theta(x)$ to replicate $q_\alpha(x)$ in addition to distinguishing between $p_{\text{data}}(x)$ and $q_\alpha(x)$, so that $p_\theta(x_i)$ is in general larger than $q_\alpha(x_i)$, and $p_\theta(\tilde{x}_i)$ is in general smaller than $q_\alpha(\tilde{x}_i)$.

(2) Minimization of Jensen–Shannon divergence for the flow model. If $p_\theta(x)$ is close to the data distribution, then the update of α is approximately minimizing the Jensen–Shannon divergence between the flow model q_α and data distribution p_{data}:

$$D_{\text{JS}}(q_\alpha \| p_{\text{data}}) = D_{\text{KL}}(p_{\text{data}} \| (p_{\text{data}} + q_\alpha)/2) + D_{\text{KL}}(q_\alpha \| (p_{\text{data}} + q_\alpha)/2).$$

$$(12.58)$$

Its gradient w.r.t. α equals the gradient of $-\mathrm{E}_{p_{\text{data}}}[\log((p_\theta + q_\alpha)/2)] + D_{\text{KL}}(q_\alpha \| (p_\theta + q_\alpha)/2)$. The gradient of the first term resembles MLE, which forces q_α to cover the modes of data distribution, and tends to lead to an over-dispersed model, which is also pointed out in [131]. The gradient of the second term is similar to reverse Kullback–Leibler divergence between q_α and p_θ, or variational approximation of p_θ by q_α, which forces q_α to chase the modes of p_θ. This may help correct the over-dispersion of MLE.

(3) Connection with GAN [81, 199]. Our parameter estimation scheme is closely related to GAN. In GAN, the discriminator D and generator G play a minimax game: $\min_G \max_D V(G, D)$,

$$V(G, D) = \mathrm{E}_{p_{\text{data}}} \left[\log D(x) \right] + \mathrm{E}_z \left[\log(1 - D(G(z_i))) \right].$$

$$(12.59)$$

The discriminator $D(x)$ is learning the probability ratio $p_{\text{data}}(x)/(p_{\text{data}}(x) + p_G(x))$, which is about the difference between p_{data} and p_G [56]. p_G is the density of the generated data. In the end, if the generator G learns to perfectly replicate p_{data}, then the discriminator D ends up with a random guess. However, in our method, the ratio is explicitly modeled by p_θ and q_α. p_θ must contain all the learned knowledge in q_α, in addition to the difference between p_{data} and q_α. In the end, we learn two explicit probability distributions p_θ and q_α as approximations to p_{data}.

12.4 Generative Latent Variable Model

Model and Origin

Both discriminative model and descriptive model are based on a bottom-up network $f_\theta(x)$. The generative model is based on top-down network with latent variables. The prototype of such a model is factor analysis. Let x be the observed example, which is a D-dimensional vector. We assume that x can be explained by a d-dimensional latent vector, each element of which is called a factor. Given z, x is generated by $x = Wz + \epsilon$, where W is a $D \times d$ matrix, sometimes called loading matrix. It is usually assumed that $z \sim N(0, I_d)$, where I_d is d-dimensional identity matrix, $\epsilon \sim N(0, \sigma^2 I_D)$, and ϵ is independent of z. The factor analysis model originated from psychometrics, where x consists of a pupil's scores on a number of subjects, and $z = (z_1, z_2)$, where z_1 is verbal intelligence and z_2 is analytical intelligence.

A recent generalization [81, 133] is to keep the prior assumption about z, but replace the linear model $x = Wz + \epsilon$ by a nonlinear model $x = g_\theta(z) + \epsilon$, where $g_\theta(z)$ is parameterized by a top-down neural network where θ collects all the weight and bias parameters. In the case of image modeling, $g_\theta(z)$ is usually a convolutional neural network, which is sometimes called deconvolutional network, due to its top-down nature. The above model leads to a conditional or generation model $p_\theta(x|z)$, such that

$$\log p_\theta(x, z) = \log[p(z)p_\theta(x|z)] \tag{12.60}$$

$$= -\frac{1}{2}\left[\|z\|^2 + \|x - g_\theta(z)\|^2/\sigma^2\right] + c, \tag{12.61}$$

where c is a constant independent of θ. σ^2 is usually treated as a tuning parameter. The model follows the manifold assumption, which assumes that the density of the D-dimensional data focuses on a lower, d-dimensional manifold.

The joint distribution of (x, z) is $p_\theta(x, z) = p(z)p_\theta(x|z)$. The marginal distribution of x is $p_\theta(x) = \int p_\theta(x, z)dz$. The marginal distribution is analytically intractable due to the integration of z. The model specifies a direct sampling method for generating x, but it does not explicitly specify the density of x.

Given x, the inference of z can be based on the posterior distribution $p_\theta(z|x) = p_\theta(x, z)/p_\theta(x)$, which is also intractable due to the intractability of the marginal $p_\theta(x)$.

The above model is often referred to as the generator network in the literature.

Generative Model with Multi-layer Latent Variables

While it is computationally convenient to have a single latent noise vector at the top layer, it does not account for the fact that patterns can appear at multiple layers of compositions or abstractions (e.g., face $\rightarrow$ (eyes, nose, mouth) $\rightarrow$ (edges, corners) $\rightarrow$ pixels), where variations and randomness occur at multiple layers. To capture such a hierarchical structure, it is desirable to introduce multiple layers of latent variables organized in a top-down architecture [183]. Specifically, we have $z = (z_l, l = 1, \ldots, L)$, where layer L is the top layer, and layer 1 is the bottom layer above x. For notational simplicity, we let $x = z_0$. We can then specify $p_\theta(z)$ as

$$p_\theta(z) = p_\theta(z_L) \prod_{l=0}^{L-1} p_\theta(z_l|z_{l+1}). \tag{12.62}$$

One concrete example is $z_L \sim N(0, I)$, $[z_l|z_{l+1}] \sim N(\mu_l(z_{l+1}), \sigma_l^2(z_{l+1}))$, $l = 0, \ldots, L-1$, where $\mu_l()$ and $\sigma_l^2()$ are the mean vector and the diagonal variance–covariance matrix of z_l, respectively, and they are functions of z_{l+1}. θ collects all the parameters in these functions. $p_\theta(x, z)$ can be obtained similarly as in Eq. (12.61).

MLE Learning and Posterior Inference

Let $p_{\text{data}}(x)$ be the data distribution that generates the example x. The learning of parameters θ of $p_\theta(x)$ can be based on $\min_\theta D_{\text{KL}}(p_{\text{data}}(x)\|p_\theta(x))$. If we observe training examples $\{x_i, i = 1, \ldots, n\} \sim p_{\text{data}}(x)$, the above minimization can be approximated by maximizing the log-likelihood

$$L(\theta) = \frac{1}{n} \sum_{i=1}^{n} \log p_\theta(x_i) \doteq E_{p_{\text{data}}}[\log p_\theta(x)], \tag{12.63}$$

which leads to MLE.

The gradient of the log-likelihood, $L'(\theta)$, can be computed according to the following identity:

$$\delta_\theta(x) = \nabla_\theta \log p_\theta(x) = \frac{1}{p_\theta(x)} \nabla_\theta p_\theta(x) \qquad (12.64)$$

$$= \frac{1}{p_\theta(x)} \int \nabla_\theta p_\theta(x, z) dz \qquad (12.65)$$

$$= \mathrm{E}_{p_\theta(z|x)} \left[\nabla_\theta \log p_\theta(x, z) \right]. \qquad (12.66)$$

Thus

$$L'(\theta) = \mathrm{E}_{p_{\text{data}}}[\delta_\theta(x)] = \mathrm{E}_{p_{\text{data}}(x)} \mathrm{E}_{p_\theta(z|x)} \left[\nabla_\theta \log p_\theta(x, z) \right] \qquad (12.67)$$

$$\doteq \frac{1}{n} \sum_{i=1}^{n} \mathrm{E}_{p_\theta(z_i|x_i)} \left[\nabla_\theta \log p_\theta(x_i, z_i) \right]. \qquad (12.68)$$

The expectation with respect to $p_\theta(z|x)$ can be approximated by Monte Carlo samples. Each learning iteration updates θ by $\theta_{t+1} = \theta_t + \eta_t L'(\theta_t)$.

Posterior Sampling

Sampling from $p_\theta(z|x)$ usually requires MCMC. One convenient MCMC is Langevin dynamics, which iterates

$$z_{t+1} = z_t + s \nabla_z \log p_\theta(z_t|x) + \sqrt{2s} e_t, \qquad (12.69)$$

where $e_t \sim \mathrm{N}(0, I)$, t indexes the time step of the Langevin dynamics, and s is the step size. The Langevin dynamics consists of a gradient descent term on $-\log p(z|x)$. In the case of generator network, it amounts to gradient descent on $\|z\|^2/2 + \|x - g_\theta(z)\|^2/2\sigma^2$, which is the penalized reconstruction error. The Langevin dynamics also consists of a white noise diffusion term $\sqrt{2s} e_t$ to create randomness for sampling from $p_\theta(z|x)$.

For small step size s, the marginal distribution of z_t will converge to $p_\theta(z|x)$ as $t \to \infty$ regardless of the initial distribution of z_0. More specifically, let $p_t(z)$ be the marginal distribution of z_t of the Langevin dynamics, and then $D_{\text{KL}}(p_t(z) \| p_\theta(z|x))$ decreases monotonically to 0, that is, by increasing t, we reduce $D_{\text{KL}}(p_t(z) \| p_\theta(z|x))$ monotonically.

Perturbation of KL-divergence

Again we understand the MLE learning algorithm by perturbing the KL-divergence for MLE. Define $D(\theta) = D_{\text{KL}}(p_{\text{data}} \| p_\theta)$. It is the objective function of MLE. Let θ_t be the estimate at iteration t. Let us consider the following perturbation of $D(\theta)$:

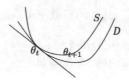

Fig. 12.10 Reprinted with permission from [95]. The surrogate S majorizes (upper bounds) D, and they touch each other at θ_t with the same tangent

$$S(\theta) = D(\theta) + D_{\text{KL}}(p_{\theta_t}(z|x) \| p_\theta(z|x)) \tag{12.70}$$

$$= D_{\text{KL}}(p_{\text{data}}(x) \| p_\theta(x)) + D_{\text{KL}}(p_{\theta_t}(z|x) \| p_\theta(z|x)) \tag{12.71}$$

$$= D_{\text{KL}}(p_{\text{data},\theta_t}(x, z) \| p_\theta(x, z)), \tag{12.72}$$

where we define $p_{\text{data},\theta_t}(x, z) = p_{\text{data}}(x)p_{\theta_t}(z|x)$. Again $S(\theta)$ is a surrogate for $D(\theta)$ at θ_t, and $S(\theta)$ is simpler than $D(\theta)$ because $S(\theta)$ is based on the joint distributions instead of the marginal distributions as in $D(\theta)$. Unlike the joint distribution $p_\theta(x, z) = p(z)p_\theta(x|z)$, the marginal distribution $p_\theta(x) = \int p_\theta(x, z)dz$ is implicit as it is an intractable integral (Fig. 12.10).

The perturbation term $D_{\text{KL}}(p_{\theta_t}(z|x) \| p_\theta(z|x))$, as a function of θ, achieves its minimum 0 at $\theta = \theta_t$, and its derivative at $\theta = \theta_t$ is zero. Thus $S(\theta)$ and $D(\theta)$ touch each other at θ_t, and they share the same gradient at θ_t.

$$-S(\theta) = E_{p_{\text{data}}(x)} E_{p_{\theta_t}(z|x)}[\log p_\theta(x, z)] - \text{entropy}(p_{\text{data},\theta_t}(x, z)). \tag{12.73}$$

$$-S'(\theta) = E_{p_{\text{data}}(x)} E_{p_{\theta_t}(z|x)}[\nabla_\theta \log p_\theta(x, z)]. \tag{12.74}$$

Thus, the learning gradient at θ_t is

$$-D'(\theta_t) = -S'(\theta_t) = E_{p_{\text{data}}}[\delta_{\theta_t}(x)] = E_{p_{\text{data}}(x)} E_{p_{\theta_t}(z|x)}[\nabla_\theta \log p_{\theta_t}(x, z)]. \tag{12.75}$$

This provides another justification for the learning algorithm.

The above perturbation of KL-divergence can be compared to that in the descriptive model, where the sign in front of the second KL-divergence is negative, in order to cancel the intractable $\log Z(\theta)$ term. For the generative model, the sign in front of the second KL-divergence is positive, in order to change the marginal distributions in the first KL-divergence, i.e., $D(\theta)$, into the joint distributions, so that $p_\theta(z, x) = p(z)p_\theta(x|z)$ is obtained in closed form.

Short-Run MCMC for Approximate Inference

We can use short-run MCMC as inference dynamics [183], with a fixed small K (e.g., $K = 25$),

$$z_0 \sim p(z), z_{k+1} = z_k + s \nabla_z \log p_\theta(z_k|x) + \sqrt{2s} e_k, \ k = 1, \ldots, K, \qquad (12.76)$$

where $p(z)$ is the prior noise distribution of z.

We can write the above short-run MCMC as

$$z_0 \sim p(z), \ z_{k+1} = z_k + s R(z_k) + \sqrt{2s} e_k, \ k = 1, \ldots, K, \qquad (12.77)$$

$R(z) = \nabla_z \log p_\theta(z|x)$, where we omit x and θ in $R(z)$ for simplicity of notation. To further simplify the notation, we may write the short-run MCMC as

$$z_0 \sim p(z), \ z_K = F(z_0, e), \qquad (12.78)$$

where $e = (e_k, k = 1, \ldots, K)$, and F composes the K steps of Langevin updates.

Let the distribution of z_K be $\tilde{p}(z)$. Recall that the distribution of z_K also depends on x and θ and step size s, so that in full notation, we may write $\tilde{p}(z)$ as $\tilde{p}_{s,\theta}(z|x)$. For each x, we define

$$\tilde{\delta}_\theta(x) = \mathrm{E}_{\tilde{p}_{s,\theta}(z|x)} \left[\nabla_\theta \log p_\theta(x, z) \right] \qquad (12.79)$$

and modify the learning algorithm to

$$\theta_{t+1} = \theta_t + \eta_t \mathrm{E}_{p_{\mathrm{data}}}[\tilde{\delta}_{\theta_t}(x)] = \theta_t + \eta_t \, \mathrm{E}_{p_{\mathrm{data}}} \mathrm{E}_{\tilde{p}_{s,\theta_t}(z|x)} \left[\nabla_\theta \log p_{\theta_t}(x, z) \right], \qquad (12.80)$$

where η_t is the learning rate and $\mathrm{E}_{\tilde{p}_{s,\theta_t}(z_i|x_i)}$ (here we use the full notation $\tilde{p}_{s,\theta}(z|x)$ instead of the abbreviated notation $q(z)$) can be approximated by sampling from $\tilde{p}_{s,\theta_t}(z_i|x_i)$ using the noise initialized K-step Langevin dynamics.

Compared to MLE learning algorithm, we replace $p_\theta(z|x)$ by $\tilde{p}_{s,\theta}(z|x)$, and fair Monte Carlo samples from $\tilde{p}_{s,\theta}(z|x)$ can be obtained by short-run MCMC.

One major advantage of the proposed method is that it is simple and automatic. For models with multiple layers of latent variables that may be organized in complex top-down architectures, the gradient computation in Langevin dynamics is automatic on modern deep learning platforms. Such dynamics naturally integrates explaining-away competitions and bottom-up and top-down interactions between multiple layers of latent variables. It thus enables researchers to explore flexible generative models without dealing with the challenging task of designing and learning the inference models. The short-run MCMC is automatic, natural, and biologically plausible as it may be related to attractor dynamics [7, 108, 198].

Objective Function and Estimating Equation

The following are justifications for learning with short-run MCMC:

(1) **Objective function**. Again we use perturbation of KL-divergence. At iteration t, with θ_t fixed, the learning algorithm follows the gradient of the following perturbation of $D(\theta)$ at $\theta = \theta_t$:

$$S(\theta) = D(\theta) + D_{\text{KL}}(\tilde{p}_{s,\theta_t}(z|x) \| p_\theta(z|x)) \tag{12.81}$$

$$- D_{\text{KL}}(p_{\text{data}}(x) \| p_\theta(x)) + D_{\text{KL}}(\tilde{p}_{s,\theta_t}(z|x) \| p_\theta(z|x)), \tag{12.02}$$

so that $\theta_{t+1} = \theta_t - \eta_t S'(\theta_t)$, where η_t is the step size, and

$$- S'(\theta) = E_{p_{\text{data}}(x)} E_{\tilde{p}_{s,\theta_t}(z|x)} [\nabla_\theta \log p_\theta(x, z)]. \tag{12.83}$$

$$- S'(\theta_t) = E_{p_{\text{data}}}[\tilde{\delta}_{\theta_t}(x)] = E_{p_{\text{data}}(x)} E_{\tilde{p}_{s,\theta_t}(z|x)} [\nabla_\theta \log p_{\theta_t}(x, z)]. \tag{12.84}$$

Compared to the perturbation of KL-divergence in MLE learning, we use $\tilde{p}_{s,\theta_t}(z|x)$ instead of $p_{\theta_t}(z|x)$. While sampling $p_{\theta_t}(z|x)$ can be impractical if it is multi-modal, sampling $\tilde{p}_{s,\theta_t}(z|x)$ is practical because it is a short-run MCMC.

(2) **Estimating equation**. The fixed point of the learning algorithm (12.80) solves the following estimating equation:

$$\frac{1}{n} \sum_{i=1}^n E_{\tilde{p}_{s,\theta}(z_i|x_i)} \left[\nabla_\theta \log p_\theta(x_i, z_i) \right] \doteq E_{p_{\text{data}}(x)} E_{\tilde{p}_{s,\theta}(z|x)} \left[\nabla_\theta \log p_\theta(x, z) \right] = 0.$$

$$\tag{12.85}$$

If we approximate $E_{\tilde{p}_{s,\theta_t}(z_i|x_i)}$ by Monte Carlo samples from $\tilde{p}_{s,\theta_t}(z_i|x_i)$, then the learning algorithm becomes Robbins–Monro algorithm for stochastic approximation [205].

The bias of the learned θ based on short-run MCMC relative to the MLE depends on the gap between $\tilde{p}_{s,\theta}(z|x)$ and $p_\theta(z|x)$. We suspect that this bias may actually be beneficial in the following sense. The learning gradient seeks to decrease $D(\theta)$ while decreasing $D_{\text{KL}}(\tilde{p}_{s,\theta_t}(z_i|x_i) \| p_\theta(z_i|x_i))$. The latter tends to bias the learned model so that its posterior distribution $p_\theta(z_i|x_i)$ is close to the short-run MCMC $\tilde{p}_{s,\theta_t}(z_i|x_i)$, i.e., our learning method may bias the model to make inference by short-run MCMC accurate.

We can optimize the step size s and other algorithmic parameters of the short-run Langevin dynamics by minimizing $D_{\text{KL}}(\tilde{p}_{s,\theta_t}(z_i|x_i) \| p_\theta(z_i|x_i))$ over s. This is a variational optimization.

12.5   Descriptive Model in Latent Space of Generative Model

Top-Down and Bottom-Up

$$
\begin{array}{cc}
\text{Top-down mapping} & \text{Bottom-up mapping} \\
\text{hidden vector } z & \text{energy } - f_\theta(x) \\
\Downarrow & \Uparrow \\
\text{signal } x \approx g_\theta(z) & \text{signal } x \\
\text{(a) Generator model} & \text{(b) Descriptive model}
\end{array}
\tag{12.86}
$$

The above diagram compares the generative model and the descriptive model. The former is based on top-down generation, whereas the latter is based on the bottom-up description.

The top-down model is a very natural representation of knowledge, with its multiple layers of latent variables representing concepts at multiple levels of abstractions. The top-down model is also called the directed acyclic graphical model. It is characterized by independence or conditional independence assumptions of the latent variables. Such assumptions limit the expressive power of the top-down model.

For the special case of the generator network, there is a latent vector z at the top layer, which generates the example x via the top-down generation mapping $g_\theta(z)$. The prior distribution of z is usually assumed to be a simple noise distribution, e.g., the Gaussian white noise distribution $z \sim N(0, I)$. The top-down $g_\theta(z)$ maps this simple isotropic unimodal prior distribution to the multi-modal data distribution p_{data}. However, the expressive power may be limited by the simple prior distribution $p(z)$ (as well as the simple Gaussian white noise distribution of ϵ in $x = g_\theta(z) + \epsilon$). The marginal distribution of $p_\theta(x) = \int p(z)p_\theta(x|z)dz$ is implicit because of the intractable integral over the latent z.

The bottom-up model only needs to specify a scalar-valued energy function $-f_\theta(x)$, instead of a vector-valued $g_\theta(z)$, while leaving the generative task to MCMC. It specifies the distribution $p_\theta(x) = \frac{1}{Z(\theta)} \exp(f_\theta(x))$ explicitly even though the normalizing constant $Z(\theta)$ is intractable. Compared to the generator model, the descriptive model tends to have stronger expressive power in terms of synthesis ability.

However, because p_{data} tends to be highly multi-modal, the learned p_θ can also be highly multi-modal. As a result, MCMC sampling cannot mix. Even though we can use short-run MCMC to learn the model and synthesize images, the model is admittedly biased. One remedy is to use more sophisticated MCMC such as parallel tempering [189] or replica exchange MCMC [226]. The other option is to move the descriptive model to the latent space.

Descriptive Energy-Based Model in Latent Space

We follow the philosophy of empirical Bayes, that is, instead of assuming a given prior distribution for the latent vector, as in the generator network, we learn a prior model from empirical observations

Specifically, we assume the latent vector follows a descriptive model or, more specifically, an energy-based correction of the isotropic Gaussian white noise prior distribution. We call this model the latent space descriptive model. Such a model adds more expressive power to the generator model. In the latent space, the descriptive model is close to unimodal as it is a correction of the unimodal Gaussian distribution, and MCMC sampling is expected to mix well.

The MLE learning of the generative model with a latent space descriptive model involves MCMC sampling of the latent vector from both the prior and posterior distributions. Parameters of the prior model can then be updated based on the statistical difference between samples from the two distributions. Parameters of the top-down network can be updated based on the samples from the posterior distribution as well as the observed data.

Let $x \in \mathbb{R}^D$ be an observed example such as an image or a piece of text, and let $z \in \mathbb{R}^d$ be the latent variables, where $D \gg d$. The joint distribution of (x, z) is

$$p_\theta(x, z) = p_\alpha(z) p_\beta(x|z), \tag{12.87}$$

where $p_\alpha(z)$ is the prior model with parameters α, $p_\beta(x|z)$ is the top-down generation model with parameters β, and $\theta = (\alpha, \beta)$.

The prior model $p_\alpha(z)$ is formulated as a descriptive model or an energy-based model

$$p_\alpha(z) = \frac{1}{Z(\alpha)} \exp(f_\alpha(z)) p_0(z), \tag{12.88}$$

where $p_0(z)$ is a reference distribution, assumed to be isotropic Gaussian white noise distribution. $f_\alpha(z)$ is the negative energy and is parameterized by a small multi-layer perceptron with parameters α. $Z(\alpha) = \int \exp(f_\alpha(z)) p_0(z) dz = E_{p_0}[\exp(f_\alpha(z))]$ is the normalizing constant or partition function.

The prior model (12.88) can be interpreted as an energy-based correction or exponential tilting of the original prior distribution p_0, which is the prior distribution in the generator model.

The generation model is the same as the top-down network in the generator model. For image modeling,

$$x = g_\beta(z) + \epsilon, \tag{12.89}$$

where $\epsilon \sim N(0, \sigma^2 I_D)$, so that $p_\beta(x|z) \sim N(g_\beta(z), \sigma^2 I_D)$. Usually, σ^2 takes an assumed value. For text modeling, let $x = (x^{(t)}, t = 1, \ldots, T)$, where each $x^{(t)}$ is a token. A commonly used model is to define $p_\beta(x|z)$ as a conditional auto-regressive model,

$$p_\beta(x|z) = \prod_{t=1}^{T} p_\beta(x^{(t)}|x^{(1)}, \dots, x^{(t-1)}, z), \qquad (12.90)$$

which is often parameterized by a recurrent network with parameters β.

In the original generator model, the top-down network g_β maps the unimodal prior distribution p_0 to be close to the usually highly multi-modal data distribution. The prior model in (12.88) refines p_0 so that g_β maps the prior model p_α to be closer to the data distribution. The prior model p_α does not need to be highly multi-modal because of the expressiveness of g_β.

The marginal distribution is $p_\theta(x) = \int p_\theta(x, z)dz = \int p_\alpha(z)p_\beta(x|z)dz$. The posterior distribution is $p_\theta(z|x) = p_\theta(x, z)/p_\theta(x) = p_\alpha(z)p_\beta(x|z)/p_\theta(x)$.

In the above model, we exponentially tilt $p_0(z)$. We can also exponentially tilt $p_0(x, z) = p_0(z)p_\beta(x|z)$ to $p_\theta(x, z) = \frac{1}{Z(\theta)}\exp(f_\alpha(x, z))p_0(x, z)$. Equivalently, we may also exponentially tilt $p_0(z, \epsilon) = p_0(z)p(\epsilon)$, as the mapping from (z, ϵ) to (z, x) is a change of variable. This leads to a descriptive model in both the latent space and data space, which makes learning and sampling more complex. Therefore, we choose to only tilt $p_0(z)$ and leave $p_\beta(x|z)$ as a directed top-down generation model.

Maximum Likelihood Learning

Suppose we observe training examples $(x_i, i = 1, \dots, n)$. The log-likelihood function is

$$L(\theta) = \frac{1}{n}\sum_{i=1}^{n} \log p_\theta(x_i) \doteq \mathrm{E}_{p_{\text{data}}}[\log p_\theta(x)]. \qquad (12.91)$$

The learning gradient can be calculated according to

$$\nabla_\theta \log p_\theta(x) = \mathrm{E}_{p_\theta(z|x)}\left[\nabla_\theta \log p_\theta(x, z)\right] = \mathrm{E}_{p_\theta(z|x)}\left[\nabla_\theta(\log p_\alpha(z) + \log p_\beta(x|z))\right]. \qquad (12.92)$$

For the prior model, $\nabla_\alpha \log p_\alpha(z) = \nabla_\alpha f_\alpha(z) - \mathrm{E}_{p_\alpha(z)}[\nabla_\alpha f_\alpha(z)]$. Thus the learning gradient for an example x is

$$\delta_\alpha(x) = \nabla_\alpha \log p_\theta(x) = \mathrm{E}_{p_\theta(z|x)}[\nabla_\alpha f_\alpha(z)] - \mathrm{E}_{p_\alpha(z)}[\nabla_\alpha f_\alpha(z)]. \qquad (12.93)$$

The above equation has an empirical Bayes nature. $p_\theta(z|x)$ is based on the empirical observation x, while p_α is the prior model. α is updated based on the difference between z inferred from empirical observation x and z sampled from the current prior.

For the generation model,

$$\delta_\beta(x) = \nabla_\beta \log p_\theta(x) = E_{p_\theta(z|x)}[\nabla_\beta \log p_\beta(x|z)], \tag{12.94}$$

where $\log p_\beta(x|z) = -\|x-g_\beta(z)\|^2/(2\sigma^2)+$constant or $\sum_{t=1}^T \log p_\beta(x^{(t)}|x^{(1)},\ldots, x^{(t-1)}, z)$ for image and text modeling, respectively, which is about the reconstruction error.

Writing $\delta_\theta(x) = (\delta_\alpha(x), \delta_\beta(x))$, we have $L'(\theta) = E_{p_{\text{data}}}[\delta_\theta(x)]$, and the learning algorithm is $\theta_{t+1} = \theta_t + \eta_t E_{p_{\text{data}}}[\delta_{\theta_t}(x)]$.

Expectations in (12.93) and (12.94) require MCMC sampling of the prior model $p_\alpha(z)$ and the posterior distribution $p_\theta(z|x)$. We can use Langevin dynamics. For a target distribution $\pi(z)$, the dynamics iterates

$$z_{k+1} = z_k + s\nabla_z \log \pi(z_k) + \sqrt{2s}e_k, \tag{12.95}$$

where k indexes the time step of the Langevin dynamics, s is a small step size, and $e_k \sim N(0, I_d)$ is the Gaussian white noise. $\pi(z)$ can be either $p_\alpha(z)$ or $p_\theta(z|x)$. In either case, $\nabla_z \log \pi(z)$ can be efficiently computed by back-propagation.

Short-Run MCMC for Synthesis and Inference

We use short-run MCMC for approximate sampling. This is in agreement with the philosophy of variational inference [133], which accepts the intractability of the target distribution and seeks to approximate it by a simpler distribution. The difference is that we adopt short-run Langevin dynamics instead of learning a separate network for approximation.

The short-run Langevin dynamics is always initialized from the fixed initial distribution p_0 and only runs a fixed number of K steps, e.g., $K = 20$,

$$z_0 \sim p_0(z), \quad z_{k+1} = z_k + s\nabla_z \log \pi(z_k) + \sqrt{2s}e_k, \quad k = 1, \ldots, K. \tag{12.96}$$

Denote the distribution of z_K to be $\tilde{\pi}(z)$. Because of fixed $p_0(z)$ and fixed K and s, the distribution $\tilde{\pi}$ is well defined. In this section, we put $\tilde{}$ sign on top of the symbols to denote distributions or quantities produced by short-run MCMC, and for simplicity, we omit the dependence on K and s in notation. The Kullback–Leibler divergence $D_{\text{KL}}(\tilde{\pi}\|\pi)$ decreases to zero monotonically as $K \to \infty$.

Specifically, denote the distribution of z_K to be $\tilde{p}_\alpha(z)$ if the target $\pi(z) = p_\alpha(z)$, and denote the distribution of z_K to be $\tilde{p}_\theta(z|x)$ if $\pi(z) = p_\theta(z|x)$. We can then

Fig. 12.11 Reprinted with permission from [191]. Generated images for CelebA ($128 \times 128 \times 3$)

replace $p_\alpha(z)$ by $\tilde{p}_\alpha(z)$ and replace $p_\theta(z|x)$ by $\tilde{p}_\theta(z|x)$ in Eqs. (12.93) and (12.94), so that the learning gradients in Eqs. (12.93) and (12.94) are modified to

$$\tilde{\delta}_\alpha(x) = \mathrm{E}_{\tilde{p}_\theta(z|x)}[\nabla_\alpha f_\alpha(z)] - \mathrm{E}_{\tilde{p}_\alpha(z)}[\nabla_\alpha f_\alpha(z)], \tag{12.97}$$

$$\tilde{\delta}_\beta(x) = \mathrm{E}_{\tilde{p}_\theta(z|x)}[\nabla_\beta \log p_\beta(x|z)]. \tag{12.98}$$

We then update α and β based on (12.97) and (12.98), where the expectations can be approximated by Monte Carlo samples. Specifically, writing $\tilde{\delta}_\theta(x) = (\tilde{\delta}_\alpha(x), \tilde{\delta}_\beta(x))$, the learning algorithm is $\theta_{t+1} = \theta_t + \eta_t \mathrm{E}_{p_{\mathrm{data}}}[\tilde{\delta}_{\theta_t}(x)]$.

The short-run MCMC sampling is always initialized from the same initial distribution $p_0(z)$ and always runs a fixed number of K steps. This is the case for both training and testing stages, which share the same short-run MCMC sampling (Figs. 12.11, 12.12, and 12.13).

Divergence Perturbation

The learning algorithm based on short-run MCMC sampling is a modification or perturbation of maximum likelihood learning, where we replace $p_\alpha(z)$ and $p_\theta(z|x)$ by $\tilde{p}_\alpha(z)$ and $\tilde{p}_\theta(z|x)$, respectively. For theoretical underpinning, we should also understand this perturbation in terms of the objective function and estimating equation:

(1) **Objective function**. In terms of objective function, the MLE loss function is $D(\theta) = D_{\mathrm{KL}}(p_{\mathrm{data}} \| p_\theta)$. At iteration t, with fixed $\theta_t = (\alpha_t, \beta_t)$, we perturb $D(\theta)$ to $S(\theta)$:

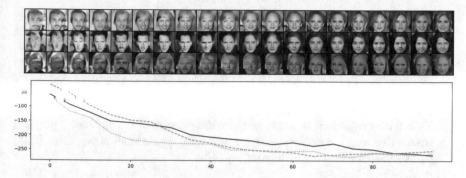

Fig. 12.12 Reprinted with permission from [191]. Transition of Markov chains initialized from $p_0(z)$ toward $\tilde{p}_\alpha(z)$ for 100 Langevin dynamics steps. *Top:* Trajectory in the CelebA data space. *Bottom:* Energy profile over time

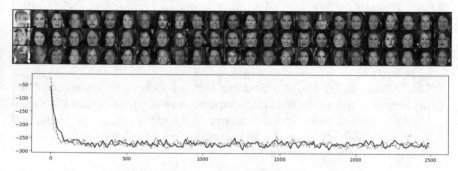

Fig. 12.13 Reprinted with permission from [191]. Transition of Markov chains initialized from $p_0(z)$ toward $\tilde{p}_\alpha(z)$ for 2500 Langevin dynamics steps. *Top:* Trajectory in the CelebA data space for every 100 steps. *Bottom:* Energy profile over time

$$S(\theta) = D_{\mathrm{KL}}(p_{\mathrm{data}} \| p_\theta) + D_{\mathrm{KL}}(\tilde{p}_{\theta_t}(z|x) \| p_\theta(z|x)) - D_{\mathrm{KL}}(\tilde{p}_{\alpha_t}(z) \| p_\alpha(z)),$$
$$(12.99)$$

where

$$D_{\mathrm{KL}}(\tilde{p}_{\theta_t}(z|x) \| p_\theta(z|x)) = \mathrm{E}_{p_{\mathrm{data}}(x)} \mathrm{E}_{\tilde{p}_{\theta_t}(z|x)} \left[\log \frac{\tilde{p}_{\theta_t}(z|x)}{p_\theta(z|x)} \right], \qquad (12.100)$$

i.e., the KL-divergence between conditional distributions of z given x also integrates over the marginal distribution x as defined before.

The learning algorithm based on short-run MCMC is $\theta_{t+1} = \theta_t - \eta_t S'(\theta_t)$ because

$$S'(\theta_t) = \mathrm{E}_{p_{\mathrm{data}}}[\tilde{\delta}_{\theta_t}(x)]. \qquad (12.101)$$

Thus the updating rule of the learning algorithm follows the stochastic gradient (i.e., Monte Carlo approximation of the gradient) of a perturbation of the log-likelihood.

$S(\theta)$ is in the form of a divergence triangle, which consists of two perturbations. $D_{KL}(\tilde{p}_{\theta_t}(z|x) \| p_\theta(z|x))$ is for inference, and $D_{KL}(\tilde{p}_{\alpha_t}(z) \| p_\alpha(z))$ is for synthesis. It is a combination of the perturbations for the descriptive model and generative model, respectively.

(2) **Estimating equation**. In terms of estimating equation, the stochastic gradient descent learning is a Robbins–Monro stochastic approximation algorithm [205] that solves the following estimating equation:

$$\frac{1}{n} \sum_{i=1}^{n} \tilde{\delta}_\alpha(x_i) = \frac{1}{n} \sum_{i=1}^{n} E_{\tilde{p}_\theta(z_i|x_i)}[\nabla_\alpha f_\alpha(z_i)] - E_{\tilde{p}_\alpha(z)}[\nabla_\alpha f_\alpha(z)] = 0,$$

$$(12.102)$$

$$\frac{1}{n} \sum_{i=1}^{n} \tilde{\delta}_\beta(x_i) = \frac{1}{n} \sum_{i=1}^{n} E_{\tilde{p}_\theta(z_i|x_i)}[\nabla_\beta \log p_\beta(x_i|z_i)] = 0. \qquad (12.103)$$

The solution to the above estimating equation defines an estimator of the parameters. The learning algorithm converges to this estimator under the usual regularity conditions of Robbins–Monro. If we replace $\tilde{p}_\alpha(z)$ by $p_\alpha(z)$, and $\tilde{p}_\theta(z|x)$ by $p_\theta(z|x)$, then the above estimating equation is the maximum likelihood estimating equation.

12.6 Variational and Adversarial Learning

From Short-Run MCMC to Learned Sampling Computations

In both the descriptive model and the generative model, we use short-run MCMC [183, 184] for the sampling computations of synthesis and inference. In this section, we shall study learning methods that replace short-run MCMC by learned computations for synthesis and inference sampling.

One popular learning method is variational auto-encoder (VAE) [133] for the generator model, where the short-run MCMC for inference is replaced by an inference model. The other popular learning method is generative adversarial networks (GAN) [81, 199], which is related to the descriptive model, where we replace short-run MCMC for synthesis by a generator model.

Both learning methods can be theoretically understood by the divergence triangle framework [95].

VAE: Learned Computation for Inference Sampling

For the generator model $p_\theta(x, z) = p(z)p_\theta(x|z)$ studied in the previous sections, the VAE [133] approximates the posterior $p_\theta(z|x)$ by a tractable $q_\phi(z|x)$, such as

$$q_\phi(z|x) \sim N(\mu_\phi(x), \mathrm{diag}(v_\phi(x))), \qquad (12.104)$$

where both μ_ϕ and v_ϕ are bottom-up networks that map x to d-dimensional vectors, with ϕ collecting all the weight and bias parameters of the bottom-up networks. For $z \sim q_\phi(z|x)$, we can write $z = \mu_\phi(x) + \mathrm{diag}(v_\phi(x))^{1/2}e$, where $e \sim N(0, I_d)$. Thus expectation with respect to $z \sim q_\phi(z|x)$ can be written as expectation with respect to e. This reparameterization trick [133] helps reduce the variance in Monte Carlo integration. We may consider $q_\phi(z|x)$ as an approximation to the iterative MCMC sampling of $p_\theta(z|x)$. In other words, $q_\phi(z|x)$ is the learned inferential computation that approximately samples from $p_\theta(z|x)$.

The VAE objective is a modification of the maximum likelihood estimation (MLE) objective $D(\theta) = D_{\mathrm{KL}}(p_{\mathrm{data}}(x)\|p_\theta(x))$:

$$S(\theta, \phi) = D(\theta) + D_{\mathrm{KL}}(q_\phi(z|x)\|p_\theta(z|x)) \qquad (12.105)$$

$$= D_{\mathrm{KL}}(p_{\mathrm{data}}(x)\|p_\theta(x)) + D_{\mathrm{KL}}(q_\phi(z|x)\|p_\theta(z|x)) \qquad (12.106)$$

$$= D_{\mathrm{KL}}(p_{\mathrm{data}}(x)q_\phi(z|x)\|p_\theta(z, x)). \qquad (12.107)$$

We define the conditional KL-divergence

$$D_{\mathrm{KL}}(q_\phi(z|x)\|p_\theta(z|x)) = \mathrm{E}_{p_{\mathrm{data}}}\mathrm{E}_{q_\phi(z|x)}\left[\log\frac{q_\phi(z|x)}{p_\theta(z|x)}\right], \qquad (12.108)$$

where we also average over p_{data}.

We estimate θ and ϕ jointly by

$$\min_\theta \min_\phi S(\theta, \phi), \qquad (12.109)$$

which can be accomplished by gradient descent.

Define $Q(z, x) = p_{\mathrm{data}}(x)q_\phi(z|x)$. Define $P(z, x) = p(z)p_\theta(x|z)$. Q is the distribution of the complete data (z, x), where $q_\phi(z|x)$ can be interpreted as an imputer that imputes the missing data z. P is the distribution of the complete-data model. Then $S(\theta, \phi) = D_{\mathrm{KL}}(Q\|P)$. The VAE is $\min_\theta \min_\phi D_{\mathrm{KL}}(Q\|P)$.

We may interpret the VAE as an alternating projection between Q and P. Figure 12.14 provides an illustration. The wake–sleep algorithm [101] is similar to the VAE, except that it updates ϕ by $\min_\phi D_{\mathrm{KL}}(P\|Q)$, where the order is flipped.

In the VAE, the model $q_\phi(z|x)$ and the parameter ϕ are shared by all the training examples x, so that $\mu_\phi(x)$ and $v_\phi(x)$ in Eq. (12.104) can be computed directly for each x given ϕ. This is different from traditional variational inference [20, 122],

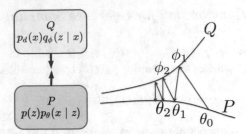

Fig. 12.14 Reprinted with permission from [95]. Variational auto-encoder as joint minimization by alternating projection. $P = p(z)p_\theta(x|z)$ is the distribution of the complete-data model, where $p(z)$ is the prior distribution of hidden vector z and $p_\theta(x|z)$ is the conditional distribution of x given z. $Q = p_d(x)q_\phi(z|x)$ is the distribution of the complete data (z, x), where $p_d(x)$ refers to the data distribution p_{data} and $q_\phi(z|x)$ is the learned inferential computation that approximately samples from the posterior distribution $p_\theta(x|z)$. (Left) Interaction between the models. (Right) Alternating projection. The two models run toward each other

where for each x, a model $q_{\mu,v}(z)$ is learned by minimizing $D_{KL}(q_{\mu,v}(z)\|p_\theta(z|x))$ with x fixed, so that (μ, v) is computed by an iterative algorithm for each x, which is an inner loop of the learning algorithm. This is similar to maximum likelihood learning, except that in maximum likelihood learning, the inner loop is an iterative algorithm that samples $p_\theta(z|x)$ instead of minimizing over (μ, v). The learned networks $\mu_\phi(x)$ and $v_\phi(d)$ in the VAE are to approximate the iterative minimization algorithm by direct mappings.

GAN: Joint Learning of Generator and Discriminator

The generator model learned by MLE or the VAE usually cannot generate very realistic images. Both MLE and the VAE target $D_{KL}(p_{data}\|p_\theta)$, though the VAE only minimizes an upper bound of $D_{KL}(p_{data}\|p_\theta)$. Consider minimizing $D_{KL}(q\|p)$ over p within a certain model class. If q is multi-modal, then p is obliged to fit all the major modes of q because $D_{KL}(q\|p)$ is an expectation with respect to q. Thus, p tends to interpolate the major modes of q if p cannot fit the modes of q closely. As a result, p_θ learned by MLE or the VAE tends to generate images that are not as sharp as the observed images.

The behavior of minimizing $D_{KL}(q\|p)$ over p is different from minimizing $D_{KL}(q\|p)$ over q. If p is multi-modal, q tends to capture some major modes of p while ignoring the other modes of p, because $D_{KL}(q\|p)$ is an expectation with respect to q. In other words, $\min_q D_{KL}(q\|p)$ encourages mode chasing, whereas $\min_p D_{KL}(q\|p)$ encourages mode covering.

Sharp synthesis can be achieved by GAN [81, 199], which pairs a generator model G with a discriminator model D. For an image x, $D(x)$ is the probability that x is an observed (real) image instead of a generated (faked) image. It can be parameterized by a bottom-up network $f_\alpha(x)$, so that $D(x) = 1/(1 + \exp(-f_\alpha(x)))$, i.e., logistic regression. We can train the pair of (G, D) by an adversarial, zero-sum game. Specifically, let $G(z) = g_\theta(z)$ be a generator. Let

$$V(D, G) = \mathrm{E}_{p_{\text{data}}}[\log D(x)] + \mathrm{E}_{z \sim p(z)}[\log(1 - D(G(z)))], \quad (12.110)$$

where $\mathrm{E}_{p_{\text{data}}}$ can be approximated by averaging over the observed examples, and E_z can be approximated by Monte Carlo average over the faked examples generated by the generator model. We learn D and G by $\min_G \max_D V(D, G)$. $V(D, G)$ is the log-likelihood for D, i.e., the log probability of the real and faked examples. However, $V(D, G)$ is not a very convincing objective for G. In practice, the training of G is usually modified into maximizing $\mathrm{E}_{z \sim p(z)}[\log D(G(z))]$ to avoid the vanishing gradient problem.

For a given θ, let p_θ be the distribution of $g_\theta(z)$ with $z \sim p(z)$. Assume a perfect discriminator. Then, according to the Bayes theorem, $D(x) = p_{\text{data}}(x)/(p_{\text{data}}(x) + p_\theta(x))$ (assuming equal numbers of real and faked examples). Then θ minimizes the Jensen–Shannon (JS) divergence

$$D_{\text{JS}}(p_{\text{data}} \| p_\theta) = D_{\text{KL}}(p_\theta \| p_{\text{mix}}) + D_{\text{KL}}(p_{\text{data}} \| p_{\text{mix}}), \quad (12.111)$$

where $p_{\text{mix}} = (p_{\text{data}} + p_\theta)/2$.

In JS divergence, the model p_θ also appears on the left-hand side of KL-divergence. This encourages p_θ to fit some major modes of p_{data} while ignoring others. As a result, GAN learning suffers from the mode collapsing problem, i.e., the learned p_θ may miss some modes of p_{data}. However, the p_θ learned by GAN tends to generate sharper images than the p_θ learned by MLE or the VAE.

Joint Learning of Descriptive and Generative Models

We can also learn the descriptive model and the generative model jointly, similar to GAN. In this joint learning scheme, we seek to learn the descriptive model by MLE, following the analysis by synthesis scheme. But we recruit the generator model as an approximate sampler, i.e., in this context, the generator model is the learned computation for synthesis sampling.

We continue to use $p_\theta(x, z) = p(z)p_\theta(x|z)$ to denote the generative model, and we denote the descriptive model by

$$\pi_\alpha(x) = \frac{1}{Z(\alpha)} \exp(f_\alpha(x)), \quad (12.112)$$

so that we will not confuse the notation.

To avoid MCMC sampling of π_α, we may approximate it by a generator model p_θ, which can generate synthesized examples directly (i.e., sampling z from $p(z)$, and transforming z to x by $x = g_\theta(z)$). We may consider p_θ an approximation to the iterative MCMC sampling of π_α. In other words, p_θ is the learned computation that approximately samples from π_α. It is an approximate direct sampler of π_α.

The MLE learning objective is $D(\alpha) = D_{KL}(p_{data}\|\pi_\alpha)$. We can learn both π_α and p_θ using the following objective function [33, 128]:

$$S(\alpha, \theta) = D(\alpha) - D_{KL}(p_\theta\|\pi_\alpha) = D_{KL}(p_{data}\|\pi_\alpha) - D_{KL}(p_\theta\|\pi_\alpha). \quad (12.113)$$

We learn α and θ by

$$\min_\alpha \max_\theta S(\alpha, \theta), \quad (12.114)$$

which defines a minimax game.

The gradient for updating α becomes

$$\nabla_\alpha S(\alpha, \theta) = \nabla_\alpha [E_{p_{data}}(f_\alpha(x)) - E_{p_\theta}(f_\alpha(x))], \quad (12.115)$$

where the intractable $\log Z(\alpha)$ term is canceled.

Because of the negative sign in front of the second KL-divergence in Eq. (12.113), we need $\max_\theta$ in Eq. (12.114), so that the learning becomes adversarial (illustrated in Fig. 12.15). Inspired by Hinton [100], Han et al. [95] called Eq. (12.114) the adversarial contrastive divergence (ACD). It underlies the work of [33, 128].

The adversarial form (Eq. (12.114) or (12.113)) defines a chasing game with the following dynamics: The generator p_θ chases the energy-based model π_α in $\min_\theta D_{KL}(p_\theta\|\pi_\alpha)$, while the energy-based model π_α seeks to get closer to p_{data} and away from p_θ. The red arrow in Fig. 12.15 illustrates this chasing game. The result is that π_α lures p_θ toward p_{data}. In the idealized case, p_θ always catches up with π_α, and then π_α will converge to the maximum likelihood estimate $\min_\alpha D_{KL}(p_{data}\|\pi_\alpha)$, and p_θ converges to π_α.

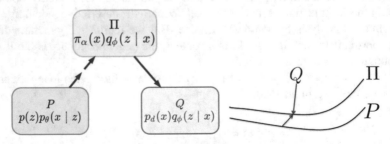

Fig. 12.15 Reprinted with permission from [95]. Adversarial contrastive divergence where the energy-based model favors real data against the generator. (Left) Interaction between the models. The red arrow indicates a chasing game, where the red arrow pointing to Π indicates that Π seeks to move away from P. The blue arrow pointing from P to Π indicates that P seeks to move close to Π. (Right) Contrastive divergence

This chasing game is different from the VAE $\min_\theta \min_\phi D_{KL}(Q\|P)$, which defines a cooperative game where q_ϕ and p_θ run toward each other.

Even though the above chasing game is adversarial, both models are running toward the data distribution. While the generator model runs after the energy-based model, the energy-based model runs toward the data distribution. As a consequence, the ~~~~~ ~ ~~~~~~ ~~~~~ ~~~~~ ~~~~ ~~~ ~~ ~~~~ ~~~ ~~~~~~ ~~~~ ~~~~~~~~~ ~~~~~~~~~ ~~~ ~~~~~ distribution. It is different from GAN [81], in which the discriminator eventually becomes confused because the generated data become similar to the real data. In the above chasing game, the energy-based model becomes close to the data distribution.

The updating of α by Eq. (12.115) is similar to Wasserstein GAN (WGAN) [8], but unlike WGAN, f_α defines a probability distribution π_α, and the learning of θ is based on $\min_\theta D_{KL}(p_\theta\|\pi_\alpha)$, which is a variational approximation to π_α. This variational approximation only requires knowing $f_\alpha(x)$, without knowing $Z(\alpha)$. However, unlike $q_\phi(z|x)$, $p_\theta(x)$ is still intractable; in particular, its entropy does not have a closed form. Thus, we can again use variational approximation, by changing the problem $\min_\theta D_{KL}(p_\theta\|\pi_\alpha)$ to

$$\min_\theta \min_\phi D_{KL}(p(z)p_\theta(x|z)\|\pi_\alpha(x)q_\phi(z|x)). \tag{12.116}$$

Define $\Pi(z,x) = \pi_\alpha(x)q_\phi(z|x)$, and then the problem is $\min_\theta \min_\phi D_{KL}(P\|\Pi)$, which is analytically tractable and underlies the work of [33]. In fact,

$$D_{KL}(P\|\Pi) = D_{KL}(p_\theta(x)\|\pi_\alpha(x)) + D_{KL}(p_\theta(z|x)\|q_\phi(z|x)). \tag{12.117}$$

Thus, we can use $\max_\alpha \min_\theta \min_\phi[D_{KL}(P\|\Pi) - D_{KL}(Q\|\Pi)]$ because $D_{KL}(Q\|\Pi) = D_{KL}(p_{\text{data}}\|\pi_\alpha)$.

Note that in the VAE (Eq. (12.107)), the objective function is in the form of KL + KL, whereas in ACD (Eq. (12.113)), it is in the form of KL − KL. In both Eqs. (12.107) and (12.113), the first KL is about maximum likelihood. The KL + KL form of the VAE makes the computation tractable by changing the marginal distribution of x to the joint distribution of (z, x). The KL − KL form of ACD makes the computation tractable by canceling the intractable $\log Z(\alpha)$ term. Because of the negative sign in Eq. (12.113), the ACD objective function becomes an adversarial one or a minimax game.

Also note that in the VAE, p_θ appears on the right-hand side of KL, whereas in ACD, p_θ appears on the left-hand side of KL. Thus in ACD, p_θ may exhibit mode chasing behavior, i.e., fitting the major modes of π_α, while ignoring other modes.

Divergence Triangle: Integrating VAE and ACD

We can combine the VAE and ACD into a divergence triangle, which involves the following three joint distributions on (z, x) defined above:

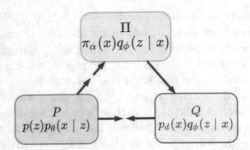

Fig. 12.16 Reprinted with permission from [95]. Divergence triangle is based on the Kullback–Leibler divergences between three joint distributions, Q, P, and Π, of (z, x). The blue arrow indicates the "running toward" behavior, and the red arrow indicates the "running away" behavior

1. Q distribution: $Q(z, x) = p_{\text{data}}(x)q_\phi(z|x)$.
2. P distribution: $P(z, x) = p(z)p_\theta(x|z)$.
3. Π distribution: $\Pi(z, x) = \pi_\alpha(x)q_\phi(z|x)$.

Han et al. [95] proposed to learn the three models p_θ, π_α, and q_ϕ by the following divergence triangle loss functional S:

$$\max_\alpha \min_\theta \min_\phi S(\alpha, \theta, \phi),$$

$$S = D_{\text{KL}}(Q\|P) + D_{\text{KL}}(P\|\Pi) - D_{\text{KL}}(Q\|\Pi). \tag{12.118}$$

See Fig. 12.16 for an illustration. The divergence triangle is based on the three KL-divergences between the three joint distributions on (z, x). It has a symmetric and anti-symmetric form, where the anti-symmetry is due to the negative sign in front of the last KL-divergence and the maximization over α. Compared to the VAE and ACD objective functions in the previous subsections, $D_{\text{KL}}(Q\|P)$ is the VAE part, and $D_{\text{KL}}(P\|\Pi) - D_{\text{KL}}(Q\|\Pi)$ is the ACD part.

The divergence triangle leads to the following dynamics between the three models: (a) Q and P seek to get close to each other. (b) P seeks to get close to Π. (c) π seeks to get close to p_{data}, but it seeks to get away from P, as indicated by the red arrow. Note that $D_{\text{KL}}(Q\|\Pi) = D_{\text{KL}}(p_{\text{data}}\|\pi_\alpha)$ because $q_\phi(z|x)$ is canceled out. The effect of (b) and (c) is that π gets close to p_{data} while inducing P to get close to p_{data} as well, or in other words, P chases π_α toward p_{data}.

[95] also employed the layer-wise training scheme of [125] to learn models by divergence triangle from the CelebA-HQ dataset [125], including 30,000 celebrity face images with resolutions of up to 1024×1024 pixels. The learning algorithm converges stably, without extra tricks, to obtain realistic results as shown in Fig. 12.17.

Figure 12.17a displays a few 1024×1024 images generated by the learned generator model with 512-dimensional latent vector. Figure 12.17b shows an example of interpolation. The two images at the two ends are generated by two different latent vectors. The images in between are generated by the vectors that

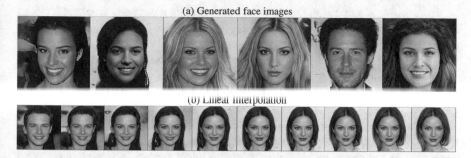

Fig. 12.17 Reprinted with permission from [95]. Learning generator model by divergence triangle from the CelebA-HQ dataset [125] that includes 30,000 high-resolution celebrity face images. (**a**) Generated face images with 1024×1024 resolution sampled from the learned generator model with 512-dimensional latent vector. (**b**) Linear interpolation of the vector representations. The images at the two ends are generated from latent vectors randomly sampled from a Gaussian distribution. Each image in the middle is obtained by first interpolating the two vectors of the two end images and then generating the image using the generator

are linear interpolations of the two vectors at the two ends. Even though the interpolation is linear in the latent vector space, the nonlinear mapping leads to a highly nonlinear interpolation in the image space. We first do a linear interpolation between the latent vectors at the two ends, i.e., $(1 - \alpha)z_0 + \alpha z_1$, where z_0 and z_1 are two latent vectors at two ends, respectively, and α is in the closed unit interval [0, 1]. The images in between are generated by mapping those interpolated vectors to image space via the learned generator. The interpolation experiment shows that the algorithm can learn a smooth generator model that traces the manifold of the data distribution.

12.7 Cooperative Learning via MCMC Teaching

Joint Training of Descriptive and Generative Models

In ACD, the generator model p_θ is used to approximate the energy-based model π_α, and we treat the examples generated by p_θ as if they are generated from π_α for the sake of updating α. The gap between p_θ and π_α can cause bias in learning. In the work of [262, 263], we proposed to bring back MCMC to bridge the gap. Instead of running MCMC from scratch, we run a finite-step MCMC toward π_α, initialized from the examples generated by p_θ. We then use the examples produced by the finite-step MCMC as the synthesized examples from π_α for updating α. Meanwhile, we update p_θ based on how the finite-step MCMC revises the initial examples generated by p_θ; in other words, the energy-based model (as a teacher) π_α distills the MCMC into the generator (as a student) p_θ. We call this scheme cooperative learning.

Specifically, we first generate $\hat{z}_i \sim N(0, I_d)$ and then generate $\hat{x}_i = g_\theta(\hat{z}_i) + \epsilon_i$, for $i = 1, \ldots, \tilde{n}$. Starting from $\{\hat{x}_i, i = 1, \ldots, \tilde{n}\}$, we run MCMC such as Langevin dynamics for a finite number of steps toward π_α to get $\{\tilde{x}_i, i = 1, \ldots, \tilde{n}\}$, which are revised versions of $\{\hat{x}_i\}$. $\{\tilde{x}_i\}$ are used as the synthesized examples from the descriptive model.

The descriptive model can teach the generator via MCMC. The key is that in the generated examples, the latent z is known. In order to update θ of the generator model, we treat $\{\tilde{x}_i, i = 1, \ldots, \tilde{n}\}$ as the training data for the generator. Since these $\{\tilde{x}_i\}$ are obtained by the Langevin dynamics initialized from $\{\hat{x}_i\}$, which are generated by the generator model with known latent factors $\{\hat{z}_i\}$, we can update θ by learning from the complete data $\{(\hat{z}_i, \tilde{x}_i); i = 1, \ldots, \tilde{n}\}$, which is a supervised learning problem, or more specifically, a nonlinear regression of $\tilde{x}_i$ on $\hat{z}_i$. At $\theta^{(t)}$, the latent factors $\hat{z}_i$ generate and thus reconstruct the initial example $\hat{x}_i$. After updating θ, we want $\hat{z}_i$ to reconstruct the revised example $\tilde{x}_i$. That is, we revise θ to absorb the MCMC transition from $\hat{x}_i$ to $\tilde{x}_i$. The left panel of diagram (12.119) illustrates the basic idea.

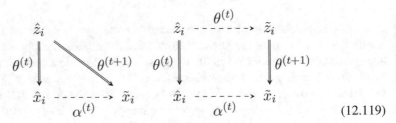

$$\tag{12.119}$$

In the two diagrams in (12.119), the double-line arrows indicate generation and reconstruction by the generator model, while the dashed-line arrows indicate Langevin dynamics for MCMC sampling and inference in the two models. The right panel of diagram (12.119) illustrates a more rigorous method, where we initialize the MCMC for inferring $\{\tilde{z}_i\}$ from the known $\{\hat{z}_i\}$ and then update θ based on $\{(\tilde{z}_i, \tilde{x}_i), i = 1, \ldots, \tilde{n}\}$.

The theoretical understanding of the cooperative learning scheme is given below:

(1) Modified contrastive divergence for the energy-based model. In the traditional contrastive divergence [100], $\hat{x}_i$ is taken to be the observed x_i. In cooperative learning, $\hat{x}_i$ is generated by $p_{\theta^{(t)}}$. Let M_α be the Markov transition kernel of finite steps of Langevin dynamics that samples π_α. Let $(M_\alpha p_\theta)(x) = \int M_\alpha(x', x) p_\theta(x') dx'$ be the marginal distribution by running M_α initialized from p_θ. Then similar to the traditional contrastive divergence, the learning gradient of the energy-based model α at iteration t is the gradient of $D_{\text{KL}}(p_{\text{data}} \| \pi_\alpha) - D_{\text{KL}}(M_{\alpha^{(t)}} p_{\theta^{(t)}} \| \pi_\alpha)$ with respect to α. In the traditional contrastive divergence, p_{data} takes the place of $p_{\theta^{(t)}}$ in the second KL-divergence.

(2) MCMC teaching of the generator model. The learning gradient of the generator θ in the right panel of diagram (12.119) is the gradient of $D_{\text{KL}}(M_{\alpha^{(t)}} p_{\theta^{(t)}} \| p_\theta)$ with respect to θ. Here $\pi^{(t+1)} = M_{\alpha^{(t)}} p_{\theta^{(t)}}$ takes the place of p_{data} as the data to

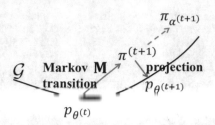

Fig. 12.18 Reprinted with permission from [264]. The MCMC teaching of the generator alternates between Markov transition and projection. The family of the generator models $\mathcal{G}$ is illustrated by the black curve, and each distribution is illustrated by a point. p_θ is a generator model, and π_α is a descriptive model

train the generator model. It is much easier to minimize $D_{\mathrm{KL}}(M_{\alpha^{(t)}} p_{\theta^{(t)}} \parallel p_\theta)$ than to minimize $D_{\mathrm{KL}}(p_{\mathrm{data}} \parallel p_\theta)$ because the latent variables are essentially known in the former, so the learning is supervised. The MCMC teaching alternates between Markov transition from $p_{\theta^{(t)}}$ to $\pi^{(t+1)}$, and projection from $\pi^{(t+1)}$ to $p_{\theta^{(t+1)}}$, as illustrated by Fig. 12.18.

Conditional Learning via Fast Thinking Initializer and Slow Thinking Solver

Xie et al. [267] extended the cooperative learning scheme to the conditional learning problem by jointly learning a conditional energy-based model and a conditional generator model. The conditional energy-based model is of the following form:

$$\pi_\alpha(x|c) = \frac{1}{Z(c, \alpha)} \exp[f_\alpha(x, c)], \qquad (12.120)$$

where x is the input signal and c is the condition. $Z(c, \alpha)$ is the normalizing constant conditioned on c. $f_\alpha(x, c)$ can be defined by a bottom-up ConvNet where α collects all the weight and bias parameters. Fixing the condition c, $f_\alpha(x, c)$ defines the value of x for the condition c, and $-f_\alpha(x, c)$ defines the conditional energy function. $\pi_\alpha(x|c)$ is also a deep generalization of conditional random fields [140]. Both the conditional generator model and the conditional energy-based model can be learned jointly by the cooperative learning scheme in Sect. 12.7.

Figure 12.19 shows some examples of pattern completion on the CMP (Center for Machine Perception) Facades dataset [238] by learning a mapping from an occluded image (256×256 pixels), where a mask of the size of 128×128 pixels is centrally placed onto the original version, to the original image. In this case, c is the observed part of the signal, and x is the unobserved part of the signal.

input   ground truth   initializer   solver   conditional GAN

Fig. 12.19 Reprinted with permission from [268]. Pattern completion by conditional learning. Each row displays one example. The first image is the testing image (256 × 256 pixels) with a hole of 128 × 128 that needs to be recovered, the second image shows the ground truth, and the third image shows the result recovered by the initializer (i.e., conditional generator model), the fourth image shows the result recovered by the solver (i.e., the MCMC sampler of the conditional energy-based model, initialized from the result of the initializer), and the last image shows the result recovered by the conditional GAN as a comparison

The cooperative learning of the conditional generator model and conditional energy-based model can be interpreted as follows. The conditional energy function defines the objective function or value function, i.e., it defines what solutions are desirable given the condition or the problem. The solutions can then be obtained by an iterative optimization or sampling algorithm such as MCMC. In other words, the conditional energy-based model leads to a solver in the form of an iterative algorithm, and this iterative algorithm is a slow thinking process. In contrast, the conditional generator model defines a direct mapping from condition or problem to solutions, and it is a fast thinking process. We can use the fast thinking generator as an initializer to generate the initial solution and then use the slow thinking solver to refine the fast thinking initialization by the iterative algorithm. The cooperative learning scheme enables us to learn both the fast thinking initializer and slow thinking solver. Unlike conditional GAN, the cooperative learning scheme has a slow thinking refining process, which can be important if the fast thinking initializer is not optimal.

In terms of inverse reinforcement learning [1, 283], the conditional energy-based model defines the reward or value function, and the iterative solver defines an optimal control or planning algorithm. The conditional generator model defines a policy. The fast thinking policy is about habitual, reflexive, or impulsive behaviors, while the slow thinking solver is about deliberation and planning. Compared with the policy, the value is usually simpler and more generalizable, because it is in general easier to specify what one wants than to specify how to produce what one wants.

Correction to: Computer Vision

Correction to:
S.-C. Zhu, Y. N. Wu, *Computer Vision,*
https://doi.org/10.1007/978-3-030-96530-3

The name of author "Ying Nian Wu" was inadvertently published as "Ying Wu".
The initially published version has now been corrected.

The updated original version for this book can be found at
https://doi.org/10.1007/978-3-030-96530-3

S.-C. Zhu, Y. N. Wu, *Computer Vision*, https://doi.org/10.1007/978-3-030-96530-3_13

Bibliography

1. Abbeel, P., & Ng, A. Y. (2004). Apprenticeship learning via inverse reinforcement learning. In *International Conference on Machine Learning* (pp. 1–8). ACM.
2. Adelson, E. H. (1995). Layered representations for vision and video. In *In Representation of Visual Scenes, Proceedings IEEE Workshop on International Conference on Computer Vision (ICCV)*, pp. 3–9.
3. Aharon, M., Elad, M., Bruckstein, A. M. (2006). The K-SVD: An algorithm for designing of overcomplete dictionaries for sparse representations. *IEEE Transactions On Signal Processing, 15*(12), 3736–3745.
4. Barbu, A. & Zhu, S.-C. (2005). Incorporating visual knowledge representation in stereo reconstruction. In *Tenth IEEE International Conference on Computer Vision (ICCV'05) Volume 1* (Vol. 1, pp. 572–579).
5. Ahuja, N., & Tuceryan, M. (1989). Extraction of early perceptual structure in dot patterns. *Computer Vision, Graphics, and Image Processing, 48*(3), 304–356.
6. Alain, G., & Bengio, Y. (2014). What regularized auto-encoders learn from the data-generating distribution. *The Journal of Machine Learning Research, 15*(1), 3563–3593.
7. Amit, D. J. (1989). Modeling brain function: The world of attractor neural networks. In *Modeling Brain Function*.
8. Arjovsky, M., Chintala, S., & Bottou, L. (2017). Wasserstein gan. arXiv preprint arXiv:1701.07875.
9. Barbu, A., & Zhu, S.-C. (2005). Generalizing swendsen-wang to sampling arbitrary posterior probabilities. *IEEE Transaction on Pattern Analysis and Machine Intelligence, 27*(8), 1239–1253.
10. Barlow, H. B. (1961). Possible principles underlying the transformation of sensory messages. *Sensory communication, 1*(01).
11. Barrett, D. G., & Dherin, B. (2020). Implicit gradient regularization. arXiv preprint arXiv:2009.11162.
12. Barrow, H. G., & Tenenbaum, J. M. (1993). Retrospective on "interpreting line drawings as three-dimensional surfaces". *Artificial Intelligence, 59*(1–2), 71–80.
13. Behrmann, J., Grathwohl, W., Chen, R. T., Duvenaud, D., & Jacobsen, J.-H. (2018). Invertible residual networks. arXiv preprint arXiv:1811.00995.
14. Belhumeur, P. N. (1996). A bayesian approach to binocular steropsis. *International Journal of Computer Vision, 19*(3), 237–260.
15. Bell, A. J., & Sejnowski, T. J. (1997). The independent components of natural scenes are edge filters. *Vision Research, 37*(23), 3327–3338.
16. Bengio, Y., Goodfellow, I. J., & Courville, A. (2015). Deep learning. Book in preparation for MIT Press.

© Springer Nature Switzerland AG 2023

S.-C. Zhu, Y. N. Wu, *Computer Vision*, https://doi.org/10.1007/978-3-030-96530-3

17. Bengio, Y., Yao, L., Alain, G., & Vincent, P. (2013). Generalized denoising auto-encoders as generative models. In *Advances in neural information processing systems*, pp. 899–907.
18. Bergen, J. R., & Adelson, E. (1991). Theories of visual texture perception. *Spatial Vision, 10*, 114–134.
19. Besag, J. (1974). Spatial interaction and the statistical analysis of lattice systems. *Journal of the Royal Statistical Society: Series B (Methodological), 36*(2), 192–225.
20. Blei, D. M., Kucukelbir, A., & McAuliffe, J. D. (2017). Variational inference: a review for statisticians. *Journal of the American Statistical Association, 112*(518), 859–877.
21. Breiman, L., Friedman, J., Stone, C. J., & Olshen, R. A. (1984). *Classification and regression trees*.
22. Bristow, H., Eriksson, A., & Lucey, S. (2013). Fast convolutional sparse coding. In *IEEE Conference on Computer Vision and Pattern Recognition* (pp. 391–398). IEEE.
23. Brooks, M. J., & Horn, B. K. (1985). *Shape and source from shading*.
24. Chandler, D. (1987). Introduction to modern statistical mechanics. In D. Chandler (Ed.) *Introduction to Modern Statistical Mechanics* (pp. 288). Foreword by D. Chandler. Oxford University Press. ISBN-10: 0195042778. ISBN-13: 9780195042771, 1.
25. Chen, H., & Zhu, S.-C. (2006). A generative sketch model for human hair analysis and synthesis. *IEEE transactions on pattern analysis and machine intelligence, 28*(7), 1025–1040.
26. Chen, S. S., Donoho, D. L., & Saunders, M. A. (1998). Atomic decomposition by basis pursuit. *SIAM Journal on Scientific Computing, 20*(1), 33–61.
27. Chiu, S. N., Stoyan, D., Kendall, W. S., & Mecke, J. (2013). *Stochastic geometry and its applications*. Wiley.
28. Cootes, T. F., Edwards, G. J., & Taylor, C. J. (2001). Active appearance models. *IEEE Transactions on Pattern Analysis and Machine Intelligence, 23*(6), 681–685.
29. Cover, T. M., & Thomas, J. A. (2012). *Elements of information theory*. Wiley.
30. Cross, G. R., & Jain, A. K. (1983). Markov random field texture models. *IEEE Transactions on Pattern Analysis and Machine Intelligence, 1*, 25–39.
31. Da Vinci, L., Kemp, M., & Walker, M. (1989). Leonardo on painting: Anthology of writings. In L. da Vinci (Ed.), *With a selection of documents relating to his career as an artist*. Yale Nota Bene.
32. Dai, J., Lu, Y., & Wu, Y. N. (2014). Generative modeling of convolutional neural networks. arXiv preprint arXiv:1412.6296.
33. Dai, Z., Almahairi, A., Bachman, P., Hovy, E., & Courville, A. (2017). Calibrating energy-based generative adversarial networks. In *International Conference on Learning Representations*.
34. Dalal, N., & Triggs, B. (2005). Histograms of oriented gradients for human detection. In *2005 IEEE computer society conference on computer vision and pattern recognition (CVPR'05)* (Vol. 1, pp. 886–893). IEEE.
35. Daugman, J. G. (1985). Uncertainty relation for resolution in space, spatial frequency, and orientation optimized by two-dimensional visual cortical filters. *JOSA A, 2*(7), 1160–1169.
36. De Bonet, J., & Viola, P. (1997). A non-parametric multi-scale statistical model for natural images. *Advances in Neural Information Processing Systems, 10*.
37. Della Pietra, S., Della Pietra, V., & Lafferty, J. (1997). Inducing features of random fields. *IEEE Transactions on Pattern Analysis and Machine Intelligence, 19*(4), 380–393.
38. Dempster, A. P., Laird, N. M., & Rubin, D. B. (1977). Maximum likelihood from incomplete data via the EM algorithm. *Journal of the Royal Statistical Society: Series B (Methodological), 39*(1), 1–38.
39. Deng, J., Dong, W., Socher, R., Li, L.-J., Li, K., & Fei-Fei, L. (2009). Imagenet: A large-scale hierarchical image database. In *Proceedings of the IEEE Conference on Computer Vision and Pattern Recognition* (pp. 248–255). IEEE.
40. Denton, E. L., Chintala, S., Fergus, R., et al. (2015). Deep generative image models using a laplacian pyramid of adversarial networks. In *Advances in neural information processing systems* (pp. 1486–1494).

41. Diaconis, P., & Freedman, D. (1981). On the statistics of vision: the julesz conjecture. *Journal of Mathematical Psychology, 24*(2), 112–138.
42. Dickinson, S., & Pizlo, Z. (2015). *Shape perception in human and computer vision.* Springer.
43. Dinh, L., Krueger, D., & Bengio, Y. (2014). NICE: Non-linear independent components estimation. arXiv preprint arXiv:1410.8516.
44. Dinh, L., Sohl-Dickstein, J., & Bengio, S. (2017). Density estimation using real NVP. In *International Conference on Learning Representations, arXiv1605.08803*
45. Donato, G., & Belongie, S. (2002). Approximate thin plate spline mappings. In *European conference on computer vision* (pp. 21–31). Springer.
46. Donoho, D. L. (2001). Sparse components of images and optimal atomic decompositions. *Constructive Approximation, 17*(3), 353–382.
47. Dosovitskiy, A., Springenberg, J. T., & Brox, T. (2015). Learning to generate chairs with convolutional neural networks. In *Proceedings of the IEEE Conference on Computer Vision and Pattern Recognition* (pp. 1538–1546).
48. Earl, D. J., & Deem, M. W. (2005). Parallel tempering: Theory, applications, and new perspectives. *Physical Chemistry Chemical Physics, 7*(23), 3910–3916.
49. Elad, M. (2010). *Sparse and redundant representations: From theory to applications in signal and image processing.* Springer.
50. Elder, J. H., Krupnik, A., & Johnston, L. A. (2002). Contour grouping with prior models. *Journal of Vision, 2*(4), 324–353.
51. Elder, J. H., Krupnik, A., & Johnston, L. A. (2003). Contour grouping with prior models. *IEEE Transaction on Pattern Analysis and Machine Intelligence, 25*(6), 661–674.
52. Elder, J. H., & Zucker, S. W. (1998). Local scale control for edge detection and blur estimation. *IEEE Transactions on Pattern Analysis and Machine Intelligence, 20*(7), 699–716.
53. Esedoglu, S., & March, R. (2003). Segmentation with depth butwithout detecting junctions. *Journal of Mathematical Imaging and Vision, 18*(1), 7–15.
54. Field, D. J. (1987). Relations between the statistics of natural images and the response properties of cortical cells. *JOSA A, 4*(12), 2379–2394.
55. Field, D. J. (1994). What is the goal of sensory coding? *Neural Computation, 6*(4), 559–601.
56. Finn, C., Christiano, P., Abbeel, P., & Levine, S. (2016). A connection between generative adversarial networks, inverse reinforcement learning, and energy-based models. arXiv preprint arXiv:1611.03852.
57. Fraley, C., & Raftery, A. E. (2002). Model-based clustering, discriminant analysis, and density estimation. *Journal of the American statistical Association, 97*(458), 611–631.
58. Frankot, R. T., & Chellappa, R. (1988). A method for enforcing integrability in shape from shading algorithms. *IEEE Transactions on Pattern Analysis and Machine Intelligence, 10*(4), 439–451.
59. Freund, Y., & Schapire, R. E. (1997). A decision-theoretic generalization of on-line learning and an application to boosting. *Journal of Computer and System Sciences, 55*(1), 119–139.
60. Frey, B. J., & Jojic, N. (1999). Transformed component analysis: joint estimation of spatial transforms and image components. In G. Corfu (Ed.) *Proceedings of the 7th ICCV.*
61. Fridman, A. (2000a). *Mixed Markov Field.*
62. Fridman, A. (2000b). *Mixed Markov Models.* PhD thesis, Division of Applied Math, Brown University.
63. Fridman, A. (2003). Mixed markov models, applied mathematics. *Proceedings of the National Academy of Sciences, 100*(14), 8092–8096.
64. Friedman, J., Hastie, T., Tibshirani, R., et al. (2000). Additive logistic regression: a statistical view of boosting (with discussion and a rejoinder by the authors). *The Annals of Statistics, 28*(2), 337–407.
65. Friedman, J. H. (1987). Exploratory projection pursuit. *Journal of the American Statistical Association, 82*(397), 249–266.
66. Friedman, J. H. (1991). Multivariate adaptive regression splines. *The Annals of Statistics, 19*(1), 1–67.

67. Friedman, J. H. (2001). Greedy function approximation: a gradient boosting machine. *Annals of Statistics, 29*(5), 1189–1232.

68. Fu, K.-S., & Bhargava, B. K. (1973). Tree systems for syntactic pattern recognition. *IEEE Transactions on Computers, 100*(12), 1087–1099.

69. Gao, R., Lu, Y., Zhou, J., Zhu, S.-C., & Wu, Y. N. (2017). Learning multi-grid generative convnets by minimal contrastive divergence. arXiv preprint arXiv:1709.08868.

70. Gao, R., Lu, Y., Zhou, J., Zhu, S.-C., & Wu, Y. N. (2018). Learning generative ConvNets via multi-grid modeling and sampling. In *Proceedings of the IEEE Conference on Computer Vision and Pattern Recognition* (pp. 9155–9164).

71. Gao, R., Nijkamp, E., Kingma, D. P., Xu, Z., Dai, A. M., & Wu, Y. N. (2020a). Flow contrastive estimation of energy-based models. In *Proceedings of the IEEE/CVF Conference on Computer Vision and Pattern Recognition* (pp. 7518–7528).

72. Gao, R., Song, Y., Poole, B., Wu, Y. N., & Kingma, D. P. (2020b). Learning energy-based models by diffusion recovery likelihood. arXiv preprint arXiv:2012.08125.

73. Gao, R.-X., Wu, T.-F., Zhu, S.-C., & Sang, N. (2007). Bayesian inference for layer representation with mixed markov random field. In *International Workshop on Energy Minimization Methods in Computer Vision and Pattern Recognition* (pp. 213–224). Springer.

74. Geman, S., & Geman, D. (1997). Stochastic relaxation, gibbs distributions, and the bayesian restoration of images. *IEEE Transactions on Pattern Analysis and Machine Intelligence, 6*(4), 380–393.

75. Geman, S., & Graffigne, C. (1986). Markov random field image models and their applications to computer vision. In *Proceedings of the International Congress of Mathematicians* (Vol. 1, p. 2).

76. Geman, S., Potter, D. F., & Chi, Z. (2002). Composition systems. *Quarterly of Applied Mathematics, 60*(4), 707–736.

77. Geyer, C. J., & Thompson, E. A. (1995). Annealing markov chain monte carlo with applications to ancestral inference. *Journal of the American Statistical Association, 90*(431), 909–920.

78. Gibbs, J. W. (1902). *Elementary principles in statistical mechanics: Developed with especial reference to the rational foundations of thermodynamics*. C. Scribner's sons.

79. Gibson, J. J. (1950). *The perception of the visual world*.

80. Girolami, M., & Calderhead, B. (2011). Riemann manifold langevin and hamiltonian monte carlo methods. *Journal of the Royal Statistical Society: Series B (Statistical Methodology), 73*(2), 123–214.

81. Goodfellow, I., Pouget-Abadie, J., Mirza, M., Xu, B., Warde-Farley, D., Ozair, S., Courville, A., & Bengio, Y. (2014). Generative adversarial nets. In *Advances in Neural Information Processing Systems* (pp. 2672–2680).

82. Grathwohl, W., Chen, R. T., Betterncourt, J., Sutskever, I., & Duvenaud, D. (2019). FFJORD: Free-form continuous dynamics for scalable reversible generative models. In *International Conference on Learning Representations*.

83. Green, P. J. (1995). Reversible jump markov chain monte carlo computation and Bayesian model determination. *Biometrika, 82*, 711–732.

84. Grenander, U. (1970). A unified approach to pattern analysis. *Advances in Computers, 10*, 175–216.

85. Grenander, U., & Miller, M. I. (2007). *Pattern theory: from representation to inference*. Oxford University Press.

86. Griffiths, R. B., & Ruelle, D. (1971). Strict convexity ("continuity") of the pressure in lattice systems. *Communications in Mathematical Physics, 23*(3), 169–175

87. Grindrod, P. (1996). *The theory and applications of reaction-diffusion equations: patterns and waves*. Clarendon Press.

88. Gu, M. G., & Kong, F. H. (1998). A stochastic approximation algorithm with mcmc method for incomplete data estimation problems. Preprint, Department of Mathematics & Statistics, McGill University.

89. Guo, C.-E., Zhu, S.-C., & Wu, Y. N. (2003a). Modeling visual patterns by integrating descriptive and generative methods. *International Journal of Computer Vision, 53*(1), 5–29.

90. Guo, C.-e., Zhu, S.-C., & Wu, Y. N. (2003b). Towards a mathematical theory of primal sketch and sketchability. In *Proceedings of the Ninth IEEE International Conference on Computer Vision, 2003* (pp. 1228–1235). IEEE.

91. Guo, C.-e., Zhu, S.-C., & Wu, Y. N. (2007). Primal sketch: Integrating structure and texture. *Computer Vision and Image Understanding, 106*(1), 5–19.

92. Gutmann, M., & Hyvarinen, A. (2010). Noise-contrastive estimation: A new estimation principle for unnormalized statistical models. In *Proceedings of the thirteenth international conference on artificial intelligence and statistics* (pp. 297–304). JMLR Workshop and Conference Proceedings.

93. Han, F., & Zhu, S.-C. (2007). A two-level generative model for cloth representation and shape from shading. *IEEE Transactions on Pattern Analysis and Machine Intelligence, 29*(7), 1230–1243.

94. Han, T., Lu, Y., Zhu, S.-C., & Wu, Y. N. (2017). Alternating back-propagation for generator network. In *Proceedings of the AAAI Conference on Artificial Intelligence* (Vol. 3, p. 13).

95. Han, T., Nijkamp, E., Fang, X., Hill, M., Zhu, S.-C., & Wu, Y. N. (2018). Divergence triangle for joint training of generator model, energy-based model, and inference model. arXiv preprint arXiv:1812.10907, pp. 8670–8679.

96. Hastie, T., Tibshirani, R., & Friedman, J. (2009). *The elements of statistical learning: data mining, inference and prediction*. Springer.

97. He, K., Zhang, X., Ren, S., & Sun, J. (2016). Deep residual learning for image recognition. In *Proceedings of the IEEE Conference on Computer Vision and Pattern Recognition* (pp. 770–778).

98. Heeger, D. J., & Bergen, J. R. (1995). Pyramid-based texture analysis/synthesis. In *Proceedings of the 22nd annual conference on Computer graphics and interactive techniques* (pp. 229–238). ACM.

99. Hill, M., Nijkamp, E., & Zhu, S.-C. (2019). Building a telescope to look into high-dimensional image spaces. *Quarterly of Applied Mathematics, 77*(2), 269–321.

100. Hinton, G. E. (2002). Training products of experts by minimizing contrastive divergence. *Neural Computation, 14*(8), 1771–1800.

101. Hinton, G. E., Dayan, P., Frey, B. J., & Neal, R. M. (1995). The wake-sleep algorithm for unsupervised neural networks. *Science, 268*(5214), 1158–1161.

102. Ho, J., Jain, A., & Abbeel, P. (2020). Denoising diffusion probabilistic models. arXiv preprint arXiv:2006.11239.

103. Hochreiter, S., & Schmidhuber, J. (1997). Long short-term memory. *Neural Computation, 9*(8), 1735–1780.

104. Hoffman, M., Sountsov, P., Dillon, J. V., Langmore, I., Tran, D., & Vasudevan, S. (2019). Neutra-lizing bad geometry in hamiltonian monte carlo using neural transport. arXiv preprint arXiv:1903.03704.

105. Hoffman, M. D., Gelman, A., et al. (2014). The No-U-Turn sampler: Adaptively setting path lengths in Hamiltonian Monte Carlo. *Journal of Machine Learning Research, 15*(1), 1593–1623.

106. Hoiem, D., Stein, A. N., Efros, A. A., & Hebert, M. (2007). Recovering occlusion boundaries from a single image. In *International Conference on Computer Vision (ICCV)* (pp. 1–8).

107. Hong, Y., Si, Z., Hu, W., Zhu, S.-C., & Wu, Y. N. (2013). Unsupervised learning of compositional sparse code for natural image representation. *Quarterly of Applied Mathematics, 72*, 373–406.

108. Hopfield, J. J. (1982). Neural networks and physical systems with emergent collective computational abilities. *Proceedings of the National Academy of Sciences, 79*(8), 2554–2558.

109. Horn, B. K. (1970). *Shape from shading: A method for obtaining the shape of a smooth opaque object from one view*.

110. Horn, B. K. (1983). The curve of least energy. *CACM Transactions on Mathematical Software, 9*(4), 441–460.

111. Horn, B. K. (1990). Height and gradient from shading. *International Journal of Computer Vision, 5*(1), 37–75.
112. Horn, B. K. P., Szeliski, R. S., & Yuille, A. L. (1993). Impossible shaded images. *IEEE Transactions on Pattern Analysis and Machine Intelligence, 15*(2), 166–170.
113. Hubel, D. H., & Wiesel, T. N. (1962). Receptive fields, binocular interaction and functional architecture in the cat's visual cortex. *The Journal of Physiology, 160*(1), 106.
114. Hyvärinen, A. (2005). Estimation of non-normalized statistical models by score matching. *Journal of Machine Learning Research, 6*(Apr), 695–709.
115. Hyvarinen, A. (2007). Connections between score matching, contrastive divergence, and pseudolikelihood for continuous-valued variables. *IEEE Transactions on Neural Networks, 18*(5), 1529–1531.
116. Hyvärinen, A., Karhunen, J., & Oja, E. (2004). *Independent component analysis* (Vol. 46). Wiley.
117. Ikeuchi, K., & Horn, B. K. (1981). Numerical shape from shading and occluding boundaries. *Artificial Intelligence, 17*(1–3), 141–184.
118. Ioffe, S., & Szegedy, C. (2015). Batch normalization: Accelerating deep network training by reducing internal covariate shift. arXiv preprint arXiv:1502.03167.
119. Bergen, J. R., & Adelson, E. H. (1988). Early vision and texture perception. *Nature, 333*, 363–364.
120. Jaynes, E. T. (1957). Information theory and statistical mechanics. *Physical Review, 106*(4), 620.
121. Jin, L., Lazarow, J., & Tu, Z. (2017). Introspective classification with convolutional nets. In *Advances in Neural Information Processing Systems* (pp. 823–833).
122. Jordan, M. I., Ghahramani, Z., Jaakkola, T. S., & Saul, L. K. (1999). An introduction to variational methods for graphical models. *Machine Learning, 37*(2), 183–233.
123. Julesz, B. (1962). Visual pattern discrimination. *IRE Transactions on Information Theory, 8*(2), 84–92.
124. Julesz, B. et al. (1981). Textons, the elements of texture perception, and their interactions. *Nature, 290*(5802), 91–97.
125. Karras, T., Aila, T., Laine, S., & Lehtinen, J. (2017). Progressive growing of gans for improved quality, stability, and variation. In *International Conference on Learning Representations*.
126. Karras, T., Laine, S., & Aila, T. (2019). A style-based generator architecture for generative adversarial networks. In *Proceedings of the IEEE conference on computer vision and pattern recognition* (pp. 4401–4410).
127. Kim, H., & Park, H. (2008). Nonnegative matrix factorization based on alternating nonnegativity constrained least squares and active set method. *SIAM Journal on Matrix Analysis and Applications, 30*(2), 713–730.
128. Kim, T., & Bengio, Y. (2016). Deep directed generative models with energy-based probability estimation. In *ICLR Workshop*.
129. Kimia, B. B., Frankel, I., & Popescu, A.-M. (2003). Euler spiral for shape completion. *International Journal of Computer Vision, 54*(1–3), 159–182.
130. Kingma, D., & Welling, M. (2014a). Efficient gradient-based inference through transformations between bayes nets and neural nets. In *International Conference on Machine Learning* (pp. 1782–1790).
131. Kingma, D. P., & Dhariwal, P. (2018). Glow: Generative flow with invertible 1×1 convolutions. In *Advances in Neural Information Processing Systems* (pp. 10215–10224).
132. Kingma, D. P., Salimans, T., Jozefowicz, R., Chen, X., Sutskever, I., & Welling, M. (2016). Improved variational inference with inverse autoregressive flow. In *Advances in Neural Information Processing Systems* (pp. 4743–4751).
133. Kingma, D. P., & Welling, M. (2014b). Auto-encoding variational bayes. In *International Conference for Learning Representations*.
134. Kipf, T. N., & Welling, M. (2016). Semi-supervised classification with graph convolutional networks. In *International Conference on Learning Representations*.

135. Kirkpatrick, S., Gelatt Jr, C. D., & Vecchi, M. P. (1983). Optimization by simulated annealing. *Science, 220*(4598), 671–680.
136. Koren, Y., Bell, R., & Volinsky, C. (2009). Matrix factorization techniques for recommender systems. *Computer, 42*(8), 30–37.
137. Kossaifi, J., Tzimiropoulos, G., & Pantic, M. (2017). Fast and exact newton and bidirectional fitting of active appearance models. *IEEE Transactions on Image Processing, 26*(2), 1040–1053.
138. Krizhevsky, A., Sutskever, I., & Hinton, G. E. (2012). Imagenet classification with deep convolutional neural networks. In *Advances in Neural Information Processing Systems* (pp. 1097–1105).
139. Kumar, M., Babaeizadeh, M., Erhan, D., Finn, C., Levine, S., Dinh, L., & Kingma, D. (2019). Videoflow: A flow-based generative model for video. arXiv preprint arXiv:1903.01434.
140. Lafferty, J., McCallum, A., & Pereira, F. C. (2001). Conditional random fields: Probabilistic models for segmenting and labeling sequence data. In *International Conference on Machine Learning* (pp. 282–289).
141. Landau, L. D., & Lifshitz, E. M. (1959). *Statistical physics, 5*, 482.
142. Lanford, O. E. (1973). Statistical mechanics and mathematical problems. In *Entropy and equilibrium states in classical statistical mechanics* (pp. 1–113).
143. Lazarow, J., Jin, L., & Tu, Z. (2017). Introspective neural networks for generative modeling. In *IEEE International Conference on Computer Vision* (pp. 2774–2783).
144. LeCun, Y., Bottou, L., Bengio, Y., Haffner, P., et al. (1998). Gradient-based learning applied to document recognition. *Proceedings of the IEEE, 86*(11), 2278–2324.
145. Lee, A. B., Mumford, D., & Huang, J. (2001). Occlusion models for natural images: A statistical study of a scale-invariant dead leaves model. *International Journal of Computer Vision, 41*(1), 35–59.
146. Lee, A. B., Mumford, D. B., & Huang, J. G. (2001). Occlusion models for natural images: A statistical study of a scale-invariant dead leaves model. *International Journal of Computer Vision, 41*(1), 35–59.
147. Lee, C.-H., & Rosenfeld, A. (1985). Improved methods of estimating shape from shading using the light source coordinate system. *Artificial Intelligence, 26*(2), 125–143.
148. Lee, D. D., & Seung, H. S. (2001). Algorithms for non-negative matrix factorization. In *Advances in Neural Information Processing Systems* (pp. 556–562).
149. Lee, K., Xu, W., Fan, F., & Tu, Z. (2018). Wasserstein introspective neural networks. In *IEEE Conference on Computer Vision and Pattern Recognition* (pp. 3702–3711).
150. Leung, T., & Malik, J. (1996). Detecting, localizing and grouping repeated scene elements from an image. In *Proceedings of the 4th European Conference on Computer Vision, UK*.
151. Leung, T., & Malik, J. (1999). Recognizing surface using three-dimensional textons. In G. Corfu (Ed.), *Proceedings of 7th International Conference on Computer Vision*.
152. Lewicki, M. S., & Olshausen, B. A. (1999). A probabilistic framework for the adaptation and comparison of image codes. *JOSA, 16*(7), 1587–1601.
153. Lin, M. H., & Tomasi, C. (2003). Surfaces with occlusions from layered stereo. In *Proceedings of the 2003 IEEE Computer Society Conference on Computer Vision and Pattern Recognition, 2003* (Vol. 1, pp. I–I). IEEE.
154. Liu, C., Rubin, D. B., & Wu, Y. N. (1998). Parameter expansion to accelerate EM: The PX-EM algorithm. *Biometrika, 85*(4), 755–770.
155. Liu, C., Zhu, S.-C., & Shum, H.-Y. (2001). Learning inhomogeneous gibbs model of faces by minimax entropy. In *International Conference on Computer Vision* (Vol. 1, pp. 281–287). IEEE.
156. Liu, F., Shen, C., & Lin, G. (2015). Deep convolutional neural fields for depth estimation from a single image. In *Proceedings of the IEEE conference on computer vision and pattern recognition*.
157. Liu, J. S. (2008). *Monte Carlo strategies in scientific computing*. Springer.
158. Lowe, D. G. (1999). Object recognition from local scale-invariant features. In *Proceedings of the seventh IEEE international conference on computer vision* (Vol. 2, pp. 1150–1157). IEEE.

159. Lowe, D. G. (2004). Distinctive image features from scale-invariant keypoints. *International Journal of Computer Vision, 60*(2), 91–110.

160. Lu, Y., Gao, R., Zhu, S.-C., & Wu, Y. N. (2018). Exploring generative perspective of convolutional neural networks by learning random field models. *Statistics and Its Interface, 11*(3), 515–529.

161. Lu, Y., Zhu, S.-C., & Wu, Y. N. (2016). Learning FRAME models using CNN filters. *Thirtieth AAAI Conference on Artificial Intelligence*.

162. Malik, J. (1986). Interpreting line drawings of curved objects. *International Journal of Computer Vision, 1*(1), 73–103.

163. Malik, J., Belongie, S., Shi, J., & Leung, T. (1999). Textons, contours and regions: Cue integration in image segmentation. In G. Corfu (Ed.), *Proceedings of the Seventh IEEE International Conference on Computer Vision*, (Vol. 2). IEEE.

164. Malik, J., & Perona, P. (1990). Preattentive texture discrimination with early vision mechanisms. *Journal of Optical Society of America A, 7*(5), 923–932.

165. Mallat, S., & Zhang, Z. (1993a). Matching pursuit in a time-frequency dictionary. *IEEE Transactions on Signal Processing, 41*, 3397–3415.

166. Mallat, S. G. (1989). A theory for multiresolution signal decomposition: The wavelet representation. *IEEE Transactions on Pattern Analysis and Machine Intelligence, 11*(7), 674–693.

167. Mallat, S. G., & Zhang, Z. (1993b). Matching pursuits with time-frequency dictionaries. *IEEE Transactions on Signal Processing, 41*(12), 3397–3415.

168. Marinari, E., & Parisi, G. (1992). Simulated tempering: a new monte carlo scheme. *EPL (Europhysics Letters), 19*(6), 451.

169. Marr, D. (2010). *Vision: A computational investigation into the human representation and processing of visual information*. MIT press.

170. Marr, D., & Nishihara, H. K. (1978). Representation and recognition of the spatial organization of three-dimensional shapes. *Proceedings of the Royal Society London B, 200*(1140), 269–294.

171. Matheron, G. (1975). *Random sets and integral geometry*. Wiley.

172. Mikolov, T., Chen, K., Corrado, G., & Dean, J. (2013). Efficient estimation of word representations in vector space. In *ICLR Workshop*.

173. Mingolla, E., & Todd, J. T. (1986). Perception of solid shape from shading. *Biological Cybernetics, 53*(3), 137–151.

174. Montufar, G. F., Pascanu, R., Cho, K., & Bengio, Y. (2014). On the number of linear regions of deep neural networks. In *Advances in Neural Information Processing Systems* (pp. 2924–2932).

175. Mumford, D. (1996). The statistical description of visual signals. *Mathematical Research, 87*, 233–256.

176. Mumford, D., & Gidas, B. (2001). Stochastic models for generic images. *Quarterly of Applied Mathematics, 59*(1), 85–111.

177. Mumford, D., & Shah, J. (1989). Optimal approximations by piecewise smooth functions and associated variational problems. *Communications on Pure and Applied Mathematics, 42*(5), 577–685.

178. Murray, J. D. (1981). A pre-pattern formation mechanism for animal coat markings. *Journal of Theoretical Biology, 88*(1), 161–199.

179. Neal, R. M. et al. (2011). MCMC using hamiltonian dynamics. *Handbook of Markov Chain Monte Carlo, 2*(11), 2.

180. Ngiam, J., Chen, Z., Koh, P. W., & Ng, A. Y. (2011). *Learning deep energy models* (pp. 1105–1112).

181. Nijkamp, E., Gao, R., Sountsov, P., Vasudevan, S., Pang, B., Zhu, S.-C., & Wu, Y. N. (2021). MCMC should mix: Learning energy-based model with neural transport latent space MCMC. In *International Conference on Learning Representations*.

182. Nijkamp, E., Hill, M., Han, T., Zhu, S.-C., & Wu, Y. N. (2019a). On the anatomy of MCMC-based maximum likelihood learning of energy-based models. arXiv.

183. Nijkamp, E., Pang, B., Han, T., Zhou, L., Zhu, S.-C., & Wu, Y. N. (2020). Learning multi-layer latent variable model via variational optimization of short run mcmc for approximate inference. In *European Conference on Computer Vision* (pp. 361–378). Springer.

184. Nijkamp, E., Zhu, S.-C., & Wu, Y. N. (2019b). Learning non-convergent short-run MCMC toward energy-based model. In *NeurIPS*.

185. Nitzberg, M., & Mumford, D. (1990). The 2.1-d sketch. In *International Conference on Computer Vision (ICCV) (pp. 138–144)*.

186. Nitzberg, M., & Shiota, T. (1992). Nonlinear image filtering with edge and corner enhancement. *IEEE Transactions on Pattern Analysis & Machine Intelligence, 14*(8), 826–833.

187. Olshausen, B. A. (2003). Learning sparse, overcomplete representations of time-varying natural images. In *Proceedings of the 2003 International Conference on Image Processing, 2003 (ICIP 2003)* (Vol. 1, pp. I–41). IEEE.

188. Olshausen, B. A., & Field, D. J. (1996a). Emergence of simple-cell receptive field properties by learning a sparse code for natural images. *Nature, 381*, 607–609.

189. Olshausen, B. A., & Field, D. J. (1996b). Emergence of simple-cell receptive field properties by learning a sparse code for natural images. *Nature, 381*(6583), 607.

190. Olshausen, B. A., & Field, D. J. (1997). Sparse coding with an overcomplete basis set: A strategy employed by V1? *Vision Research, 37*(23), 3311–3325.

191. Pang, B., Han, T., Nijkamp, E., Zhu, S.-C., & Wu, Y. N. (2020). Learning latent space energy-based prior model. arXiv preprint arXiv:2006.08205.

192. Pascanu, R., Montufar, G., & Bengio, Y. (2013). On the number of response regions of deep feed forward networks with piece-wise linear activations. arXiv preprint arXiv:1312.6098.

193. Pennington, J., Socher, R., & Manning, C. (2014). Glove: Global vectors for word representation. In *Conference on Empirical Methods in Natural Language Processing* (pp. 1532–1543).

194. Pentland, A. P. (1984). Local shading analysis. *IEEE Transactions on Pattern Analysis and Machine Intelligence, 2*, 170–187.

195. Perona, P., & Malik, J. (1990). Scale-space and edge detection using anisotropic diffusion. *IEEE Transactions on Pattern Analysis and Machine Intelligence, 12*(7), 629–639.

196. Ping-Sing, T., & Shah, M. (1994). Shape from shading using linear approximation. *Image and Vision Computing, 12*(8), 487–498.

197. Portilla, J., & Simoncelli, E. P. (2000). A parametric texture model based on joint statistics of complex wavelet coefficients. *International Journal of Computer Vision, 40*, 1.

198. Poucet, B., & Save, E. (2005). Attractors in memory. *Science, 308*(5723), 799–800.

199. Radford, A., Metz, L., & Chintala, S. (2015). Unsupervised representation learning with deep convolutional generative adversarial networks. arXiv preprint arXiv:1511.06434.

200. Ren, X., Fowlkes, C. C., & Malik, J. (2005). Scale-invariant contour completion using conditional random fields. In *International Conference on Computer Vision (ICCV)* (pp. 1214–1221).

201. Ren, X., Fowlkes, C. C., & Malik, J. (2006). Figure/ground assignment in natural images. In *In European Conference on Computer Vision (ECCV)* (pp. 614–627).

202. Rezende, D. J., & Mohamed, S. (2015). Variational inference with normalizing flows. In *International Conference on Machine Learning*.

203. Rhodes, B., Xu, K., & Gutmann, M. U. (2020). Telescoping density-ratio estimation. *Advances in Neural Information Processing Systems, 33*, 4905–4916.

204. Richardson, M., & Domingos, P. (2006). Markov logic networks. *Machine Learning, 62*(1), 107–136.

205. Robbins, H., & Monro, S. (1951). A stochastic approximation method. *The Annals of Mathematical Statistics*, pp. 400–407.

206. Rosset, S., Zhu, J., & Hastie, T. (2004). Boosting as a regularized path to a maximum margin classifier. *The Journal of Machine Learning Research, 5*, 941–973.

207. Roth, S., & Black, M. J. (2005). Fields of experts: A framework for learning image priors. In *Proceedings of the IEEE Conference on Computer Vision and Pattern Recognition* (Vol. 2, pp. 860–867). IEEE.
208. Roweis, S., & Ghahramani, Z. (1999). A unifying review of linear gaussian models. *Neural Computation, 11*, 2.
209. Rubin, D. B., & Thayer, D. T. (1982). EM algorithms for ML factor analysis. *Psychometrika, 47*(1), 69–76.
210. Ruderman, D. L. (1994). The statistics of natural images. *Network: Computation in Neural Systems, 5*(4), 517–548.
211. Saund, E. (1999a). Perceptual organization of occluding contours generated by opaque surfaces. In *Computer Vision and Pattern Recognition (CVPR)* (pp. 624–630).
212. Saund, E. (1999b). Perceptual organization of occluding contours of opaque surfaces. *Computer Vision and Image Understanding, 76*(1), 70–82.
213. Scharstein, D., & Szeliski, R. (2003). High-accuracy stereo depth maps using structured light. In *Proceedings of the 2003 IEEE Computer Society Conference on Computer Vision and Pattern Recognition, 2003* (Vol. 1, pp. I–I). IEEE.
214. Serre, T., Wolf, L., Bileschi, S., Riesenhuber, M., & Poggio, T. (2007). Robust object recognition with cortex-like mechanisms. *IEEE Transactions on Pattern Analysis and Machine Intelligence, 29*(3), 411–426.
215. Shannon, C. E. (1948). A mathematical theory of communication. *The Bell System Technical Journal, 27*(3), 379–423.
216. Shannon, C. E. (2001). A mathematical theory of communication. *ACM SIGMOBILE Mobile Computing and Communications Review, 5*(1), 3–55.
217. Si, Z., & Zhu, S.-C. (2011). Learning hybrid image templates (hit) by information projection. *IEEE Transactions on Pattern Analysis and Machine Intelligence, 99*(7), 1354–1367.
218. Simoncelli, E. P., Freeman, W. T., Adelson, E. H., & Heeger, D. J. (1992). Shiftable multiscale transforms. *IEEE Transactions on Information Theory, 38*(2), 587–607.
219. Simonyan, K., & Zisserman, A. (2015). Very deep convolutional networks for large-scale image recognition. In *ICLR*.
220. Skiena, S. (1990). *Combinatorics and graph theory with mathematica.*
221. Smith, S. L., Dherin, B., Barrett, D. G., & De, S. (2021). On the origin of implicit regularization in stochastic gradient descent. arXiv preprint arXiv:2101.12176.
222. Sohl-Dickstein, J., Weiss, E. A., Maheswaranathan, N., & Ganguli, S. (2015). Deep unsupervised learning using nonequilibrium thermodynamics. arXiv preprint arXiv:1503.03585.
223. Song, Y., Sohl-Dickstein, J., Kingma, D. P., Kumar, A., Ermon, S., & Poole, B. (2020). Score-based generative modeling through stochastic differential equations. arXiv preprint arXiv:2011.13456.
224. Zhu, S.-C., Liu, X., & Wu, Y. N. (2000). Exploring julesz texture ensemble by effective markov chain monte carlo. *PAMI, 22*, 6.
225. Yu, S. X., Lee, T. S., & Kanade, T. (2002). A hierarchical markov random field model for figure-ground segregation. In *Proceedings of the 4th International Conference on Energy Minimization Methods in Computer Vision and Pattern Recognition (EMMCVPR)* (pp. 118–131).
226. Sugita, Y., & Okamoto, Y. (1999). Replica-exchange molecular dynamics method for protein folding. *Chemical Physics Letters, 314*(1–2), 141–151.
227. Swersky, K., Ranzato, M., Buchman, D., Marlin, B., & Freitas, N. (2011). On autoencoders and score matching for energy based models. In Getoor, L., & Scheffer, T. (Eds.) *Proceedings of the 28th international conference on machine learning (ICML-11)* (pp. 1201–1208). ACM.
228. Szeliski, R. (1991). Fast shape from shading. *Computer Vision, Graphics, and Image Processing: Image Understanding, 53*(2), 129–153.
229. Tanner, M. A. (1996). *Tools for Statistical Inference.* Springer.
230. Tibshirani, R. (1996). Regression shrinkage and selection via the Lasso. *Journal of the Royal Statistical Society: Series B (Methodological), 58*(1), 267–288.

231. Tieleman, T. (2008). Training restricted Boltzmann machines using approximations to the likelihood gradient. In *International Conference on Machine Learning* (pp. 1064–1071). ACM.

232. Torralba, A., & Oliva, A. (2002). Depth estimation from image structure. *IEEE Transaction on Pattern Analysis and Machine Intelligence, 24*(9), 1226–1238.

233. Tran, D., Vafa, K., Agrawal, K. K., Dinh, L., & Poole, B. (2019). Discrete flows: Invertible generative models of discrete data. arXiv preprint arXiv:1905.10347.

234. Tu, Z. (2007). Learning generative models via discriminative approaches. In *IEEE Conference on Computer Vision and Pattern Recognition* (pp. 1–8). IEEE.

235. Tu, Z., & Zhu, S.-C. (2002). Image segmentation by data-driven markov chain monte carlo. *IEEE Transactions on Pattern Analysis and Machine Intelligence, 24*(5), 657–673.

236. Tu, Z., & Zhu, S.-C. (2006). Parsing images into regions, curves and curve groups. *International Journal of Computer Vision, 69*(2), 223–249.

237. Turing, A. M. (1990). The chemical basis of morphogenesis. *Bulletin of Mathematical Biology, 52*(1–2), 153–197.

238. Tyleček, R., & Šára, R. (2013). Spatial pattern templates for recognition of objects with regular structure. In *German Conference on Pattern Recognition* (pp. 364–374). Springer.

239. Vaswani, A., Shazeer, N., Parmar, N., Uszkoreit, J., Jones, L., Gomez, A. N., Kaiser, Ł., & Polosukhin, I. (2017). Attention is all you need. In *Advances in Neural Information Processing Systems* (pp. 5998–6008).

240. Vega, O. E., & Yang, Y.-H. (1993). Shading logic: A heuristic approach to recover shape from shading. *IEEE Transactions on Pattern Analysis and Machine Intelligence, 15*(6), 592–597.

241. Vincent, P. (2010). A connection between score matching and denoising autoencoders. *Neural Computation, 23*(7), 1661–1674.

242. Vincent, P., Larochelle, H., Bengio, Y., & Manzagol, P.-A. (2008). Extracting and composing robust features with denoising autoencoders. In *International Conference on Machine Learning* (pp. 1096–1103). ACM.

243. Waltz, D. (1975). Understanding line drawings of scenes with shadows.

244. Wang, J., Betke, M., & Gu, E. (2005). Mosaicshape: Stochastic region grouping with shape prior. In *Computer Vision and Pattern Recognition (CVPR)* (pp. 902–908).

245. Wang, J. Y., & Adelson, E. H. (1993). Layered representation for motion analysis. In *Computer Vision and Pattern Recognition (CVPR)* (pp. 361–366).

246. Wang, J. Y., & Adelson, E. H. (1994). Representing moving images with layers. *IEEE Transactions on Image Processing, 3*(5), 625–638.

247. Wang, Y., & Zhu, S.-C. (2004). Analysis and synthesis of textured motion: Particles and waves. *IEEE Transactions on Pattern Analysis and Machine Intelligence, 26*(10), 1348–1363.

248. Wang, J., & Zhu, S.-C. (2008). Perceptual scale-space and its applications. *International Journal of Computer Vision, 80*(1), 143–165.

249. Welling, M. (2009). Herding dynamical weights to learn. In *International Conference on Machine Learning* (pp. 1121–1128). ACM.

250. Welling, M., Zemel, R. S., & Hinton, G. E. (2002). Self supervised boosting. In *Advances in Neural Information Processing Systems* (pp. 665–672).

251. Willianms, L. R., & Hanson, A. R. (1996). Perceptual completion of occluded surfaces. *Computer Vision and Image Understanding (CVIU), 64*(1), 1–20.

252. Wu, T.-F., Xia, G.-S., & Zhu, S.-C. (2007). Compositional boosting for computing hierarchical image structures. In *IEEE conf. on Computer Vision and Pattern Recognition (CVPR)* (pp. 1–8).

253. Wu, Y. N., Gao, R., Han, T., & Zhu, S.-C. (2019). A tale of three probabilistic families: Discriminative, descriptive and generative models. *Quarterly of Applied Mathematics, 77*(2), 423–465.

254. Wu, Y. N., Si, Z., Gong, H., & Zhu, S.-C. (2010). Learning active basis model for object detection and recognitio. *International Journal of Computer Vision, 90*, 198–235.

255. Wu, Y. N., Zhu, S.-C., & Liu, X. (1999a). *Equivalence of Julesz and Gibbs Ensembles*. ICCV.

256. Wu, Y. N., Xie, J., Lu, Y., & Zhu, S.-C. (2018). Sparse and deep generalizations of the frame model. *Annals of Mathematical Sciences and Applications, 3*(1), 211–254.

257. Wu, Y. N., Zhu, S.-C., & Guo, C.-E. (2008). From information scaling of natural images to regimes of statistical models. *Quarterly of Applied Mathematics, 66*, 81–122.

258. Wu, Y. N., Zhu, S.-C., & Liu, X. (1999b). Equivalence of julesz and gibbs texture ensembles. In *Proceedings of the Seventh IEEE International Conference on Computer Vision* (Vol. 2, pp. 1025–1032). IEEE.

259. Wu, Y. N., Zhu, S. C., & Liu, X. (2000). Equivalence of julesz ensembles and frame models. *International Journal of Computer Vision, 38*(3), 247–265.

260. Xie, J., Gao, R., Zheng, Z., Zhu, S.-C., & Wu, Y. N. (2019a). Learning dynamic generator model by alternating back-propagation through time. In *Proceedings of the AAAI Conference on Artificial Intelligence* (Vol. 33(01)).

261. Xie, J., Hu, W., Zhu, S.-C., & Wu, Y. N. (2014). Learning sparse frame models for natural image patterns. *International Journal of Computer Vision, 114*, 1–22.

262. Xie, J., Lu, Y., Gao, R., & Wu, Y. N. (2018a). Cooperative learning of energy-based model and latent variable model via MCMC teaching. In *The AAAI Conference on Artificial Intelligence*.

263. Xie, J., Lu, Y., Gao, R., Zhu, S.-C., & Wu, Y. N. (2018b). Cooperative training of descriptor and generator networks. *IEEE Transactions on Pattern Analysis and Machine Intelligence*, (preprints).

264. Xie, J., Lu, Y., Gao, R., Zhu, S.-C., & Wu, Y. N. (2018c). Cooperative training of descriptor and generator networks. *IEEE Transactions on Pattern Analysis and Machine Intelligence, 42*(1), 27–45.

265. Xie, J., Lu, Y., Zhu, S.-C., & Wu, Y. N. (2016a). Inducing wavelets into random fields via generative boosting. *Journal of Applied and Computational Harmonic Analysis, 41*, 4–25.

266. Xie, J., Lu, Y., Zhu, S.-C., & Wu, Y. N. (2016b). A theory of generative ConvNet. In *International Conference on Machine Learning* (pp. 2635–2644).

267. Xie, J., Zheng, Z., Fang, X., Zhu, S.-C., & Wu, Y. N. (2019b). Multimodal conditional learning with fast thinking policy-like model and slow thinking planner-like model. arXiv preprint arXiv:1902.02812.

268. Xie, J., Zheng, Z., Fang, X., Zhu, S.-C., & Wu, Y. N. (2021). Cooperative training of fast thinking initializer and slow thinking solver for conditional learning. In *IEEE Transactions on Pattern Analysis and Machine Intelligence*.

269. Xie, J., Zheng, Z., Gao, R., Wang, W., Zhu, S.-C., & Wu, Y. N. (2018d). *Learning descriptor networks for 3D shape synthesis and analysis* (pp. 8629–8638).

270. Xie, J., Zhu, S.-C., & Nian Wu, Y. (2017). *Synthesizing dynamic patterns by spatial-temporal generative ConvNet* (pp. 7093–7101).

271. Xing, X., Gao, R., Han, T., Zhu, S.-C., & Wu, Y. N. (2020). Deformable generator networks: Unsupervised disentanglement of appearance and geometry. In *IEEE Transactions on Pattern Analysis and Machine Intelligence* (pp. 1–1).

272. Xing, X., Han, T., Gao, R., Zhu, S.-C., & Wu, Y. N. (2019). Unsupervised disentangling of appearance and geometry by deformable generator network. In *The IEEE Conference on Computer Vision and Pattern Recognition (CVPR)* (pp. 10354–10363).

273. Yao, B., Yang, X., & Zhu, S.-C. (2007). An integrated image annotation tool and large scale ground truth database. In *Proceeding of the 6th International Conference on Energy Minimization Methods in Computer Vision and Pattern Recognition (EMMCVPR)* (pp. 169–183).

274. Younes, L. (1999). On the convergence of markovian stochastic algorithms with rapidly decreasing ergodicity rates. *Stochastics: An International Journal of Probability and Stochastic Processes, 65*(3-4), 177–228.

275. Zheng, Q., & Chellappa, R. (1991). Estimation of illumination direction, albedo, and shape from shading. In *IEEE Transactions on PAMI*.

276. Zhu, S. C. (1999). Embedding gestalt laws in markov random fields. *IEEE Transactions on Pattern Analysis and Machine Intelligence, 21*(11), 1170–1187.

277. Zhu, S.-C., Guo, C.-E., Wang, Y., & Xu, Z. (2005). What are textons? *International Journal of Computer Vision, 62*(1–2), 121–143.
278. Zhu, S.-C., & Mumford, D. (1998). Grade: Gibbs reaction and diffusion equations. In *Sixth International Conference on Computer Vision (ICCV)* (pp. 847–854). IEEE.
279. Zhu, S.-C., & Mumford, D. B. (1997). Prior learning and gibbs reaction-diffusion. *IEEE Transactions on Pattern Analysis and Machine Intelligence, 19*(11), 1236–1250.
280. Zhu, S. C., Mumford, D., et al. (2007). A stochastic grammar of images. *Foundations and Trends® in Computer Graphics and Vision, 2*(4), 259–362.
281. Zhu, S.-C., Wu, Y. N., & Mumford, D. (1997). Minimax entropy principle and its application to texture modeling. *Neural Computation, 9*(8), 1627–1660.
282. Zhu, S.-C., Wu, Y. N., & Mumford, D. (1998). Filters, random fields and maximum entropy (frame): Toward a unified theory for texture modeling. *International Journal of Computer Vision, 27*(2), 107–126.
283. Ziebart, B. D., Maas, A. L., Bagnell, J. A., & Dey, A. K. (2008). Maximum entropy inverse reinforcement learning. In *The AAAI Conference on Artificial Intelligence*.

Printed in the United States
by Baker & Taylor Publisher Services